# Chemistry of the Main Group Elements

# Chemistry of the Main Group Elements

Andrew R. Barron

MiDAS Green Innovations
2020

First Printing: 2020

ISBN 978-1-8380085-1-2

ISBN 978-1-83800-851-2 90000

9 781838 008512

MiDAS Green Innovation, Ltd
Swansea, SA1 8RD, UK

www.midasgreeninnovation.com

# Dedication

To my wife, Merrie, for all your support, friendship and love.

You are the 'light of my life'.

# Contents

# Acknowledgements

I would like to thank my Secondary School Chemistry teacher, Mr. Roberts of Heath End School, who instilled in me his love of chemistry as well as a fascination with dramatic demonstrations.

Without Nobel Laureate Prof. Sir Geoffrey Wilkinson, I would have never embarked on an academic career. His creation of a "certain well-known text-book" was the inspiration for this work.

I would like to thank all the members of the Barron Research Group (post-doctoral fellows, PhD and undergraduate students) past and present, for their work in making me look good.

The myriad collaborators of whom I have had the pleasure of learning from over the years are to be thanked for ensuring that my research has never become stale and new challenges have always been on the horizon.

Last, but not least, I would like to thank my wife, Merrie, for putting up with me during the Editing of this book while 'social distancing' during the COVID-19 pandemic.

# Preface

The main group (s- and p-block) elements are among the most diverse in the Periodic Table. Ranging from non-metallic gases (e.g., hydrogen and fluorine), through semi-metals (e.g., metalloids such as silicon) to highly reactive metals (e.g., sodium and potassium). The study of the main group elements is important for a number of reasons. On an academic level they exemplify the trends and predictions in structure and reactivity that are the key to the Periodic Table. They represent the diversity of inorganic chemistry, and the fundamental aspects of structure and bonding that are also present for the transition metal, lanthanide and actinide elements.

The main group elements represent the most prevalent elements in the Earth's crust, as well as most of the key elements of life, and have enormous industrial, economic, and environmental importance. In this regard an understanding of the chemistry of the main group elements is vital for students within science, engineering, and medicine; however, it is hoped that those who make political and economic decisions would make better ones (or at least more responsible ones) if they had a fraction of the knowledge of the world around them.

Since the position of the main group elements within the Periodic Table defines their properties (and vice versa) this represents a logical organization of the topics. Prior to introducing the elements, a series of general and background topics are covered to provide the basis for further discussion. The subsequent organization is based upon a Chapter for each Group of the s- and p-block elements; however, hydrogen is given its own chapter due to its importance as an element. Although the Group 12 elements are often omitted from a discussion of main group elements they have been included herein.

Some chapters are organized with regard to individual elements (e.g., carbon, silicon, etc.) and others are arranged with regard the types of compounds (e.g., oxides, halides, etc.). This is based upon particular interest or importance of an element. An effort has been made to ensure that topics are not covered twice (unless necessary) and so in general a particular subject is covered in the Group chapter associated with the lower Group number. For example, the halides of boron are described in the Chapter on the Group 13 elements rather than Group 17 elements.

In addition to the basic synthesis, structure, properties, and reactivity of the elements and their compounds, sections describing some industrial use, as well as historical or social perspective have been added. These sections were as a result of attempts to put the chemistry into a context outside of the classroom. It is important that the discovery and use of elements be understood to be a human endeavor rather than a series of abstract concepts or facts. It is only by an appreciation of the past that we can advance the future.

# Chapter 1: General Concepts and Trends

## Oxidation State

The oxidation state of an element is defined as *the formal charge on the atom if all bonds were assumed to be fully ionic.*

In an ionic compound the oxidation state is equal to the charge on the ion, e.g., in NaCl the charge on the sodium is +1 and the oxidation state is also +1. In contrast, the charge on an atom in a covalent compound never approaches the charge implied from the oxidation number, e.g., in $CCl_4$, the oxidation state of carbon is +4 but the charge on the carbon atom is significantly less. Thus, while oxidation state is a simple formulism, it is also very useful in a number of ways:

- Classification of compounds of elements, especially those of the transition metals.
- Enables the balance of reduction-oxidation (redox) reactions.

To determine the oxidation state of an element within a particular compound the following steps act as a guideline:

1. The oxidation state of any atom of any element in its elemental form is zero.
2. The oxidation state of a monatomic ion is equal to its charge.
3. Any homoleptic X-X bonds (e.g., the C-C bond in $H_3C$-$CH_3$) are assumed to be non-ionic and do not contribute to the oxidation state of X.
4. Fluorine always has oxidation state of -1.
5. Elements of Group 1 (IA) (except hydrogen) have an oxidation state of +1 in all compounds.
6. Elements of Group 2 (IIA) have an oxidation state of +2 in compounds.
7. Elements of Group 17 (VIIA) have an oxidation state of -1 when they combine with elements below or to the left of their position in the periodic table.
8. Oxygen is usually assigned the oxidation state of -2, except in compounds with fluorine, where oxygen has a positive oxidation number.
9. Hydrogen is assigned the oxidation state of +1 with non-metals and 1 with metals.

10. In any compound the sum of the oxidation states (oxidation numbers) is equal to the overall charge.

In writing the oxidation state of an element within a compound it is common to use a Roman numeral, rather than the charge, i.e., Al(III) rather than $Al^{3+}$.

## Ionization Energy

The ionization energy (IE), or more properly the ionization enthalpy, is defined as the energy required to lose an electron from a gaseous atom or ion:

$$M^{n+}_{(g)} \rightarrow M^{(n+1)+}_{(g)} + e^-$$

Each subsequent ionization energy is greater than the previous one because of the increase in charge on the ion. For example, for any given atom or ion the 1st ionization energy is less than the 2nd ionization energy, and so on. This is shown in Table 1.1.

| Ionization | Ionization energy (kJ/mol) |
|---|---|
| $Al^0 \rightarrow Al^+ + e^-$ | 548 |
| $Al^+ \rightarrow Al^{2+} + e^-$ | 1823 |
| $Al^{2+} \rightarrow Al^{3+} + e^-$ | 2751 |

**Table 1.1: The first three ionization energies for aluminum.**

How does the ionization energy vary with elements in the Periodic Table? If we consider the 1st ionization potential of the elements in a particular group of the Periodic Table, we note that there is a decrease in ionization potential as you go down the Group. For example, Table 1.2 shows the 1st ionization potential for the Group 1 alkali metals. The reason for this trend is due to the increased shielding of the outer shell electrons ($ns^1$) by the completed (filled) inner shells. The $ns^1$ electron thus exhibits a lower effective nuclear charge and makes it easier to remove.

In contrast, to individual Groups, moving across a particular Period results in a general increase in the ionization potential as is shown in Figure 1.1. The lack of additional screening of filled shells across the Period means that the ionization energy for the outer shell electrons is dominated by the increase in nuclear charge (number of protons) with increased atomic number.

| Element | Ionization energy (kJ/mol) |
|---|---|
| Li | 526 |
| Na | 502 |
| K | 425 |
| Rb | 409 |
| Cs | 382 |

**Table 1.2: Variation of the first ionization potential ($M^0 \rightarrow M^+$) for the Group 1 (IA) elements.**

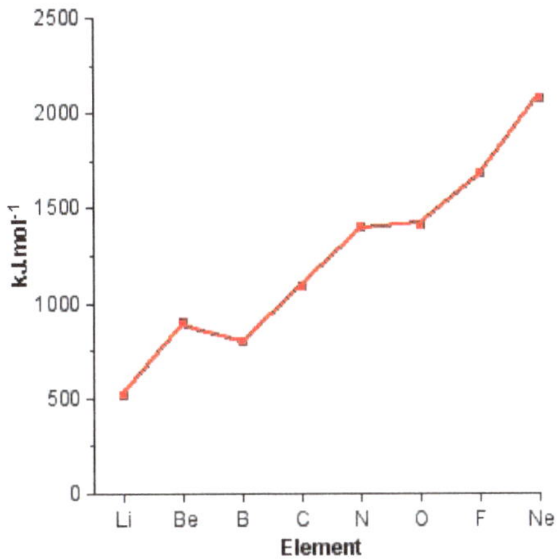


**Figure 1.1: Plot of the first ionization potential for the elements, Li to Ne.**

From Figure 1.1 it can be seen that the Periodic trend is not linear, there are significant steps in the plot. Boron, for example, has a lower first ionization potential than beryllium, why? A consideration of the electron configuration for the elements provides in answer (Figure 1.2). Beryllium has a $2s^2$ outer shell configuration, while boron has a $2s^22p^1$ configuration. The $2p^1$ electron is easy to remove because it exhibits increased shielding from the nucleus due to the filled 2s orbital.

As we move from, boron to nitrogen, the 2p shell is filled (Figure 1.2) without additional shielding and the effect of the increased nuclear charge dominates. Finally, the $2p^4$ configuration for oxygen (Figure 1.2) results in an electron pair, which repel each other, thus making it easier to remove an electron (lower ionization potential) that expected from the increased nuclear charge.

From Figure 1.1 we can see that the effect of electron pairing is less than that of a filled shell.

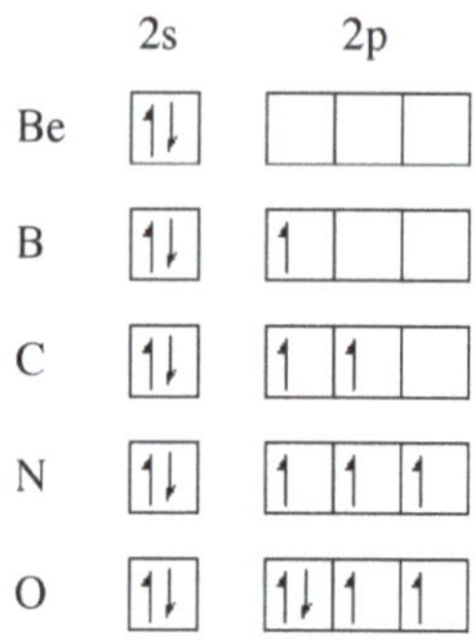


**Figure 1.2: Outer shell electron configuration of beryllium through oxygen.**

The trend for the 2nd ionization potential is similar, but different, to that of the 1st ionization potential. As may be seen in Figure 1.3 the steps observed for the 1st ionization energy plot (i.e., between Be/B and N/O have moved one element to the right. A view of the electron configuration for the $E^+$ ions (Figure 1.4) shows that the rational for the trend in the 1st ionization potential trends still applies but to the ion of the element to the right in the Periodic table. Now $B^+$ has a $2s^2$ outer shell configuration, while $C^+$ has a $2s^22p^1$ configuration. A similar plot for the 3rd ionization energy would move the steps another element to the right.

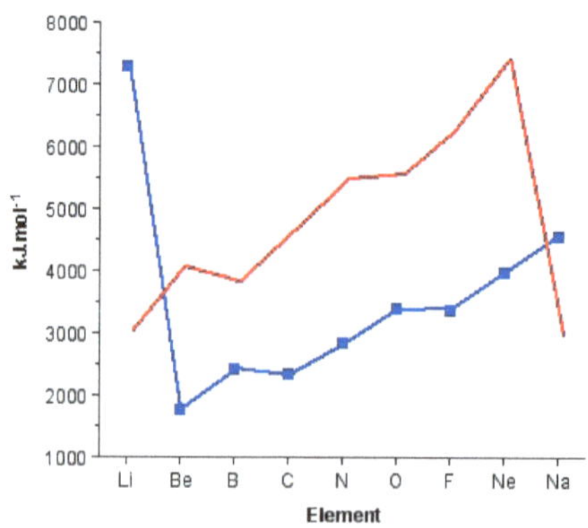


**Figure 1.3: A plot of 2nd ionization energy (enthalpy) for elements Li to Na (blue with square data points) showing the relative trend to the 1st ionization potential (red line not to scale).**

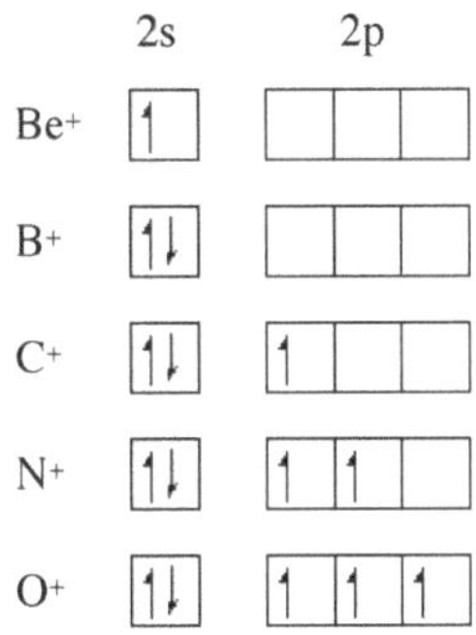


**Figure 1.4: The outer shell electron configuration of E for beryllium through oxygen.**

The other observation to be made from Figure 1.3 is the very large 2nd ionization potential for lithium associated with the ionization of $Li^+$ to $Li^{2+}$. The large increase is due to the removal of an electron from the filled $1s^2$ shell.

## Electron Affinity

The electron affinity (EA) of an element is defined as the energy given off when a neutral atom in the gas phase gains an extra electron to form a negatively charged ion:

$$X^{n-}{}_{(g)} + e^- \rightarrow X^{(n+1)-}{}_{(g)}$$

Electron affinities are more difficult to measure than ionization energies and are usually less accurately known. Electron affinities are large and negative for elements such as fluorine and oxygen, and small and positive for metals.

Electron affinities generally become smaller as you go down a Group of the periodic table (Table 1.3). This is because the electron being added to the atom is placed in a larger orbital, where it spends less time near the nucleus of the atom, and also the number of electrons on an atom increases as we go down a column, so the force of repulsion between the electron being added and the electrons already present on a neutral atom becomes larger. Electron affinities are further complicated since the repulsion between the electron being added to the atom and the electrons already present on the atom depends on the volume of the atom. Thus, for the nonmetals in Groups 6 (VIA) and 7 (VIIA), this force of repulsion is largest for the very smallest atoms in these columns: oxygen and fluorine. As a result, these elements have a smaller electron affinity than the elements below them in these columns as shown in Table 1.3.

| Element | Electron affinity (kJ/mol) |
|---|---|
| F | -322 |
| Cl | -349 |
| Br | -325 |
| I | -295 |

**Table 1.3: The electron affinity for the non-metallic halogens.**

Although there is a general trend that for Group 1 (IA) to Group 17 (VIIA) elements the electron affinity increases across the Periodic table from left to right, the details of the trend are more complex. As may be seen from Figure 1.5, there is a cyclic trend.

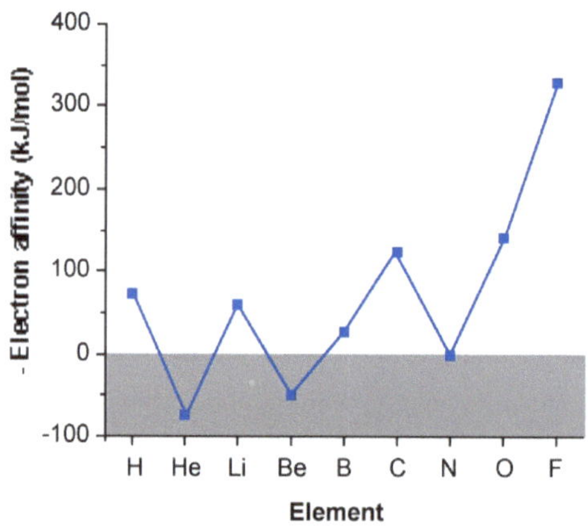


**Figure 1.5: Plot of the negative electron affinity (i.e., -EA) for the elements, hydrogen to fluorine. Values for helium, beryllium, nitrogen, and neon are not known with any accuracy but are all positive.**

The explanation of this is a consequence of the unusually stable electron configurations exhibited by atoms with filled or half-filled shells, i.e., helium, beryllium, nitrogen and neon (Table 1.4). These configurations are so stable that it actually takes energy to force one of these elements to pick up an extra electron to form a negative ion.

| Element | Electron affinity (kJ/mol) | Electron configuration |
|---|---|---|
| H | -72.8 | $1s^1$ |
| He | +ve | $1s^2$ |
| Li | -59.8 | [He] $2s^1$ |
| Be | +ve | [He] $2s^2$ |
| B | -27 | [He] $2s^22p^1$ |
| C | -122.3 | [He] $2s^22p^2$ |
| N | +ve | [He] $2s^22p^3$ |
| O | -141.1 | [He] $2s^22p^4$ |
| F | -328.0 | [He] $2s^22p^5$ |
| Ne | +ve | [He] $2s^22p^6$ |

**Table 1.4: Electron affinities of the elements: hydrogen to neon. Values for helium, beryllium, nitrogen, and neon are not known with any accuracy but are all positive.**

## Electronegativity

An issue with ionization potential and electron affinity is that they are defined and measured as reactions of isolated atoms in the gas phase. Although values have been determined for molecular fragments it is still difficult to correlate with reaction trends in solution. To overcome this issue the concept of electronegativity was developed.

**Figure 1.6: American chemist Linus Carl Pauling (1901 - 1994).**

Electronegativity is defined as *the tendency of an atom in a molecule to attract electrons to itself.* Although several electronegativity scales have been developed, that by Linus Pauling (Figure 1.6) is the most often used. Table 1.5 provides selected Pauling electronegativity values (unit less).

| Element | Pauling scale |
|---|---|
| F | 4.0 |
| O | 3.5 |
| Cl | 3.0 |
| N | 3.0 |
| S | 2.5 |
| C | 2.5 |
| H | 2.1 |
| B | 2.0 |
| Na | 0.9 |

**Table 1.5: Selected Pauling electronegativity values.**

The advantage of the Pauling electronegativity scale is that it allows the prediction of general behavior. For example, the larger the difference in electronegativity between two elements the more ionic character or more polar the bonding interaction. Thus, a H-O bond (3.5 - 2.1 = 1.4) is more polar than a H-S bond (2.5 - 2.1 = 0.4).

# Chapter 2: Structure and Bonding

## Valence shell electron pair repulsion (VSEPR) theory

The idea of a correlation between molecular geometry and the number of valence electrons was first presented in 1940 by Sidgwick and Powell; however, in 1957, Ronald Gillespie (Figure 2.1) and Sir Ronald Nyholm (Figure 2.2) refined this concept to build a more detailed theory. It is their work that provides the basis of the *valence shell electron pair repulsion* (VSEPR) theory, and as such it is also known as the Gillespie-Nyholm theory.

**Figure 2.1: English chemist Ronald J. Gillespie FRS (1924 -).**

**Figure 2.2: Australian chemist Sir Ronald Sydney Nyholm (1917 - 1971).**

One attribute of VSEPR is that with the ability to predict the shape of a molecule for a compound comes the ability to predict some of the physical and chemical properties of that compound.

The formal definition that is the basis for VSEPR is as follows: *Pairs of electrons in the valence shell of a central atom of a molecule repel each other and take up positions as far apart as possible.*

Within this definition it is implicitly assumed that the core shells are not polarized and therefore take no part in bonding, and therefore can be ignored. Since the maximum repulsion of the electron pairs (be they associated with a bonding interaction or a lone pair) control the shape of a molecule, for each number of electron pairs we can define the geometric optimum position that maximizes the distance between the electron pairs (Table 2.1). The number of electron pairs surrounding an atom, both bonding and non-bonding, is called its steric number.

| Number of central atom electron pairs | Bonding pairs | Non-bonding pairs | Shape | Example |
|---|---|---|---|---|
| 2 | 2 | 0 | Linear | $BeCl_2$ |
| 3 | 3 | 0 | Triangular | $BF_3$ |
| 3 | 2 | 1 | Bent | $SnCl_2$ |
| 4 | 4 | 0 | Tetrahedral | $CCl_4$ |
| 4 | 3 | 1 | Pyramidal | $NH_3$ |
| 4 | 2 | 2 | Bent | $H_2O$ |
| 5 | 5 | 0 | Trigonal bipyramidal (tbp) | $PF_5$ |
| 5 | 4 | 1 | Pseudo-tdp | $BF_4^-$ |
| 5 | 3 | 2 | T-shaped | $BrF_3$ |
| 5 | 2 | 3 | Linear | $XeF_2$ |
| 6 | 6 | 0 | Octahedral | $SF_6$, $PF_6^-$ |
| 6 | 5 | 1 | Square pyramidal | $IF_5$ |
| 6 | 4 | 2 | Square planar | $XeF_4$, $IF_4^-$ |

**Table 2.1: Shapes of molecules and ions.**

The steps for defining the molecular shape are as follows:

1. Draw a simple Lewis structure including single, double, and triple bonds where appropriate.
2. Count the number of electrons on the central atom assuming it is neutral.
3. Add one electron for each σ-bond.
4. Subtract one electron for each π-bond.
5. Subtract one electron for each positive (+) charge.

6. Add one electron for each negative (-) charge.
7. Divide the number of electrons by two to give the number of electron pairs.
8. Use the list in Table 2.1 to predict the structure of the molecule.

For example, the shape of $NH_4^+$ can be predicted as follows:

1. Nitrogen has 5 valence electrons.
2. Add 4 electrons for the four σ-bonds: 5 + 4 = 9.
3. Subtract one electron for the positive (+) charge: 9 1 = 8.
4. Divide the number of electrons by two to give the number of electron pairs: 8/2 = 4.
5. Four bonding pairs and no lone pairs = tetrahedral geometry.

## Lone pairs versus bonding pairs

The prediction of the detailed molecular structure (including bond angles) is not as simple as shown in Table 2.1. In molecules with either lone pair electrons or multiple (double or triple) bonds the angles about the central atom are distorted due to the increased electron repulsion (Figure 2.3).

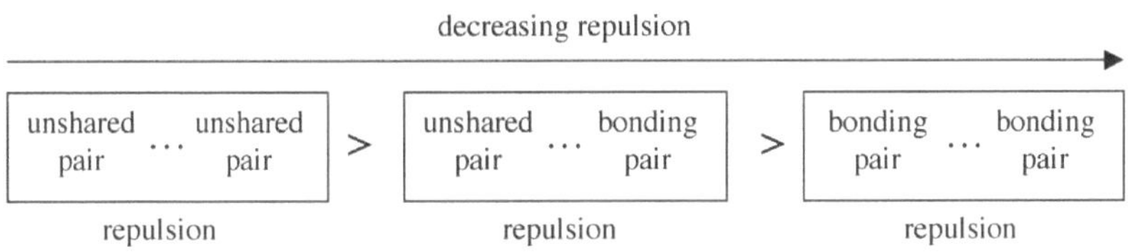


**Figure 2.3: The order of decreasing repulsion between non-bonding (unshared) and bonding electron pairs.**

The differences in repulsion caused by a lone pair or a bonding pair may be rationalized in a simple manner by a lone pair taking up more space than a bonding pair (Figure 2.4).

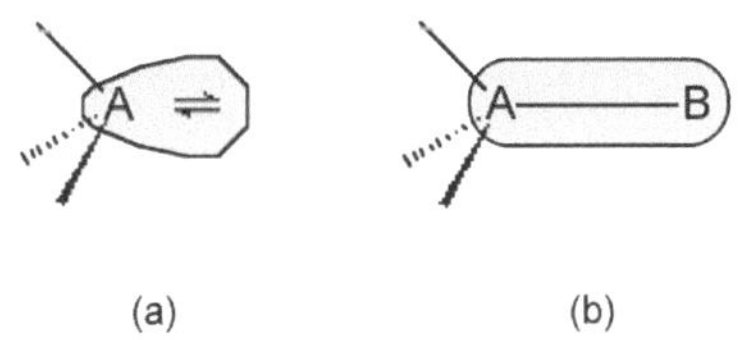


**Figure 2.4: Representation of the relative space taken up by (a) a lone pair of electrons and (b) a bonding pair of electrons.**

Water is one of the classic cases in considering the issue of non-bonding (unshared) electron pairs:

1. Oxygen has 6 valence electrons.
2. Add 2 electrons for the two σ-bonds: 6 + 2 = 8.
3. Divide the number of electrons by two to give the number of electron pairs: 8/2 = 4.
4. Two bonding pairs and two lone pairs = tetrahedral geometry.

From this an idealized tetrahedral geometry would give the H-O-H angle as 109.5°; however, from Figure 2.3, we know that the lone pair···lone pair repulsion is greater than the lone pair···bonding pair repulsion which is greater than the bonding pair···bonding pair repulsion, and thus, the H-O-H angle should be decreased from the ideal tetrahedral. The experimentally determined H-O-H angle in water is in fact 104.5°.

Ethylene is a good case in considering the issue of multiple bonds. Ethylene contains both σ-bond and π-bond between the carbon atoms. This combination can be thought of as a super bond, and as such its effect is similar to a lone pair:

1. Carbon has 4 valence electrons.
2. Add 3 electrons for the two σ-bonds: 4 + 3 = 7.
3. Subtract one electron for each π-bond: 7 1 = 6.
4. Divide the number of electrons by two to give the number of electron pairs: 6/2 = 3.
5. Three bonding pairs and no lone pairs = triangular geometry.

From this an idealized tetrahedral geometry would give the H-C-H angle as 120°, however, the π-bond repulsion is greater than the σ-bond repulsion, and thus, the H-C-H angle should be decreased from the ideal tetrahedral. The experimentally determined H-C-H angle in ethylene is in fact 118.3°.

## Resonance structures

If a molecule or ion has two or more resonance forms, it is necessary to consider each form before angles are predicted. For example, the carbonate anion, $CO_3^{2-}$, can be drawn as a single structure from which it would be predicted that two groups of O-C-O angles would result (Figure 2.5); however, $CO_3^{2-}$ should actually be drawn in each of its resonance forms as in Figure 2.6. From Figure 2.6 it is clear that the real structure will be an average of the three resonance forms, and hence there will be a single O-C-O angle = 120° (Figure 2.7).

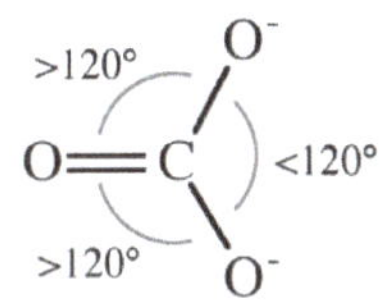


**Figure 2.5: VSEPR predicted structure of a single resonance form of $CO_3^{2-}$.**

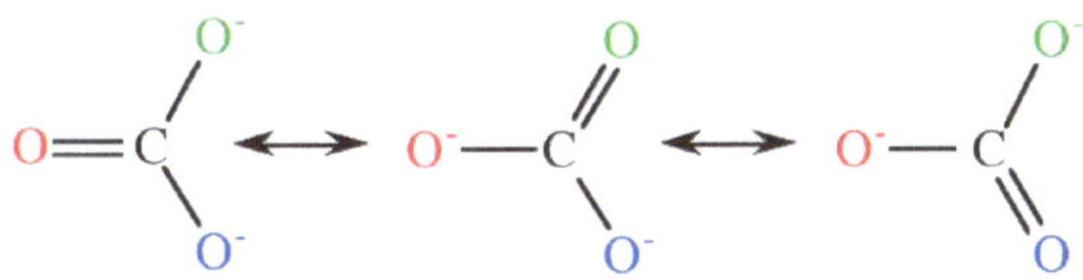

**Figure 2.6: Resonance forms of $CO_3^{2-}$.**

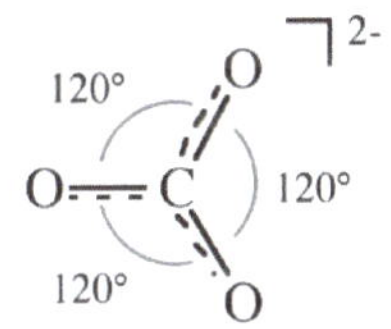


**Figure 2.7: Structure of $CO_3^{2-}$.**

## Atom electronegativities

In an A-X bond where the atom electronegativities are very different the bonding pair is assumed to occupy less space than in bond between two atoms of similar electronegativities. As the bonding pair occupies less space it will repel neighboring electron pairs less.

For example, based on the above, a comparison of the tetrahedral compounds, $H_2O$ and $F_2O$ would suggest that the F-O-F angle be smaller than the H-O-H angle since fluorine has a higher electronegativity than hydrogen (4.0 and 2.1, see Table 1.5). This is indeed observed (Figure 2.8).

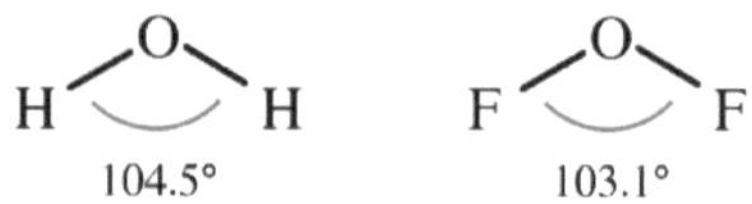

**Figure 2.8: Structures of $H_2O$ and $F_2O$.**

In many cases, inter-ligand steric interactions can also be used to explain the difference in angle. For example, while fluorine is more electronegative than chlorine (4.0 versus 3.0, see Table 1.5) it is also significantly smaller (Table 2.2). Thus, a larger Cl-X-Cl angle than a F-X-F angle in the homologous compound, can be attributed to a greater steric interaction rather than difference in electronegativities.

| Ligand | Covalent radius (Å) | Van der Waal radius (Å) |
|---|---|---|
| F | 0.57 | 1.35 |
| Cl | 1.02 | 1.80 |

**Table 2.2: Comparison of size between fluoride and chloride ligands.**

## Bent's rule

There is the potential for more than one isomer for molecules that adopt structures in which there is a symmetry difference between at least two of the ligand positions. For example, a trigional bipyramidal compound of the formula $PXY_4$ has two possible structures. One where the X occupies an axial position (Figure 2.9a) and the other where it occupies an equatorial position (Figure 2.9b).

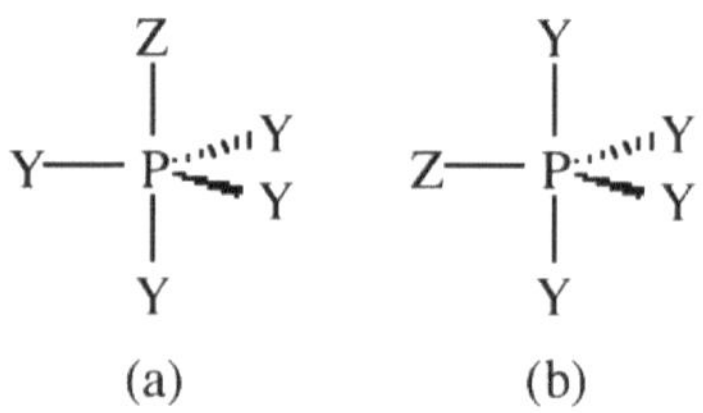

**Figure 2.9: The two possible structures of $PXY_4$.**

Through the consideration of structures Henry Bent (Figure 2.10) suggested a rule: *More electronegative substituents prefer hybrid orbitals with less s character, and conversely, more electropositive substituents prefer hybrid orbitals with greater s character.*

For example, in $PFCl_5$ the fluorine is the most electronegative substituent and will therefore occupy the axial (p-character only) position (i.e., Figure 2.9a.

**Figure 2.10: American chemist Henry A. Bent (1927 - 2015).**

# Chapter 3: Crystal Structure

## Introduction

In any sort of discussion of crystalline materials, it is useful to begin with a discussion of crystallography: the study of the formation, structure, and properties of crystals. A crystal structure is defined as the particular repeating arrangement of atoms (molecules or ions) throughout a crystal. Structure refers to the internal arrangement of particles and not the external appearance of the crystal. However, these are not entirely independent since the external appearance of a crystal is often related to the internal arrangement. For example, crystals of cubic rock salt (NaCl) are physically cubic in appearance. Only a few of the possible crystal structures are of concern with respect to simple inorganic salts and these will be discussed in detail, however, it is important to understand the nomenclature of crystallography.

## Crystallography

### Bravais lattice

The Bravais lattice is the basic building block from which all crystals can be constructed. The concept originated as a topological problem of finding the number of different ways to arrange points in space where each point would have an identical atmosphere. That is each point would be surrounded by an identical set of points as any other point, so that all points would be indistinguishable from each other. Mathematician Auguste Bravais (Figure 3.1) discovered that there were 14 different collections of the groups of points, which are known as Bravais lattices.

**Figure 3.1: French physicist Auguste Bravais (1811 - 1863).**

These lattices fall into seven different "crystal systems, as differentiated by the relationship between the angles between sides of the unit cell and the distance between points in the unit cell.

- The unit cell is the smallest group of atoms, ions or molecules that, when repeated at regular intervals in three dimensions, will produce the lattice of a crystal system.
- The lattice parameter is the length between two points on the corners of a unit cell.
- Each of the various lattice parameters are designated by the letters *a*, *b*, and *c*. If two sides are equal, such as in a tetragonal lattice, then the lengths of the two lattice parameters are designated *a* and *c*, with *b* omitted.
- The angles are designated by the Greek letters *α*, *β*, and *γ*, such that an angle with a specific Greek letter is not subtended by the axis with its Roman equivalent. For example, *α* is the included angle between the *b* and *c* axis.

Table 3.1 shows the various crystal systems, while Figure 3.2 shows the 14 Bravais lattices. It is important to distinguish the characteristics of each of the individual systems. An example of a material that takes on each of the Bravais lattices is shown in Table 3.2.

| System | Axial lengths and angles | Unit cell geometry |
|---|---|---|
| Cubic | $a = b = c, \alpha = \beta = \gamma = 90°$ | 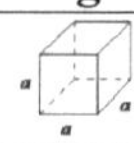 |
| Tetragonal | $a = b \neq c, \alpha = \beta = \gamma = 90°$ | 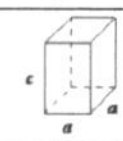 |
| Orthorhombic | $a \neq b \neq c, \alpha = \beta = \gamma = 90°$ | 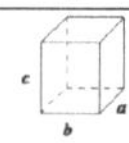 |
| Rhombohedral | $a = b = c, \alpha = \beta = \gamma \neq 90°$ | 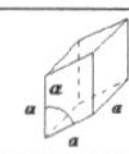 |
| Hexagonal | $a = b \neq c, \alpha = \beta = 90°, \gamma = 120°$ | 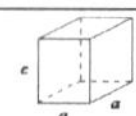 |
| Monoclinic | $a \neq b \neq c, \alpha = \gamma = 90°, \beta \neq 90°$ | 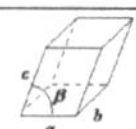 |
| Triclinic | $a \neq b \neq c, \alpha \neq \beta \neq \gamma$ | 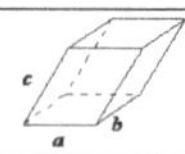 |

**Table 3.1: Geometrical characteristics of the seven crystal systems.**

| Crystal system | Example |
|---|---|
| triclinic | $K_2S_2O_8$ |
| monoclinic | $As_4S_4$, $KNO_2$ |
| rhombohedral | Hg, Sb |
| hexagonal | Zn, Co, NiAs |
| orthorhombic | Ga, $Fe_3C$ |
| tetragonal | In, $TiO_2$ |
| cubic | Au, Si, NaCl |

**Table 3.2: Examples of elements and compounds that adopt each of the crystal systems.**

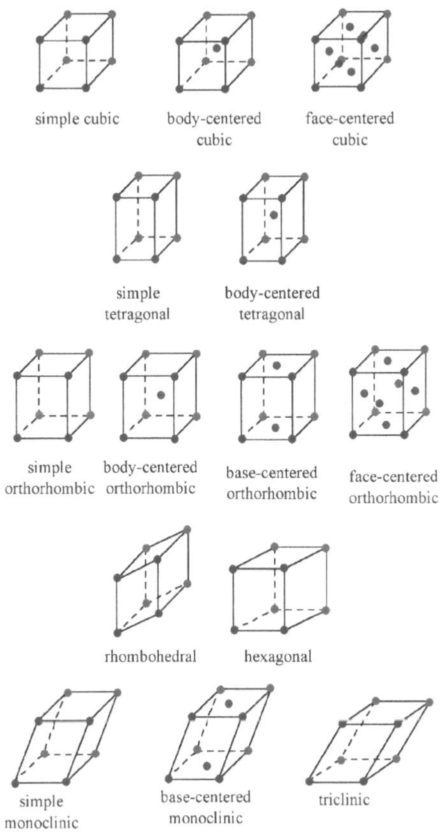


**Figure 3.2: Bravais lattices.**

The cubic lattice is the most symmetrical of the systems. All the angles are equal to 90°, and all the sides are of the same length (i.e., $a = b = c$). Only the length of one of the sides ($a$) is required to describe this system completely. In addition to simple cubic, the cubic lattice also includes *body-centered cubic* and *face-centered cubic* (Figure 3.2). Body-centered cubic results from the presence of an atom (or ion) in the center of a cube, in addition to the atoms (ions) positioned at the vertices of the cube. In a similar manner, a face-centered cubic requires, in addition to the atoms (ions) positioned at the vertices of the cube, the presence of atoms (ions) in the center of each of the cubes face.

The tetragonal lattice has all of its angles equal to 90° and has two out of the three sides of equal length ($a = b$). The system also includes body-centered tetragonal (Figure 3.2).

In an orthorhombic lattice all of the angles are equal to 90°, while all of its sides are of unequal length. The system needs only to be described by three lattice parameters. This system also includes body-centered orthorhombic, base-centered orthorhombic, and face-centered orthorhombic (Figure 3.2). A base-centered lattice has, in addition to the atoms (ions) positioned at the vertices of the orthorhombic lattice, atoms (ions) positioned on just two opposing faces.

The rhombohedral lattice is also known as trigonal, and has no angles equal to 90°, but all sides are of equal length ($a = b = c$), thus requiring only by one lattice parameter, and all three angles are equal ($\alpha = \beta = \gamma$).

A hexagonal crystal structure has two angles equal to 90°, with the other angle ($\gamma$) equal to 120°. For this to happen, the two sides surrounding the 120° angle must be equal ($a = b$), while the third side ($c$) is at 90° to the other sides and can be of any length.

The monoclinic lattice has no sides of equal length, but two of the angles are equal to 90°, with the other angle (usually defined as $\beta$) being something other than 90°. It is a tilted parallelogram prism with rectangular bases. This system also includes base-centered monoclinic (Figure 3.2).

In the triclinic lattice none of the sides of the unit cell are equal, and none of the angles within the unit cell are equal to 90°. The triclinic lattice is chosen such that all the internal angles are either acute or obtuse. This crystal system has the lowest symmetry and must be described by 3 lattice parameters (*a*, *b*, and *c*) and the 3 angles ($\alpha$, $\beta$, and $\gamma$).

## Atom positions, crystal directions and Miller indices

*Atom positions and crystal axes*

The structure of a crystal is defined with respect to a unit cell. As the entire crystal consists of repeating unit cells, this definition is sufficient to represent the entire crystal. Within the unit cell, the atomic arrangement is expressed using coordinates. There are two systems of coordinates commonly in use, which can cause some confusion. Both use a corner of the unit cell as their origin.

The first, less-commonly seen system is that of Cartesian or orthogonal coordinates (X, Y, Z). These usually have the units of Angstroms and relate to the distance in each direction between the origin of the cell and the atom. These coordinates may be manipulated in the same fashion are used with two- or three-dimensional graphs. It is very simple, therefore, to calculate inter-atomic distances and angles given the Cartesian coordinates of the atoms. Unfortunately, the repeating nature of a crystal cannot be expressed easily using such coordinates. For example, consider a cubic cell of dimension 3.52 Å. Pretend that this cell contains an atom that has the coordinates (1.5, 2.1, 2.4). That is, the atom is 1.5 Å away from the origin in the *x* direction (which coincides with the *a*-axis), 2.1 Å in the *y* (which coincides with the *b* cell axis) and 2.4 Å in the *z* (which coincides with the *c* cell axis). There will be an equivalent atom in the next unit cell along the *x*-direction, which will have the coordinates (1.5 + 3.52, 2.1, 2.4) or (5.02, 2.1, 2.4). This was a rather simple calculation, as the cell has very high symmetry and so the cell axes, *a*, *b* and *c*, coincide with the Cartesian axes, X, Y and Z. However, consider lower symmetry cells such as triclinic or monoclinic in which the cell axes are not mutually orthogonal. In such cases, expressing the repeating nature of the crystal is much more difficult to accomplish.

Accordingly, atomic coordinates are usually expressed in terms of fractional coordinates, ($x$, $y$, $z$). This coordinate system is coincident with the cell axes ($a$, $b$, $c$) and relates to the position of the atom in terms of the fraction along each axis. Consider the atom in the cubic cell discussion above. The atom was 1.5 Å in the *a*-direction away from the origin. As the *a*-axis is 3.52 Å long, the atom is ($^{1.5}/_{3.52}$) or 0.43 of the axis, away from the origin. Similarly, it is ($^{2.1}/_{3.52}$) or 0.60 of the *b*-axis and ($^{2.4}/_{3.5}$) or 0.68 of the *c*-axis. The fractional coordinates of this atom are, therefore, (0.43, 0.60, 0.68). The coordinates of the equivalent atom in the next cell over in the *a*-direction; however, are easily calculated as this atom is simply 1 unit cell away in *a*. Thus, all one has to do is add 1 to the *x* coordinate: (1.43, 0.60, 0.68). Such transformations can be performed regardless of the shape of the unit cell. Fractional coordinates, therefore, are used to retain and manipulate crystal information.

*Crystal directions*

The designation of the individual vectors within any given crystal lattice is accomplished by the use of whole number multipliers of the lattice parameter of the point at which the vector exits the unit cell. The vector is indicated by the notation [*hkl*], where *h*, *k*, and *l* are reciprocals of the point at which the vector exits the unit cell. The origination of all vectors is assumed defined as [000].

For example, the direction along the *a*-axis according to this scheme would be [100] because this has a component only in the *a*-direction and no component along either the *b* or *c* axial direction. A vector diagonally along the face defined by the *a* and *b* axis would be [110], while going from one corner of the unit cell to the opposite corner would be in the [111] direction. Figure 3.3 shows some examples of the various directions in the unit cell. The crystal direction notation is made up of the lowest combination of integers and represents unit distances rather than actual distances. A [222] direction is identical to a [111], so [111] is used. Fractions are not used. For example, a vector that intercepts the center of the top face of the unit cell has the coordinates $x = {}^{1}/_{2}$, $y = {}^{1}/_{2}$, $z = 1$. All have to be inversed to convert to the lowest combination of integers (whole numbers); i.e., [221] in Figure 3.3. Finally, all parallel vectors have the same crystal direction, e.g., the four vertical edges of the cell shown in Figure 3.3 all have the crystal direction [*hkl*] = [001].

Crystal directions may be grouped in families. To avoid confusion there exists a convention in the choice of brackets surrounding the three numbers to differentiate a crystal direction from a family of direction. For a direction, square brackets [*hkl*] are used to indicate an individual direction. Angle brackets <*hkl*> indicate a family of directions. A family of directions includes any directions that are equivalent in length and types of atoms encountered. For example, in a cubic lattice, the [100], [010], and [001] directions all belong to the <100> family of planes because they are equivalent. If the cubic lattice were rotated 90°, the *a*, *b*, and *c* directions would remain indistinguishable, and there would be no way of telling on which crystallographic positions the atoms are situated, so the family of directions is the same. In a hexagonal crystal, however, this is not the case, so the [100] and [010] would both be <100> directions, but the [001] direction would be distinct. Finally, negative directions are identified with a bar over the negative number instead of a minus sign.

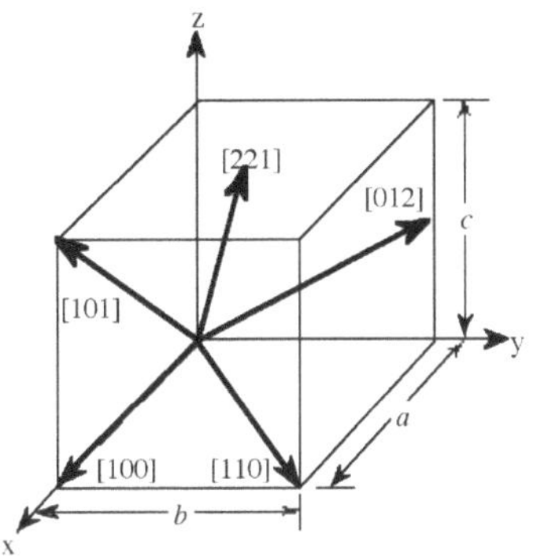


**Figure 3.3: Some common directions in a cubic unit cell.**

*Crystal planes*

Planes in a crystal can be specified using a notation called Miller indices. The Miller index is indicated by the notation [*hkl*] where *h*, *k*, and *l* are reciprocals of the plane with the *x*, *y*, and *z* axes. To obtain the Miller indices of a given plane requires the following steps:

1. The plane in question is placed on a unit cell.
2. Its intercepts with each of the crystal axes are then found.
3. The reciprocal of the intercepts is taken.
4. These are multiplied by a scalar to insure that is in the simple ratio of whole numbers.

For example, the face of a lattice that does not intersect the *y* or *z* axis would be (100), while a plane along the body diagonal would be the (111) plane. An illustration of this along with the (111) and (110) planes is given in Figure 3.4.

As with crystal directions, Miller indices directions may be grouped in families. Individual Miller indices are given in parentheses (*hkl*), while braces {*hkl*} are placed around the indices of a family of planes. For example, (001), (100), and (010) are all in the {100} family of planes, for a cubic lattice.

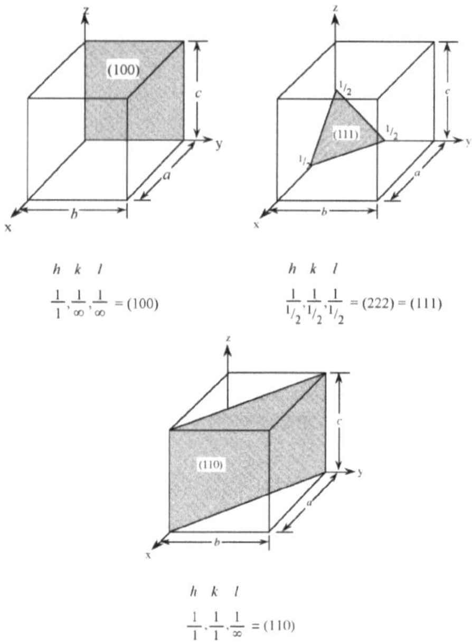


**Figure 3.4: Examples of Miller indices notation for crystal planes.**

## Description of crystal structures

Crystal structures may be described in a number of ways. The most common manner is to refer to the size and shape of the unit cell and the positions of the atoms (or ions) within the cell. However, this information is sometimes insufficient to allow for an understanding of the true structure in three dimensions. Consideration of several unit cells, the arrangement of the atoms with respect to each other, the number of other atoms they in contact with, and the distances to neighboring atoms, often will provide a better understanding. A number of methods are available to describe extended solid-state structures. The most applicable with regard to elemental and compound semiconductor, metals and the majority of insulators is the close packing approach.

*Close packed structures: hexagonal close packing and cubic close packing*

Many crystal structures can be described using the concept of close packing. This concept requires that the atoms (ions) are arranged so as to have the maximum density. In order to understand close packing in three dimensions, the most efficient way for equal sized spheres to be packed in two dimensions must be considered.

The most efficient way for equal sized spheres to be packed in two dimensions is shown in Figure 3.5, in which it can be seen that each sphere (the dark gray shaded sphere) is surrounded by, and is in contact with, six other spheres (the light gray spheres in Figure 3.5). It should be noted that contact with six other spheres the maximum possible is the spheres are the same size, although lower density packing is possible. Close packed layers are formed by repetition to an infinite sheet. Within these close packed layers, three close packed rows are present, shown by the dashed lines in Figure 3.5.

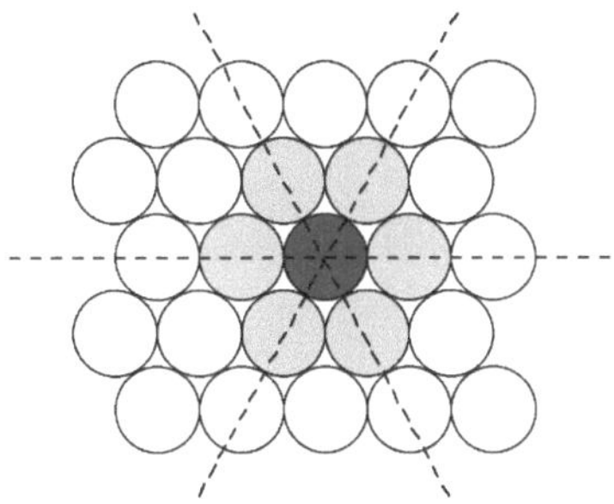

**Figure 3.5: Schematic representation of a close packed layer of equal sized spheres. The close packed rows (directions) are shown by the dashed lines.**

The most efficient way for equal sized spheres to be packed in three dimensions is to stack close packed layers on top of each other to give a close packed structure. There are two simple ways in which this can be done, resulting in either a hexagonal or cubic close packed structure.

If two close packed layers A and B are placed in contact with each other so as to maximize the density, then the spheres of layer B will rest in the hollow (vacancy) between three of the spheres in layer A. This is demonstrated in Figure 3.6. Atoms in the second layer, B (shaded light gray), may occupy one of two possible positions (Figure 3.6a or 3.6b) but neither both together or a mixture of each. If a third layer is placed on top of layer B such that it exactly covers layer A, subsequent placement of layers will result in the sequence ···ABABAB···. This is known as hexagonal close packing or *hcp*.

The hexagonal close packed cell is a derivative of the hexagonal Bravais lattice system (Figure 3.2) with the addition of an atom inside the unit cell at the coordinates ($^1/_3$, $^2/_3$, $^1/_2$). The basal plane of the unit cell coincides with the close packed layers (Figure 3.7). In other words, the close packed layer makes-up the {001} family of crystal planes.

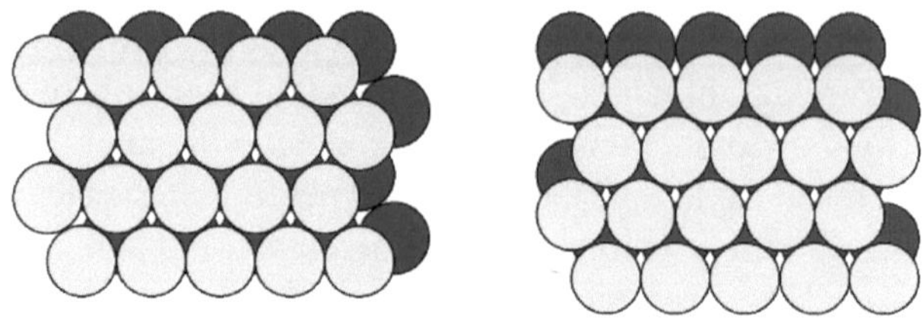


**Figure 3.6: Schematic representation of two close packed layers arranged in A (dark grey) and B (light grey) positions. The alternative stacking of the B layer is shown in (a) and (b).**

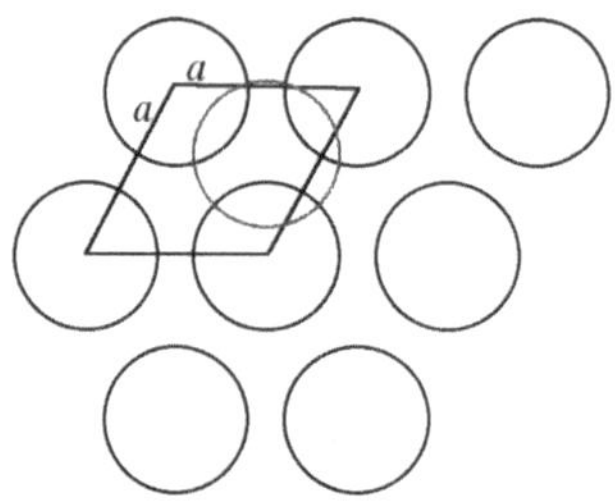


**Figure 3.7: A schematic projection of the basal plane of the hcp unit cell on the close packed layers.**

The packing fraction in a hexagonal close packed cell is 74.05%; that is 74.05% of the total volume is occupied. The packing fraction or density is derived by assuming that each atom is a hard sphere in contact with its nearest neighbors. Determination of the packing fraction is accomplished by calculating the number of whole spheres per unit cell (2 in *hcp*), the volume occupied by these spheres, and a comparison with the total volume of a unit cell. The number gives an idea of how open or filled a structure is. By comparison, the packing fraction for body-centered cubic (Figure 3.2) is 68% and for diamond cubic (an important semiconductor structure to be described later) is it 34%.

In a similar manner to the generation of the hexagonal close packed structure, two close packed layers are stacked (Figure 3.5); however, the third layer (C) is placed such that it does not exactly cover layer A, while sitting in a set of troughs in layer B (Figure 3.8), then upon repetition the packing sequence will be ···ABCABCABC···. This is known as cubic close packing or *ccp*.

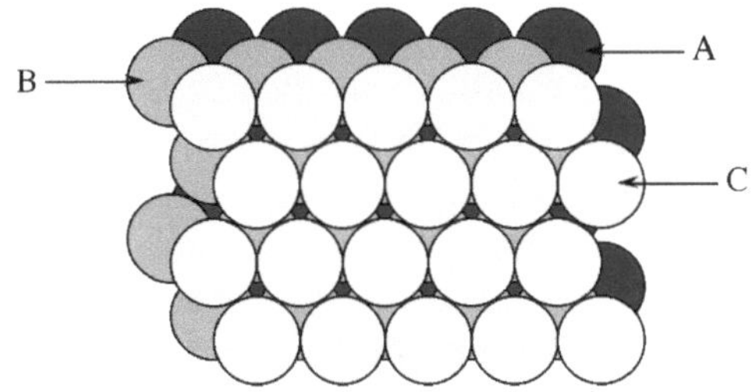


**Figure 3.8: Schematic representation of the three close packed layers in a cubic close packed arrangement: A (dark grey), B (medium grey), and C (light grey).**

The unit cell of cubic close packed structure is actually that of a face-centered cubic (*fcc*) Bravais lattice. In the *fcc* lattice the close packed layers constitute the {111} planes. As with the *hcp* lattice packing fraction in a cubic close packed (*fcc*) cell is 74.05%. Since face centered cubic or *fcc* is more commonly used in preference to cubic close packed (*ccp*) in describing the structures, the former will be used throughout this text.

*Coordination number*

The coordination number of an atom or ion within an extended structure is defined as the number of nearest neighbor atoms (ions of opposite charge) that are in contact with it. A slightly different definition is often used for atoms within individual molecules: the number of donor atoms associated with the central atom or ion. However, this distinction is rather artificial, and both can be employed.

The coordination numbers for metal atoms in a molecule or complex are commonly 4, 5, and 6, but all values from 2 to 9 are known and a few examples of higher coordination numbers have been reported. In contrast, common coordination numbers in the solid state are 3, 4, 6, 8, and 12. For example, the atom in the center of body-centered cubic lattice has a coordination number of 8, because it touches the eight atoms at the corners of the unit cell, while an atom in a simple cubic structure would have a coordination number of 6. In both *fcc* and *hcp* lattices each of the atoms have a coordination number of 12.

*Octahedral and tetrahedral vacancies*

As was mentioned above, the packing fraction in both *fcc* and *hcp* cells is 74.05%, leaving 25.95% of the volume unfilled. The unfilled lattice sites (interstices) between the atoms in a cell are called interstitial sites or vacancies.

The shape and relative size of these sites is important in controlling the position of additional atoms. In both *fcc* and hcp cells most of the space within these atoms lies within two different sites known as octahedral sites and tetrahedral sites. The difference between the two lies in their coordination number, or the number of atoms surrounding each site. Tetrahedral sites (vacancies) are surrounded by four atoms arranged at the corners of a tetrahedron. Similarly, octahedral sites are surrounded by six atoms which make-up the apices of an octahedron. For a given close packed lattice an octahedral vacancy will be larger than a tetrahedral vacancy.

Within a face centered cubic lattice, the eight tetrahedral sites are positioned within the cell, at the general fractional coordinate of ($^n/_4$,$^n/_4$,$^n/_4$) where n = 1 or 3, e.g., ($^1/_4$,$^1/_4$,$^1/_4$,), ($^1/_4$,$^1/_4$,$^3/_4$), etc. The octahedral sites are located at the center of the unit cell ($^1/_2$,$^1/_2$,$^1/_2$), as well as at each of the edges of the cell, e.g., ($^1/_2$,0,0). In the hexagonal close packed system, the tetrahedral sites are at (0,0,$^3/_8$) and ($^1/_3$,$^2/_3$,$^7/_8$), and the octahedral sites are at ($^1/_3$,$^1/_3$,$^1/_4$) and all symmetry equivalent positions.

## Important structure types

The majority of crystalline materials do not have a structure that fits into the one atom per site simple Bravais lattice. A number of other important crystal structures are found, however, only a few of these crystal structures are those of which occur for the elemental and compound semiconductors and the majority of these are derived from *fcc* or *hcp* lattices. Each structural type is generally defined by an archetype, a material (often a naturally occurring mineral), which has the structure in question and to which all the similar materials are related. With regard to commonly used elemental and compound semiconductors the important structures are diamond, zinc blende, wurtzite, and to a lesser extent chalcopyrite; however, rock salt, β-tin, cinnabar and caesium chloride are observed as high pressure or high temperature phases and are therefore also discussed. The following provides a summary of these structures.

### *Diamond cubic*

The diamond cubic structure consists of two interpenetrating face-centered cubic lattices, with one off set $^1/_4$ of a cube along the cube diagonal. It may also be described as face centered cubic lattice in which half of the tetrahedral sites are filled while all the octahedral sites remain vacant. The diamond cubic unit cell is shown in Figure 3.9. Each of the atoms (e.g., C) is four-coordinate, and the shortest interatomic distance (C-C) may be determined from the unit cell parameter ($a$):

$$\text{C-C} = \frac{a\sqrt{3}}{4} \approx 0.422\,a$$

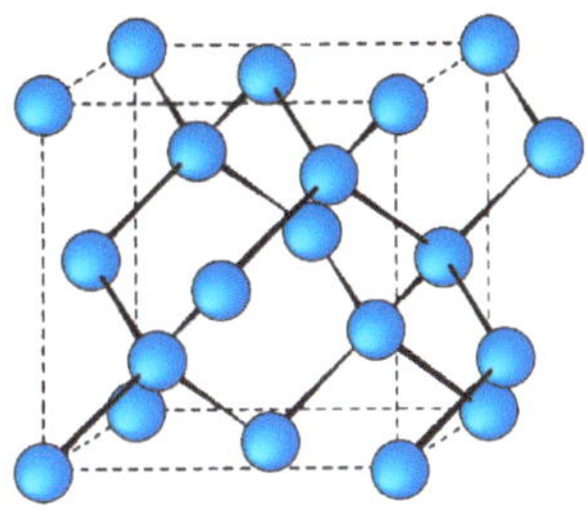

**Figure 3.9: Unit cell structure of a diamond cubic lattice showing the two-interpenetrating face-centered cubic lattices.**

*Zinc blende*

This is a binary phase (ME) and is named after its archetype, a common mineral form of zinc sulfide (ZnS). As with the diamond lattice, zinc blende consists of the two interpenetrating *fcc* lattices. However, in zinc blende one lattice consists of one of the types of atoms (Zn in ZnS), and the other lattice is of the second type of atom (S in ZnS). It may also be described as face centered cubic lattice of S atoms in which half of the tetrahedral sites are filled with Zn atoms. All the atoms in a zinc blende structure are 4-coordinate. The zinc blende unit cell is shown in Figure 3.10.

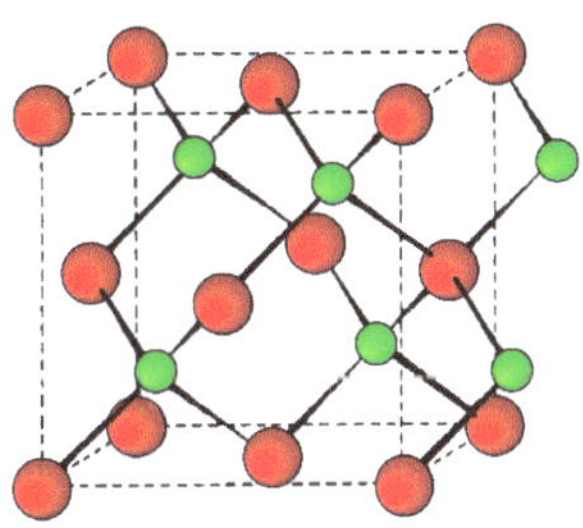

**Figure 3.10: Unit cell structure of a zinc blende (ZnS) lattice. Zinc atoms are shown in green (small), sulfur atoms shown in red (large), and the dashed lines show the unit cell.**

A number of inter-atomic distances may be calculated for any material with a zinc blende unit cell using the lattice parameter ($a$):

$$\text{Zn-S} = \frac{a\sqrt{3}}{4} \approx 0.422\,a$$

$$\text{Zn-Zn} = \text{S-S} = \frac{a}{\sqrt{2}} \approx 0.707\,a$$

*Chalcopyrite*

The mineral chalcopyrite $CuFeS_2$ is the archetype of this structure. The structure is tetragonal ($a = b \neq c$, $\alpha = \beta = \gamma = 90°$), and is essentially a superlattice on that of zinc blende. Thus, is easiest to imagine that the chalcopyrite lattice is made-up of a lattice of sulfur atoms in which the tetrahedral sites are filled in layers, ···FeCuCuFe···, etc. (Figure 3.11). In such an idealized structure $c = 2a$, however, this is not true of all materials with chalcopyrite structures.

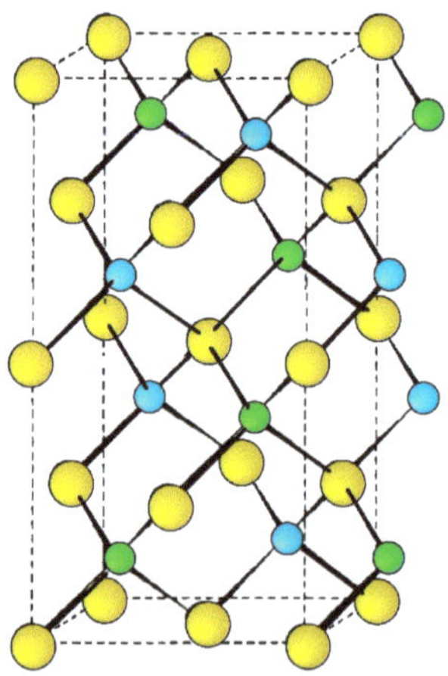

**Figure 3.11: Unit cell structure of a chalcopyrite lattice. Copper atoms are shown in blue; iron atoms are shown in green; and sulfur atoms are shown in yellow. The dashed lines show the unit cell.**

*Rock salt*

As its name implies the archetypal rock salt structure is NaCl (table salt). In common with the zinc blende structure, rock salt consists of two interpenetrating face-centered cubic lattices. However, the second lattice is off set $^1/_2a$

along the unit cell axis. It may also be described as face centered cubic lattice in which all of the octahedral sites are filled, while all the tetrahedral sites remain vacant, and thus each of the atoms in the rock salt structure are 6-coordinate. The rock salt unit cell is shown in Figure 3.12.

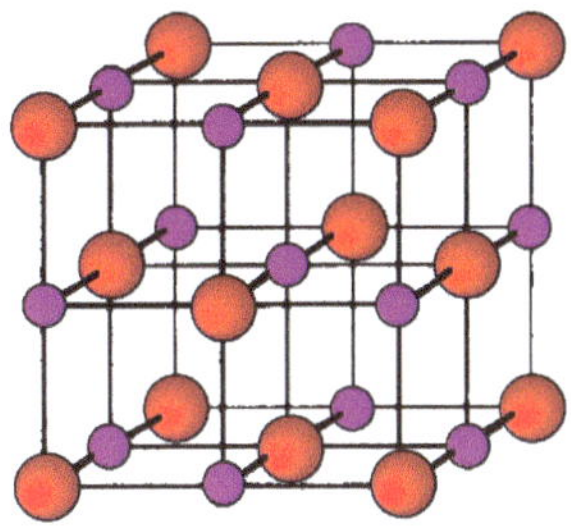

**Figure 3.12: Unit cell structure of a rock salt lattice. Sodium ions are shown in purple (small spheres) and chloride ions are shown in red (large spheres).**

A number of inter-atomic distances may be calculated for any material with a rock salt structure using the lattice parameter (*a*):

$$\text{Na-Cl} = \frac{a}{2} \approx 0.5\,a$$

$$\text{Na-Na} = \text{Cl-Cl} = \frac{a}{\sqrt{2}} \approx 0.707\,a$$

*Cinnabar*

Cinnabar, named after the archetype mercury sulfide, HgS, is a distorted rock salt structure in which the resulting cell is rhombohedral (trigonal) with each atom having a coordination number of six.

*Wurtzite*

This is a hexagonal form of the zinc sulfide. It is identical in the number of and types of atoms, but it is built from two interpenetrating hcp lattices as opposed to the *fcc* lattices in zinc blende. As with zinc blende all the atoms in a wurtzite structure are 4-coordinate. The wurtzite unit cell is shown in Figure 3.13. A number of inter atomic distances may be calculated for any material with a wurtzite cell using the lattice parameter (*a*):

$$\text{Zn-S} = a\sqrt{3/8} = 0.612\,a = \frac{3c}{8} = 0.375\,c$$

$$\text{Zn-Zn} = \text{S-S} = a = 1.632\,c$$

However, it should be noted that these formulae do not necessarily apply when the ratio *a*/*c* is different from the ideal value of 1.632.

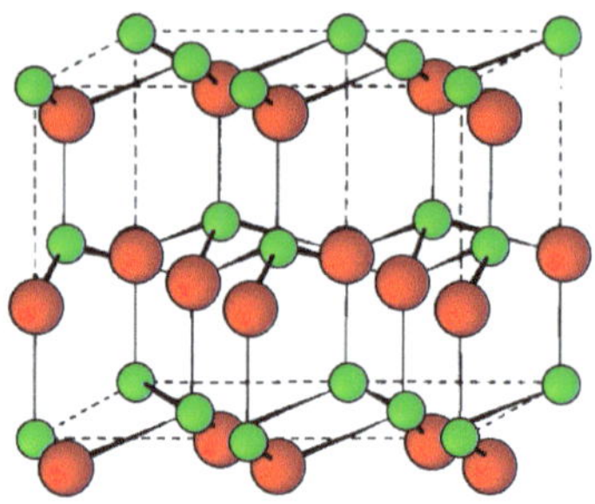

**Figure 3.13: Unit cell structure of a wurtzite lattice. Zinc atoms are shown in green (small spheres), sulfur atoms shown in red (large spheres), and the dashed lines show the unit cell.**

### *Cesium chloride*

The cesium chloride structure is found in materials with large cations and relatively small anions. It has a simple (primitive) cubic cell (Figure 3.2) with a chloride ion at the corners of the cube and the cesium ion at the body center. The coordination numbers of both $Cs^+$ and $Cl^-$, with the inner atomic distances determined from the cell lattice constant (*a*):

$$\text{Cs-Cl} = \frac{a\sqrt{3}}{2} \approx 0.866\,a$$

$$\text{Cs-Cs} = \text{Cl-Cl} = a$$

### *β-Tin*

The room temperature allotrope of tin is β-tin or white tin. It has a tetragonal structure, in which each tin atom has four nearest neighbors (Sn-Sn = 3.016 Å) arranged in a very flattened tetrahedron, and two next nearest neighbors

(Sn-Sn = 3.175 Å). The overall structure of β-tin consists of fused hexagons, each being linked to its neighbor via a four-membered $Sn_4$ ring.

# Defects in crystalline solids

Up to this point we have only been concerned with ideal structures for crystalline solids in which each atom occupies a designated point in the crystal lattice. Unfortunately, defects ordinarily exist in equilibrium between the crystal lattice and its environment (Figure 3.14). These defects are of two general types: point defects and extended defects. As their names imply, point defects are associated with a single crystal lattice site, while extended defects occur over a greater range.

## Point defects: too many or too few or just plain wrong

Point defects have a significant effect on the properties of a semiconductor, so it is important to understand the classes of point defects and the characteristics of each type. Figure 3.14 summarizes various classes of native point defects; however, they may be divided into two general classes; defects with the wrong number of atoms (deficiency or surplus) and defects where the identity of the atoms is incorrect.

*Interstitial impurity*

An interstitial impurity occurs when an extra atom is positioned in a lattice site that should be vacant in an ideal structure (Figure 3.14b). Since all the adjacent lattice sites are filled the additional atom will have to squeeze itself into the interstitial site, resulting in distortion of the lattice and alteration in the local behavior of the structure. Small atoms, such as carbon, will prefer to occupy these interstitial sites. Interstitial impurities readily di use through the lattice via interstitial diffusion, which can result in a change of the properties of a material as a function of time. Oxygen impurities in silicon generally are located as interstitials.

*Vacancies*

The converse of an interstitial impurity is when there are not enough atoms in a particular area of the lattice. These are called vacancies. Vacancies exist in any material above absolute zero and increase in concentration with temperature. In the case of compound semiconductors, vacancies can be either cation vacancies (Figure 3.14c) or anion vacancies (Figure 3.14d), depending on what type of atom are missing.

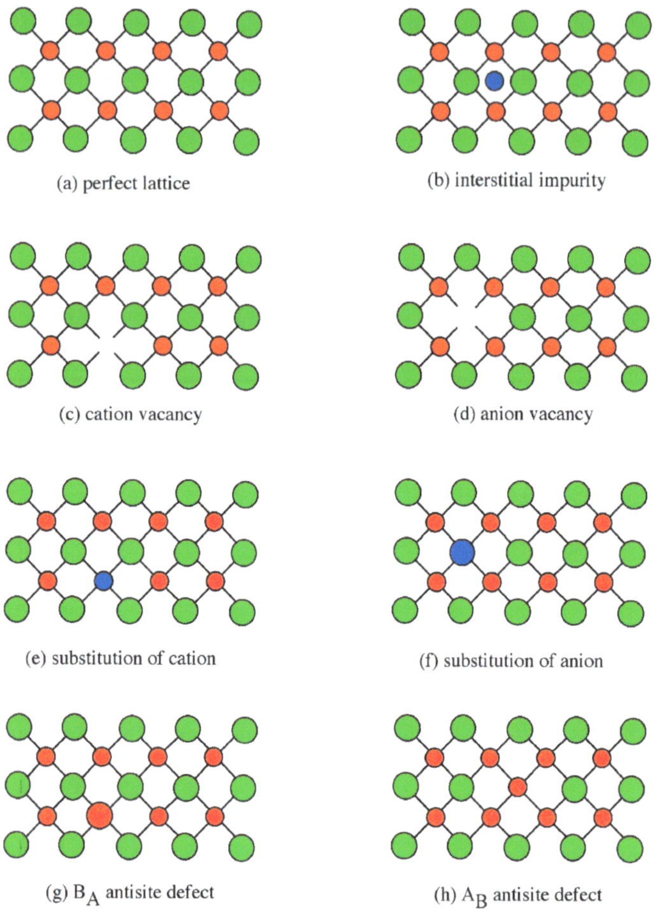


**Figure 3.14: Point defects in a crystal lattice.**

### *Substitution*

Substitution of various atoms into the normal lattice structure is common and used to change the electronic properties of both compound and elemental semiconductors. Any impurity element that is incorporated during crystal growth can occupy a lattice site. Depending on the impurity, substitution defects can greatly distort the lattice and/or alter the electronic structure. In general, cations will try to occupy cation lattice sites (Figure 3.14e), and anion will occupy the anion site (Figure 3.14f). For example, a zinc impurity in GaAs will occupy a gallium site, if possible, while a sulfur, selenium and tellurium atoms would all try to substitute for an arsenic. Some impurities will occupy either site indiscriminately, e.g., Si and Sn occupy both Ga and As sites in GaAs.

*Antisite defects*

Antisite defects are a particular form of substitution defect and are unique to compound semiconductors. An antisite defect occurs when a cation is misplaced on an anion lattice site or vice versa (Figure 3.14g and h). Dependant on the arrangement these are designated as either $A_B$ antisite defects or $B_A$ antisite defects. For example, if an arsenic atom is on a gallium lattice site the defect would be an $As_{Ga}$ defect. Antisite defects involve fitting lattice site atoms of a different size, than the rest of the lattice, and therefore this often results in a localized distortion of the lattice. In addition, cations and anions will have a different number of electrons in their valence shells, so this substitution will alter the local electron concentration and the electronic properties of this area of the semiconductor.

## Extended defects: dislocations in a crystal lattice

Extended defects may be created either during crystal growth or as a consequence of stress in the crystal lattice. The plastic deformation of crystalline solids does not occur such that all bonds along a plane are broken and reformed simultaneously. Instead, the deformation occurs through a dislocation in the crystal lattice. Figure 3.15 shows a schematic representation of a dislocation in a crystal lattice. Two features of this type of dislocation are the presence of an extra crystal plane, and a large void at the dislocation core. Impurities tend to segregate to the dislocation core in order to relieve strain from their presence.

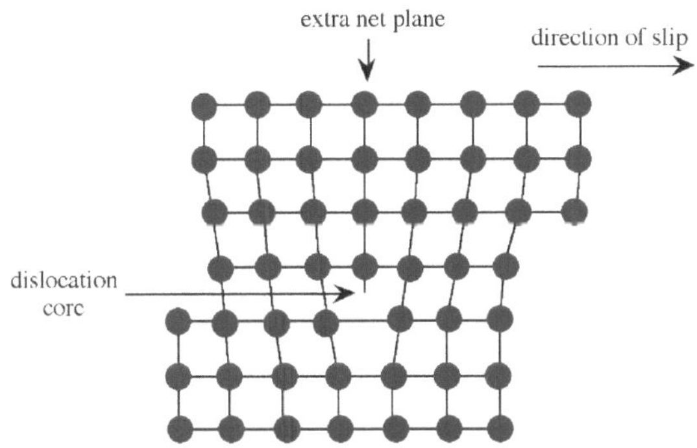


**Figure 3.15: Dislocation in a crystal lattice.**

# Epitaxy

Epitaxy, is a transliteration of two Greek words epi, meaning "upon", and taxis, meaning "ordered". With respect to crystal growth it applies to the process of growing thin crystalline layers on a crystal substrate. In epitaxial growth, there is a precise crystal orientation of the film in relation to the substrate. The growth of epitaxial films can be done by a number of methods including molecular beam epitaxy, atomic layer epitaxy, and chemical vapor deposition, all of which will be described later.

Epitaxy of the same material, such as a gallium arsenide film on a gallium arsenide substrate, is called homoepitaxy, while epitaxy where the film and substrate material are different is called heteroepitaxy. Clearly, in homoepitaxy, the substrate and film will have the identical structure, however, in heteroepitaxy, it is important to employ where possible a substrate with the same structure and similar lattice parameters. For example, zinc selenide (zinc blende, $a$ = 5.668 Å) is readily grown on gallium arsenide (zinc blende, $a$ = 5.653 Å). Alternatively, epitaxial crystal growth can occur where there exists a simple relationship between the structures of the substrate and crystal layer, such as is observed between $Al_2O_3$ (100) on Si (100). Whichever route is chosen a close match in the lattice parameters is required, otherwise, the strains induced by the lattice mismatch results in distortion of the film and formation of dislocations. If the mismatch is significant epitaxial growth is not energetically favorable, causing a textured film or polycrystalline untextured film to be grown. As a general rule of thumb, epitaxy can be achieved if the lattice parameters of the two materials are within about 5% of each other. For good quality epitaxy, this should be less than 1%. The larger the mismatch, the larger the strain in the film. As the film gets thicker and thicker, it will try to relieve the strain in the film, which could include the loss of epitaxy of the growth of dislocations. It is important to note that the <100> directions of a film must be parallel to the <100> direction of the substrate. In some cases, such as Fe on MgO, the [111] direction is parallel to the substrate [100]. The epitaxial relationship is specified by giving first the plane in the film that is parallel to the substrate [100].

# Chapter 4: Stereochemistry

Stereo isomers have the same empirical formula or molecular formula but different structural formulas. A typical example is butane, $C_4H_{10}$, which can have two different possible structures: n-butane (Figure 4.1a) and iso-butane also known as 1-methylpropane (Figure 4.1b).

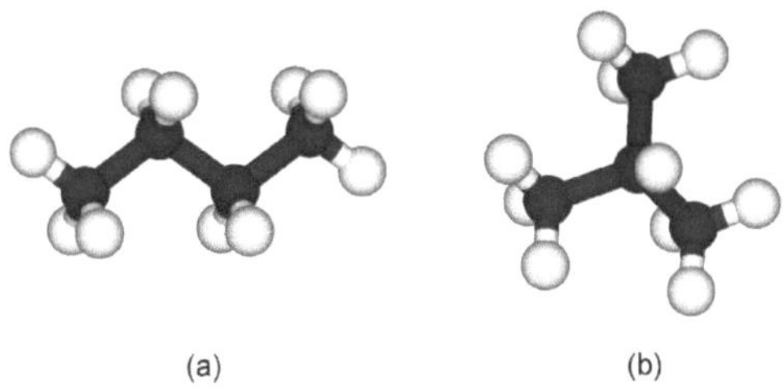


**Figure 4.1: The two stereo isomers of butane: (a) n-butane and (b) iso-butane.**

## Geometric isomers

Geometric isomers have the same empirical formula or molecular formula and also the same structural formula but have a different relative arrangement of the substituent groups. For example, the two geometric isomers of 1,2-dichloroethene (Figure 4.2) have the molecular formula of $C_2H_2Cl_2$, and the same structural formula of Cl(H)C=C(H)Cl, but the relative position of the two chlorine atoms can either be the same side of the C=C double bond (i.e., *cis*, see Figure 4.2a) or on opposite sides of the C=C double bond (i.e., *trans*, see Figure 4.2b). The use of *cis* and *trans* is not limited to organic compounds such as olefins, but can also be used in metal complexes, e.g., Figure 4.3.

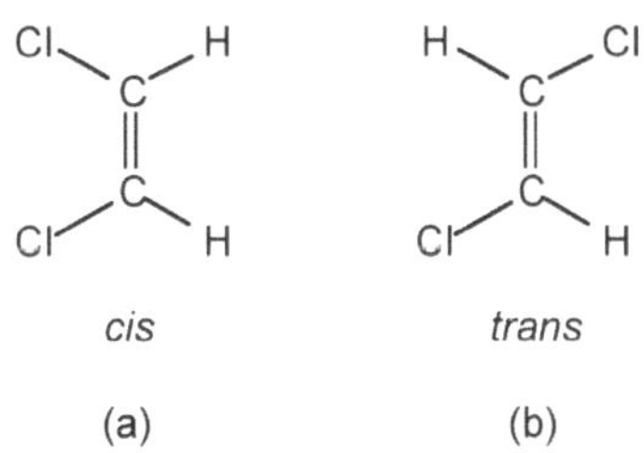


**Figure 4.2: The two geometric isomers of 1,2-dichloroethene.**

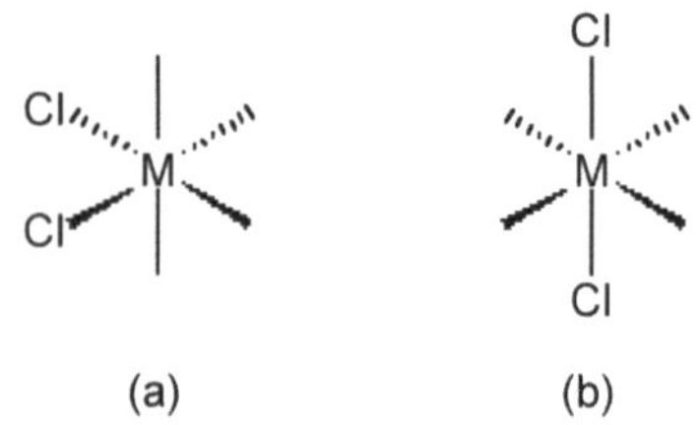


**Figure 4.3: Examples of (a) *cis* and (b) *trans* geometric isomers for metal complexes.**

When it is not possible to describe geometric isomers by the terms *cis* or *trans*, the terms *facial* (*fac*, Figure 4.4a) or *meridinal* (*mer*, Figure 4.4b) are commonly employed.

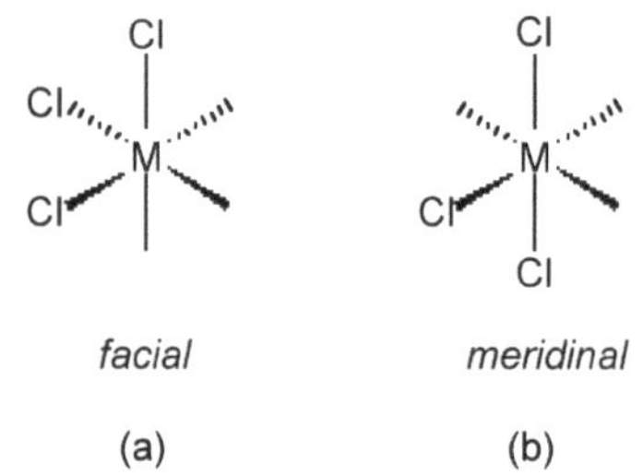


**Figure 4.4: Examples of (a) *fac* and (b) *mer* geometric isomers for metal complexes.**

## Optical isomers

If the mirror object and mirror image of a molecule are not the same (i.e., they are not superimposable) they are known as enantiomers. If they also have optical activity, they are called *chiral* and are described as having chirality. Enantiomers of a particular compound have the same general properties with two exceptions:

1. Their behavior to polarized light.
2. Their reaction with other chiral molecules.

It is possible to determine if a molecule is chiral or not from its symmetry. Chiral molecules will have no symmetry or axis of rotation. The optical activity of a chiral molecule will turn the plane of polarized light either to the right (+) or the left (-). The former is known as dextrorotatory (D), while the latter is known as levorotatory (L).

## Configuration

The configuration of a chiral molecule may be represented in a number of ways. Lactic acid ($C_3H_6O_3$) provides a suitable and simple example. The two optical isomers of lactic acid can be represented in 3-dimensional (3D) form as shown in Figure 4.5; however, it is also possible to draw the same molecules in 2D using the Fischer projection (Figure 4.6). While the Fisher projection does not appear to define the geometry about the central carbon atom, the convention is that the side chains are out of the plane of the page (towards the observer), while the top and bottom groups are into the plane of the page (away from the observer). Thus, Figure 4.6a is a representation of Figure 4.5a, and Figure 4.6b is a representation of Figure 4.5b.

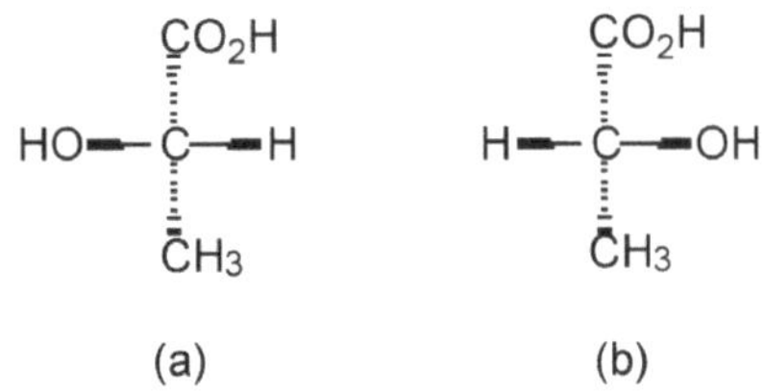


**Figure 4.5: 3D representations of lactic acid.**

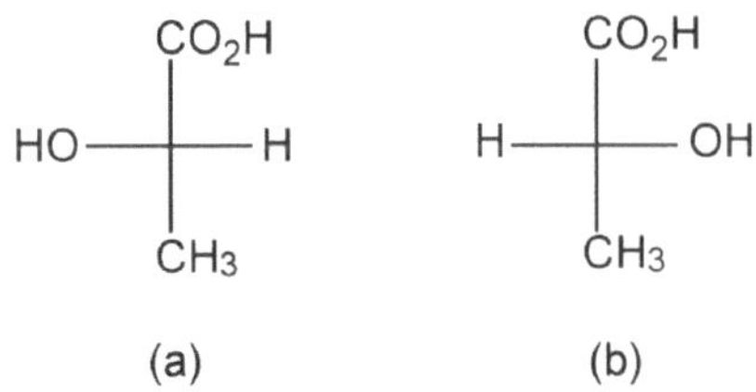


**Figure 4.6: 2D Fischer representations of lactic acid.**

## Fischer-Roseno convention

The Fischer-Roseno convention defines chirality of a molecule using the Fischer projection. The convention is based upon *D*-(+)-glyceraldehydes (Figure 4.7). To determine the chiral label for a molecule, the Fischer projection is drawn with the longest carbon chain pointing away from the observer, i.e., into the plain of the page. The number 1 carbon (that of the highest substitution) is positioned at the top of the diagram. If the functional group is on the right of the diagram the isomer is given the label *D* (from the dextrorotatory enantiomer of glyceraldehydes rather than the molecule being dextrorotatory per se), i.e., Figure 4.7. On the other hand, if the functional group is on the left of the diagram the isomer is given the label *L* (from the levorotatory enantiomer of glyceraldehyde rather than the molecule being levorotatory per se).

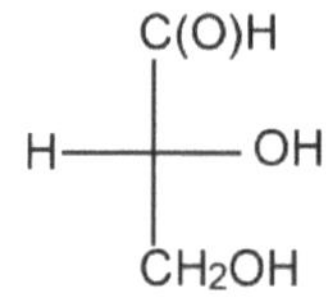


**Figure 4.7: The Fischer projection of *D*-(+)-glyceraldehyde.**

*Rules*

Where molecules are large and complex a number of additional rules are useful:

1. Make any alcohol substituent to be nearest to the top, e.g., a butan-2-ol (Figure 4.8a) rather than a butan-3-ol (Figure 4.8b).
2. Aldehydes and ketones are the number 1 carbon atom.
3. If there are both sets of substituents, Rule 1 has precedence over Rule 2.

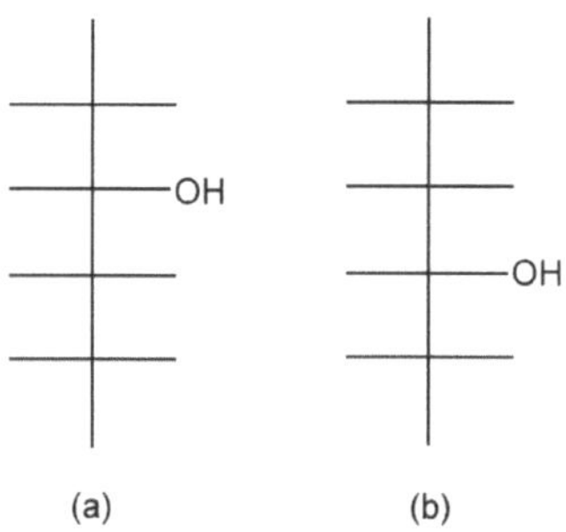


**Figure 4.8: Preferred positioning of an alcohol substituent to be nearest the top (a) rather than at the bottom (b), according to the rules of the Fischer-Roseno convention.**

## Nomenclature in terms of *R* and *S*

The *R*/*S* nomenclature is a more widely used alternative to the *D*/*L* nomenclature and is based upon the hierarchical order of the substituents. From the Fischer projection, the groups are ordered with the largest atom first, as shown in Figure 4.9. The lowest numbered substituent is then oriented away from the observer (i.e., H in Figure 4.9). The order of the remaining substituents 1, 2, 3 is then traced: if it goes clockwise the molecule is labeled *R*, if it goes anti-clockwise (anti clockwise) the molecule is labeled *S*.

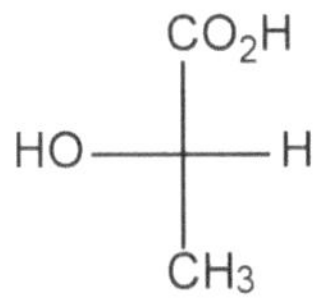


$OH > CO_2H > CH_3 > H$
1 2 3 4

**Figure 4.9: The structure of *S*-lactic acid, and the ordering of the substituents.**

The advantage of the R/S methodology is that it can also be used for geometric isomers. Thus, while cis and trans are useful when the groups either side of a C=C double bond are the same (e.g., *cis*-1,2-dichloroethene, Figure 4.10a); however, when four different groups are present (e.g., A, B, X, and Y as seen in Figure 4.10b) then an alternative must be used. By taking either group A/B or X/Y and placing in order of precedence, the relative order defines the isomer. Thus, if in Figure 4.10b, A>B and X>Y the label *Z* (*zusammen*, from the German for together) is used. Conversely, if in Figure 4.10b A<B and X>Y the label *E* (*entgegen*, from the German for opposite) is used.

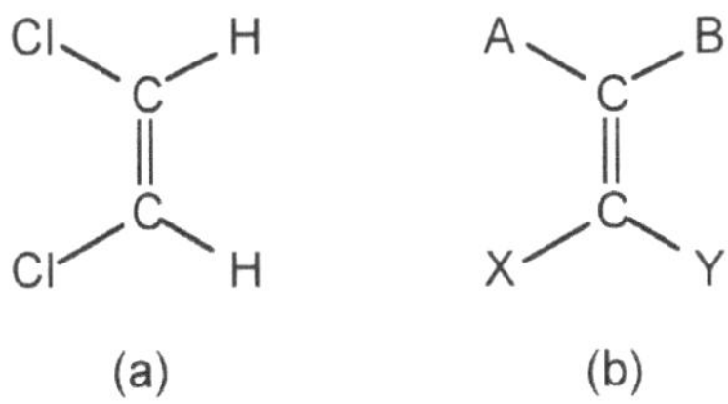


**Figure 4.10: The structures of (a) *cis*-1,2-dichloroethene and (b) a generic olefin with four different substituent groups.**

## More complex molecules

As molecules become more complex with multiple functional groups and multiple chiral centers the *D*/*L* nomenclature has the potential to become confusing. Erythrose has two chiral centers, and based upon the Fischer projections it may be seen that since the alcohol substituents in the 2 and 3 positions are on the same side of the backbone in the Fischer projection irrespective if it is *D*-(-)-erythrose or *L*-(+)-erythrose (Figure 4.11a and b). Thus, the naming is simplified despite the presence of multiple side groups. However, in threose (Figure 4.11c) the substituents are on different sides of the Fischer projection raising the issue as to whether this should be *L* or *D*. However, in such cases the substituent at the bottom of the side chain takes priority, thus Figure 4.11c is *D*-(-)-threose.

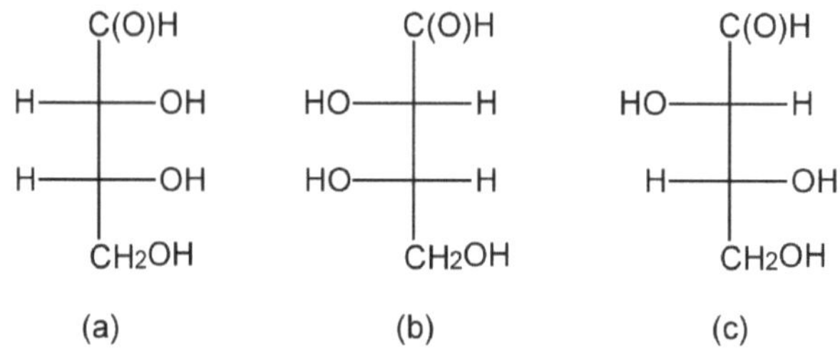


**Figure 4.11: The Fischer projections of (a) *D*-(-)-erythrose, (b) *L*-(+)-erythrose, and (c) *D*-(-)-threose.**

It should be noted that *D*-(-)-erythrose and *L*-(+)-erythrose are enantiomers, and *D*-(+)-threose and *L*-(+)-threose are also enantiomers, but threose and erythrose are diasteriomers. Enantiomers have the same physical properties (just different behavior to polarized light), while diasteriomers may have different physical properties such as melting points and solubility.

*Amino acids*

Amino acids (or more properly α-amino acids) are compounds containing both an amine ($NH_2$) group and a carboxylic acid ($CO_2H$) attached to the same carbon atom. The presence of both acid and base groups can result in the formation of the zwitterionic form with ammonium ($NH_3^+$) and carboxylate ($CO_2^-$) groups.

In defining the labels for amino acids, the basis is *D*-(-)-allothreonine (Figure 4.12), whereas for an amino acid such as threonine the top substituent is used for a label, e.g., Figure 4.13.

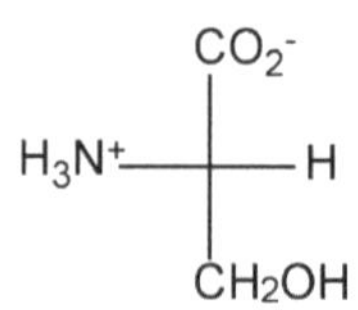


**Figure 4.12: The Fischer projection of *D*-(-)-allothreonine.**

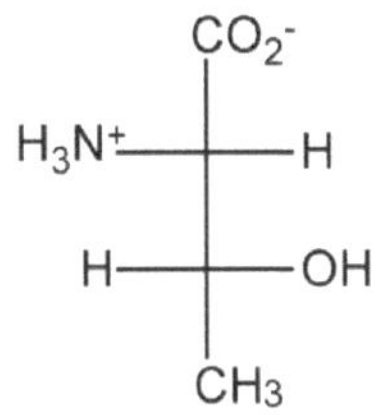

**Figure 4.13: The Fischer projection of *D*-(-)-threonine.**

Just because a compound has more than one chiral center does not mean that it is optically active. A consideration of erythritol shows the presence of a mirror plane of symmetry (Figure 4.14). As such, erythritol is not optically active, i.e., there is no effect on polarized light. In contrast, while threitol has the same molecular and geometric formula, the lack of a mirror plane of symmetry (Figure 4.15) means that the *L*- and *D*+ forms are optically active.

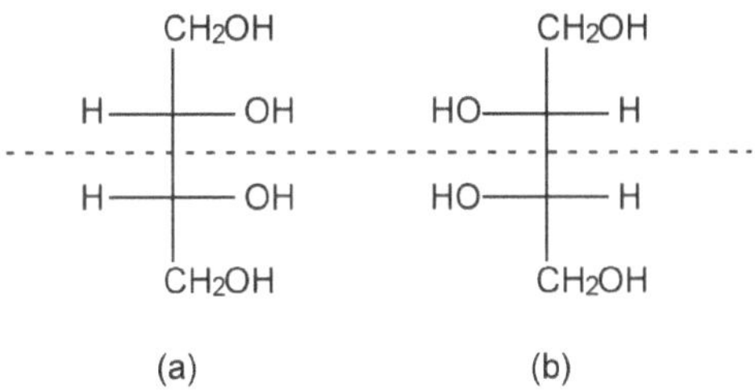

**Figure 4.14: The Fischer projections of D and L erythritol showing the presence of a mirror plane of symmetry. The presence of a mirror plane means that erythritol is not optically active and hence there is no need to label with + or -.**

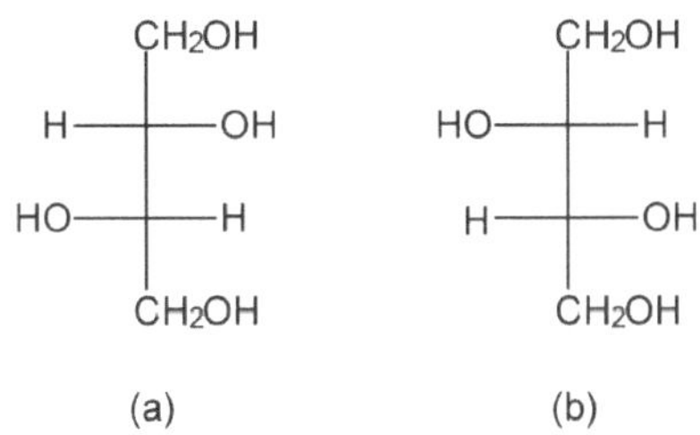

**Figure 4.15: The Fischer projections of (a) *L*-(-)-threitol and (b) *D*-(+)-threitol.**

*Similar chiral centers*

Compounds with a *D*:*L* ratio of 1:1 are called racemic compounds and are totally optically inactive. However, it is also possible for a racemic solution of a compound to crystallize to form crystals of pure *D* or *L*, which may be manually separated.

## Optical isomers that are not tetrahedral

*Pyramidal molecules*

Pyramidal molecules can also exhibit chirality when the three substituents are distinct, e.g., $PR_1R_2R_3$ (Figure 4.16). Unfortunately, most pyramidal compounds undergo an inversion of their isomers (in a similar manner to turning an umbrella inside out), such that the chiral forms interconvert rapidly. In such a case, it is not possible to resolve (separate) the different forms. Phosphines may generally be resolved because the barrier to their inversion is sufficiently large (ca. 132 kJ/mol).

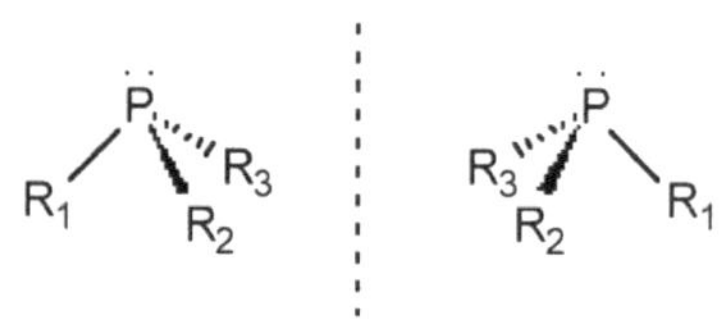


**Figure 4.16: An example of a chiral phosphine where $R_1 \neq R_2 \neq R_3$.**

The notation used for pyramidal molecules is the same as that for tetrahedral molecules, in that the individual substituents are ordered with the largest atom first. The lone pair is defined as having the lowest numbering and thus is oriented away from the observer. The order of the remaining substituents 1, 2, and 3 is then traced: if it goes clockwise the molecule is labeled *R*, if it goes anti-clockwise (counterclockwise) the molecule is labeled *S*.

*Chirality in octahedral complexes*

The presence of a chelate ligand on an octahedral complex can induce chirality in the complex. A chelate ligand is a molecule or ion that is bonded to at least two points of a central atom or ion. The term chelate is from the Greek *chelè*, meaning claw.

Unlike tetrahedral based compounds, chiral octahedral compounds have their nomenclature based on the structure of a helix. For example, in the case of a *bis*-chelate complex (Figure 4.17), with one of the chelate ligands pointing straight up the helix, the direction of the other defines the chirality of the helix, i.e., if it is left- or right-hand helix. If the ligand points up to the left the complex is assigned the symbol shown in Figure 4.17a, while if the ligand points up to the right the complex is assigned the symbol shown in Figure 4.17b.

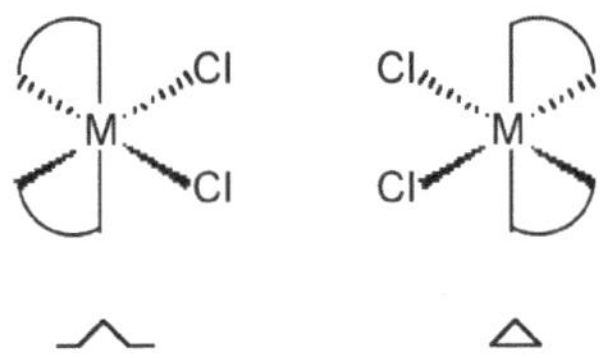


(a) (b)

**Figure 4.17: The two optical isomers of a *bis*-chelate metal complex.**

In complexes with three chelate ligands (i.e., a *tris*-chelate complex, Figure 4.18), the same methodology applies, in that any two ligands can be chosen and the same rules applied.

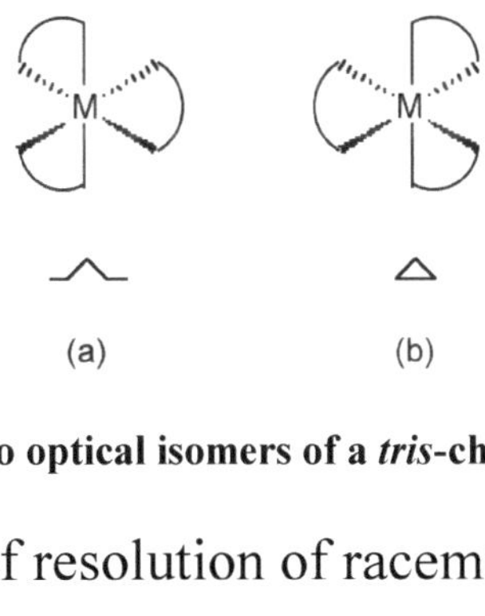


(a) (b)

**Figure 4.18: The two optical isomers of a *tris*-chelate metal complex.**

## Methods of resolution of racemic mixtures

A racemic mixture, or racemate, is one that has equal amounts of left- and right-handed enantiomers of a chiral molecule. The first known racemic mixture was "racemic acid," which Louis Pasteur (Figure 4.19) found to be a mixture of the two enantiomeric isomers of tartaric acid (Figure 4.20).

In nature, it is common that only one of the possible optical isomers is naturally produced, but in laboratory synthesis it is more common that both isomers are made in equal amounts. The separation of a racemate into its components (the pure enantiomers) is called a chiral resolution.

**Figure 4.19: French microbiologist and chemist Louis Pasteur (1822 - 1895) who is remembered for his remarkable breakthroughs in the causes and preventions of disease.**

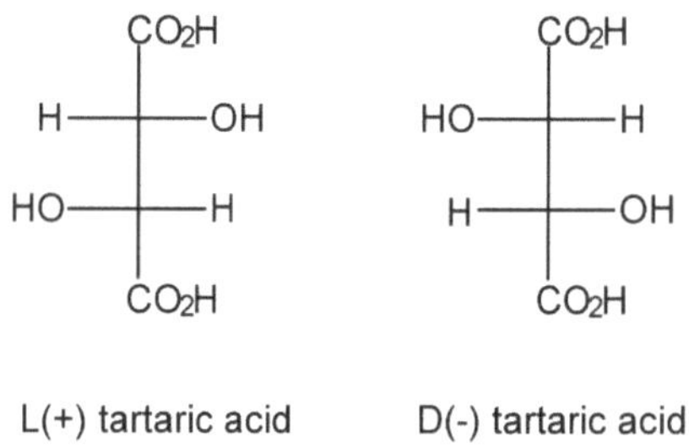


**Figure 4.20: The Fischer projections of the two enantiomeric isomers of chiral tartaric acid.**

*Mechanical separation*

The crystals of enantiomerically pure compounds often have different appearances. Thus, just as Pasteur did in 1848, it is often possible to look under a microscope and physically separate the two different enantiomers.

*Resolution by formation of diasteriomers*

One of the differentiating properties of a chiral molecule is that each enantiomer reacts with another chiral molecule to form a diastereomeric pair of compounds. For example, a racemic acid, (+)HA and (-)HA, will react with a chiral base, (-)B, to form a mixture of diasteromers, e.g., $[(-)BH^{+}.(+)A^{-}]$ and $[(-)BH^{+}.(-)A^{-}]$. Because diasteriomers have different physical properties they can be separated by recrystallization. Once separated by recrystallization, the addition of excess acid will liberate the enantiomerically pure compound, i.e., (+)HA or (-)HA. A typical chiral base would be a natural alkaloid base to ensure that it is pure + or -. If it is a base that is needed to be separated, then (-)malic acid is a suitable acid.

*Resolution by chromatography*

In a related method to resolution by the formation of diasteriomers, a chiral column material will allow for the chromatographic separation of an enantiomeric mixture. Thus, the retention rate for (+)X will be different from (-)X on a column in which the stationary phase is chiral, i.e., (-)A. Many common chiral stationary phases are based on oligosaccharides such as cellulose or cyclodextrin (in particular with β-cyclodextrin).

## Labile stereo isomers and racemization

It is only possible to separate (resolve) racemic mixtures if the molecule stays as one form for a long time. In other words, if there is a mechanism by which the two forms are interconverted, then resolution cannot be achieved.

*Intramolecular rearrangement*

Intramolecular rearrangements involve no bonds being broken. The classic example is the inversion of an amine, which has a low energy barrier (24.7 kJ/mol):

Another example of an intramolecular rearrangement is the conversion of a square-based pyramidal geometry via a trigonal bipyramidal geometry to the isomeric square-based pyramidal geometry, or the alternative isomerization of one trigonal bipyramidal geometry to another via a square-based pyramidal transition state. Such a process is known as a Berry rotation. The Berry mechanism is a pseudorotation process for simultaneously interchanging two equatorial groups with the two axial groups, while the third equatorial group (called the pivot group) remains an equatorial group (Figure 4.21).

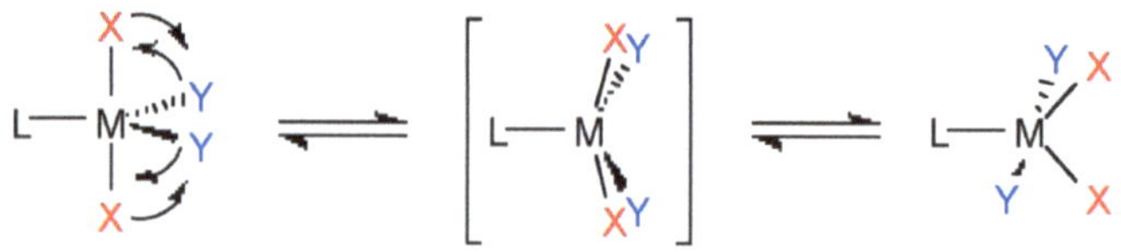


**Figure 4.21: The Berry pseudorotation.**

*Intermolecular processes*

Intermolecular processes involve bond breaking (and bond formation). For example, a tetrahedral compound could lose one ligand, which creates a labile pyramidal compound. Rapid inversion of the pyramidal compound is followed by re-attachment of the ligand. Such a process is often so fast that resolution cannot be achieved.

*Intramolecular processes*

Intramolecular processes with bond breaking (and making) also lead to racemization. For example, in an octahedral complex with chelate ligands, if one of the ends of a chelate ligand detaches, it can reattach in the same configuration, or attach differently to change the chirality. In this manner the chirality can be changed from one form to another.

## Conformation

The conformation of a molecule arises from the rotation of a single bond (Figure 4.22). However, even though there is rotation about the bond, there are energy barriers due to steric interactions of the substituents. In order to understand (and predict) these interactions, it is necessary to visualize the molecule in such a manner as to highlight the across-bond interactions; this is done by using a Newman projection.

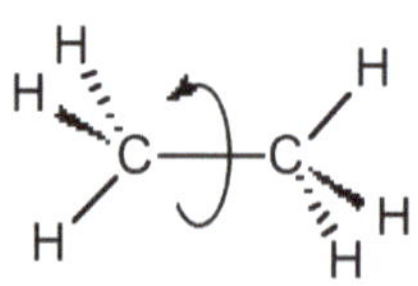


**Figure 4.22: The rotation about the C-C single bond in ethane.**

A Newman projection, useful in alkane stereochemistry, visualizes chemical conformations of a carbon- carbon chemical bond from front to back, with the front carbon represented by a dot and the back carbon as a circle (Figure 4.23).

The front carbon atom is called proximal, while the back atom is called distal. This type of representation is useful for assessing the torsional angle between bonds. Using ethane as an example, the Newman projection along the C-C single bond results in two basic conformations: *eclipsed* (Figure 4.23a) and *staggered* (Figure 4.23b). The staggered conformation will be energetically favored since the substituents are the most distant from each other. Conversely, the staggered will be the highest in energy. The difference in energy between the two conformations will define the barrier to rotation. In the case of ethane this is very small (12.5 kJ/mol).

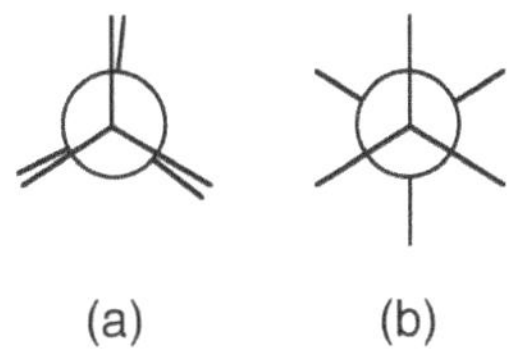


**Figure 4.23: The Newman projection of ethane showing (a) *eclipsed* (*E*) and (b) *staggered* (*S*) conformations.**

Although we often only consider the two extreme conformations, in reality there is a continuum around 360° rotation of the C-C bond. Figure 4.24 shows the relative energy as a function of the dihedral angle for ethane. Since each carbon in ethane has equivalent substitution (i.e., three H atoms) the energy for each staggered conformation is the same. This is not true for more complex molecules such as butane (Figure 4.25).

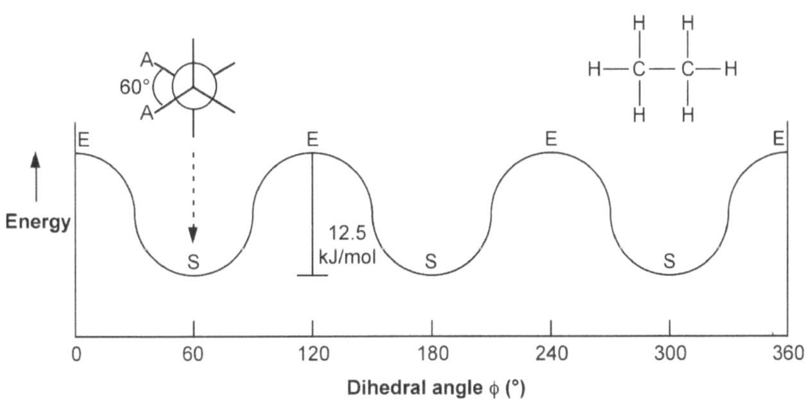


**Figure 4.24: The relative energy as a function of the dihedral angle for ethane.**

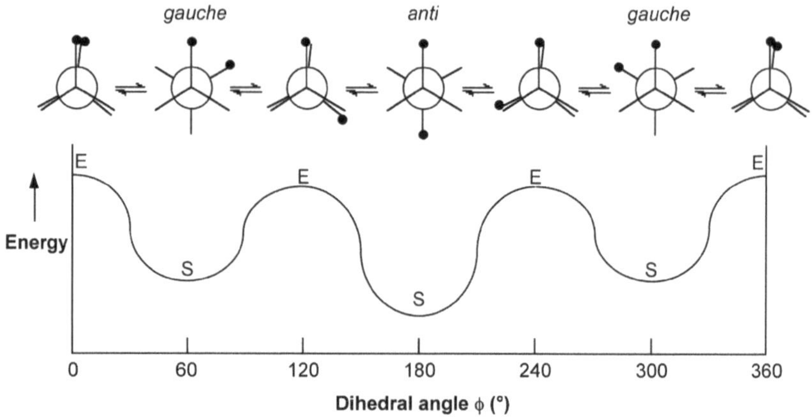


**Figure 4.25: The relative energy as a function of the dihedral angle for butane.**

As may be seen from Figure 4.25, the staggered conformation in which the two methyl groups (represented by the black circles) are as far away from each other (*anti*) is the most energetically favored. The other two staggered conformations (*gauche*) are mirror images of each other and are hence conformation enantiomers. It should also be noted from Figure 4.25 that the eclipsed conformations vary in energy since the presence of methyl···methyl near neighbors is clearly less energetically favorable than methyl···H near neighbors.

Generally, the rotation about C-C bonds has a low barrier to rotation, however, if substituents are sufficiently bulky the molecule will not twist around the bond, e.g., substituted bi-phenyl with sterically bulky substituents (Figure 4.26).

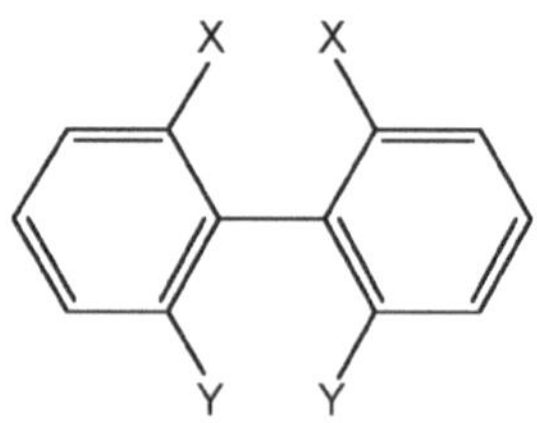


**Figure 4.26: The ease of rotation about the C-C bond of a substituted bi-phenyl will depend on the steric bulk of X and Y.**

Free rotation about a C-C bond is not fully possible when you have a ring system, e.g., in a cyclic compound such as cyclohexane, $C_6H_{12}$. The limited rotation about the C-C bonds results in the flipping of the ring conformation from the chair form (Figure 4.27a) to the boat form (Figure 4.27b). Since the boat form has steric hindrance between the hydrogen atoms, the chair form is the more stable.

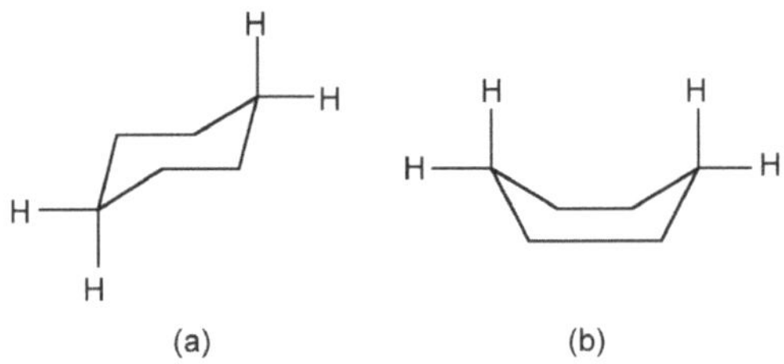


**Figure 4.27: The chair (a) and boat (b) conformations of cyclohexane, $C_6H_{12}$.**

*Conformation of compounds with lone pairs*

Lone pairs often behave in a different manner to substituents in regard to conformations. Thus, methylamine ($CH_3NH_2$) would be predicted to have the staggered conformation shown in Figure 4.28a. However, the lower steric bulk of the lone pair results in the nitrogen being 0.09 Å away from the true center of the $CH_3$ projection (Figure 4.28b).

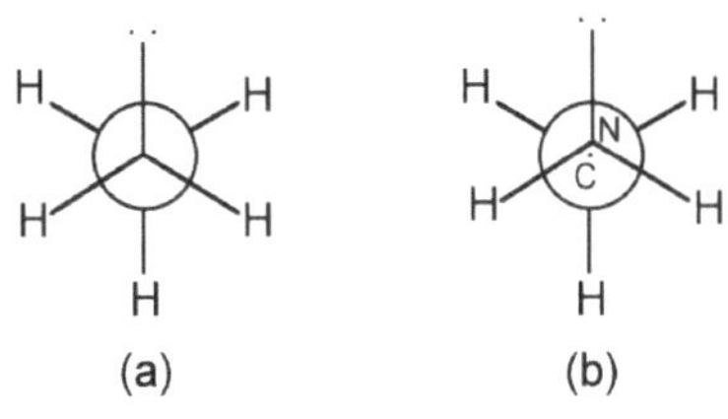


**Figure 4.28: The ideal (a) and observed (b) conformations of methylamine, $CH_3NH_2$.**

In compounds with more than one lone pair, the lowest energy form is not always the anti-conformation. For example, at 20 °C hydrazine ($H_2NNH_2$) is 100% gauche (Figure 4.29a); but for substituted hydrazines (i.e., a diamine, $R_2NNR_2$) if the substituent groups are sufficiently large then the *anti*-conformation will dominate (Figure 4.29b).

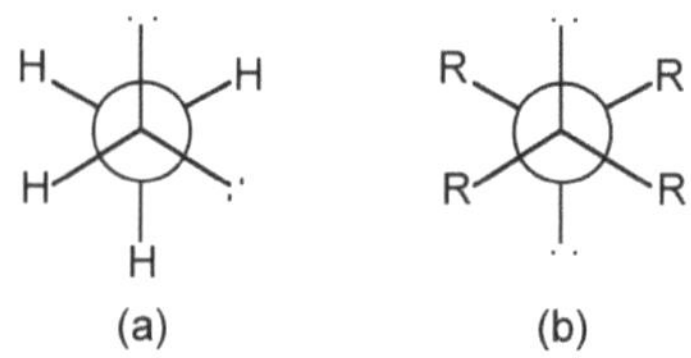


**Figure 4.29: The stable conformation of hydrazine ($H_2NNH_2$) at 20 °C (a) as compared to the *anti*-conformation (b) observed for diamines ($R_2NNR_2$).**

The conformation of hydrogen peroxide ($H_2O_2$) is dominated by the lone pairs rather than the hydrogen atoms. Instead of the expected anti conformation (c.f., Figure 4.25 where the black circles would represent the hydrogen atoms) in the free state the dihedral angle is 94° Figure 4.30). The conformation of hydrogen peroxide is therefore neither eclipsed nor staggered but an intermediate structure. When in the solid state, hydrogen bonding will cause the shape and angles to change.

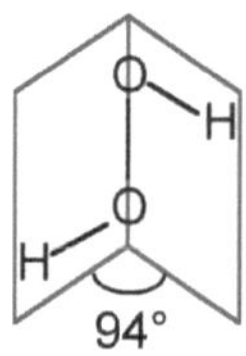


**Figure 4.30: The structure of hydrogen peroxide in the free state.**

# Chapter 5: Choosing a Solvent

The choice of solvent is an important parameter for any chemical reaction. The following provides a guide to some of the consideration to be made in choosing a solvent to ensure the desired reaction occurs.

## Solvation

Solvation may be defined as the interaction between the solvent and the solute; however, two general classes of solvation have different consequences to the stability of either reagents or products in a chemical reaction, and hence the potential of a reaction to occur.

Specific solvation is where the solvent interacts with one of the ions (or molecules) in solution via a covalent interaction. Furthermore, there will be a specific number of solvent molecules bound to each ion (or molecule), e.g., $[Cu(NH_3)_4]^{2+}$ and $[Mg(H_2O)_6]^{2+}$ (Figure 5.1a).

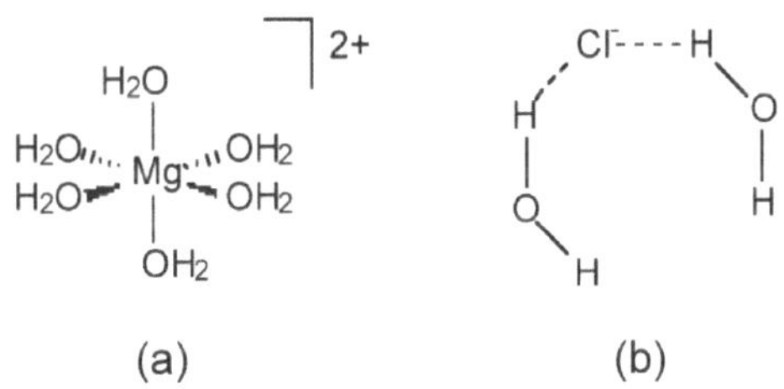

**Figure 5.1: Examples of (a) specific and (b) non-specific solvation.**

Non-specific solvation is as a result of van der Waals or dipole-dipole forces between the solvent and an ion (or molecule). There will be no defined number of interactions and the solvent···ion interaction will be highly fluxional, e.g., while water solvates the chloride ion (Figure 5.1b) the number of water molecules around each anion is not fixed.

Table 5.1 shows the ability of three solvents to act with specific and non-specific solvation. The relative solvation ability of each solvent results in three different products from the dissolution of iron(III) chloride ($FeCl_3$).

| Solvation | DMSO ($Me_2SO$) | Pyridine ($C_5H_5N$) | Acetonitrile (MeCN) |
|---|---|---|---|
| Specific | Good | Very good | Poor |
| Non-specific | Good | Poor | Moderately good |

**Table 5.1: The ability for specific and non-specific solvation.**

Dissolution of $FeCl_3$ in DMSO results in the dissociation of a chloride ligand due to both the specific solvation of the $FeCl_2^+$ cation and the non-specific solvation of the $Cl^-$ anion:

$$FeCl_3 \xrightleftharpoons{\text{DMSO}} [FeCl_2(DMSO)_4]^+ + Cl^-_{(solv)}$$

In fact, the good solvation properties of DMSO means that depending on the concentration (and temperature) a series of dissociations may occur:

$$[FeCl_2(DMSO)_4]^+ \xrightleftharpoons[-\,Cl^-_{(solv)}]{+\,DMSO} [FeCl(DMSO)_5]^{2+} \xrightleftharpoons[-\,Cl^-_{(solv)}]{+\,DMSO} [Fe(DMSO)_6]^{3+}$$

In contrast, if $FeCl_3$ is dissolved in pyridine (py) the neutral Lewis acid-base complex is formed:

$$FeCl_3 + py \rightarrow FeCl_3(py)$$

This is because, while pyridine is a very good at specific solvation (Table 5.1), it is poor at solvating the chloride anion.

In a similar manner, $FeCl_3(MeCN)_3$ will be formed by the dissolution in acetonitrile, because although it is not good at specific solvation, it is not sufficiently good at non-specific solvation to stabilize the chloride anion. However, since the $FeCl_4^-$ anion has a lower charge density that $Cl^-$, it can be supported by the non-specific solvation of acetonitrile and thus a disproportionation reaction occurs:

$$2\,FeCl_3(MeCN)_3 \rightleftharpoons [FeCl_2(MeCN)_4]^+ + FeCl_4^-$$

# Interference by the solvent

Rather than solvating a molecule or ion, the solvent can take an active and detrimental role in the synthesis of a desired compound.

## Solvolysis

The archetypal solvolysis reaction is the reaction with water, i.e., hydrolysis,

$$SO_2Cl_2 + 4\,H_2O \rightarrow SO_2(OH)_2 + 2\,H_3O^+ + 2\,Cl^-$$

However, solvolysis is a general reaction, involving bond breaking by the solvent. Thus, the reaction with ammonia is ammonolysis,

$$SO_2Cl_2 + 4\,NH_3 \rightarrow SO_2(NH_2)_2 + 2\,NH_4^+ + 2\,Cl^-$$

the reaction with acetic acid is acetolysis,

$$SO_2Cl_2 + 4\,MeCO_2H \rightarrow SO_2(O_2CMe)_2 + 2\,MeCO_2H_2^+ + 2\,Cl^-$$

and the reaction with an alcohol is alcoholysis,

$$SO_2Cl_2 + 4\,EtOH \rightarrow SO_2(OEt)_2 + 2\,EtOH_2^+ + 2\,Cl^-$$

where Et = $C_2H_5$. In each case the same general reaction takes place yielding the cation associated with the solvent.

## Competition reactions

Where more than one reaction could occur, the reaction involving the solvent can often compete with the desired reaction. If it is desired to synthesis the Lewis acid-base complex between diethylether ($Et_2O$) and boron trifluoride ($BF_3$) it is important to choose a solvent that will not compete with the complex formation. For example, pyridine is a poor choice because the nitrogen donor is a stronger Lewis base than the diethylether,

$$BF_3(Et_2O) + py \rightarrow BF_3(py) + Et_2O$$

and thus, no reaction would occur between diethylether and boron trifluoride. In contrast, since acetonitrile (MeCN) is a poor Lewis base, then the reaction will occur.

If the synthesis of $GeCl_6^{2-}$ from germanium tetrachloride ($GeCl_4$) and a source of chloride anion, then water would be a poor choice of solvent since hydrolysis of $GeCl_4$ would result. Liquid hydrogen chloride would be equally poor solvent since strong $Cl^{-}\cdots H\text{-}Cl$ hydrogen bonding would stabilize the chloride anion and preclude reaction. In contrast, nitromethane ($CH_3NO_2$) would be a polar enough solvent to solvate the $GeCl_4$, but it will be displaced by the chloride anion, which would be only weakly solvated.

## Salt formation

The formation of a salt via a double displacement reaction, can be promoted by the choice of solvent by shifting the equilibrium by stabilization of one or more reagent/product.

$$MX + M'X' \rightleftharpoons MX' + M'X$$

### Salt stabilization through relative acidity

The attempted formation of nitronium perchlorate from nitric acid and perchloric acid,

$$HNO_3 + HClO_4 \rightarrow NO_2^+ + ClO_4^- + H_2O$$

in water will result in the decomposition of the $NO_2^+$ cation,

$$NO_2^+ + 2\,H_2O \rightarrow HNO_3 + H_3O^+$$

However, if the reaction is carried out in a stronger acid, i.e., sulfuric acid, the $NO_2^+$ cation is stable, and the resulting salt can be recrystallized.

In a similar manner, the weak basic character of water means the equilibrium reaction, has a very small equilibrium constant, $K_{eq}$,

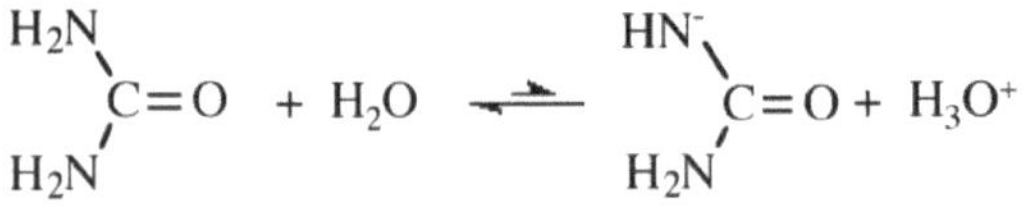

However, if the reaction is carried out in a strongly basic solvent such as ammonia the uride anion is stabilized and can be precipitated by cation exchange.

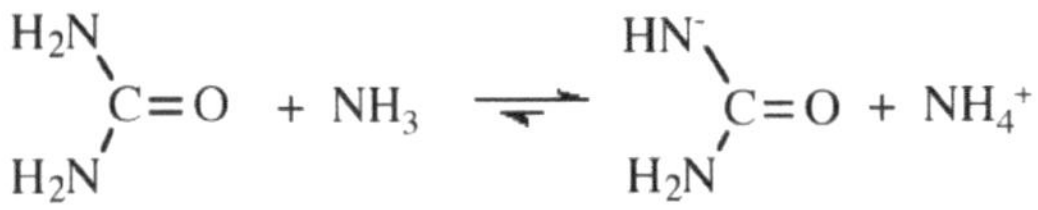

## Salt stabilization through solvation

The following observations may be explained by a consideration of the solvation ability of the solvent.

1. The reaction of silver nitrate with barium chloride in water yields silver chloride and barium nitrate, e.g.,

$$2\ AgNO_3 + BaCl_2 \xrightarrow{H_2O} 2\ AgCl\downarrow + Ba^{2+} + 2\ NO_3^-$$

2. The reaction of barium nitrate with silver chloride in ammonia yields barium chloride and silver nitrate, e.g.,

$$2\ AgCl + Ba(NO_3)_2 \xrightarrow{NH_3} BaCl_2\downarrow + 2\ Ag^+ + 2\ NO_3^-$$

Since silver nitrate and barium nitrate are soluble in both solvents, the disparity must be due to differences in the solubility of the chlorides in each solvent. A consideration of the relative stability of solid silver chloride versus the solvated species (Figure 5.2) shows that the enthalpy of solvation in water is less than the lattice energy. Thus, if silver chloride were present as $Ag^+$ and $Cl^-$ in water it would spontaneously precipitate. In contrast, the enthalpy of solvation in ammonia is greater than the lattice energy, thus solid AgCl will dissolve readily in liquid ammonia. The reason for the extra stabilization from the specific solvation of the silver cation by the ammonia, i.e., the formation of the covalent complex $[Ag(NH_3)_2]^+$.

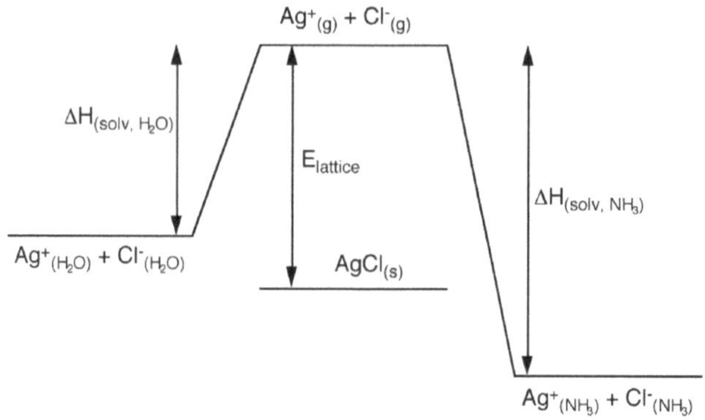


**Figure 5.2: Enthalpy of solvation of silver chloride in water and ammonia in comparison to the lattice energy.**

As may be seen from Figure 5.3, the opposite effect occurs for barium chloride. Here the enthalpy of solvation in ammonia is less than the lattice energy. Thus, if barium chloride were present as $Ba^{2+}$ and $Cl^{-}$ in ammonia it would spontaneously precipitate. In contrast, the enthalpy of solvation in water is greater than the lattice energy, thus solid $BaCl_2$ will dissolve readily in water. The stabilization of $Ba^{2+}$(aq) occurs because water will have a larger sphere of non-specific solvation as a consequence of having two lone pairs, allowing interaction with the $Ba^{2+}$ as well as other water molecules (Figure 5.4).

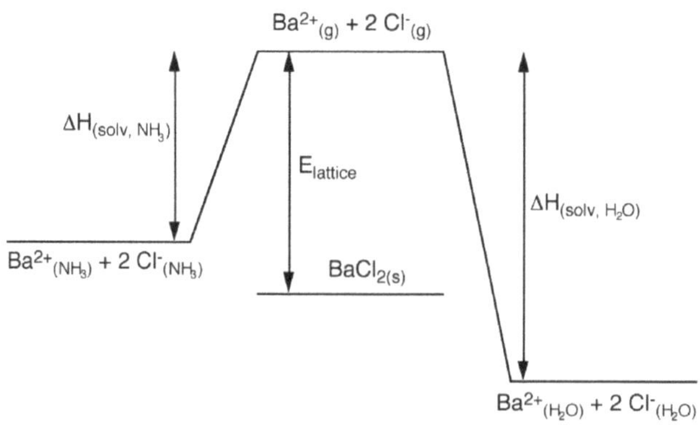


**Figure 5,3: Enthalpy of solvation of barium chloride in water and ammonia in comparison to the lattice energy.**

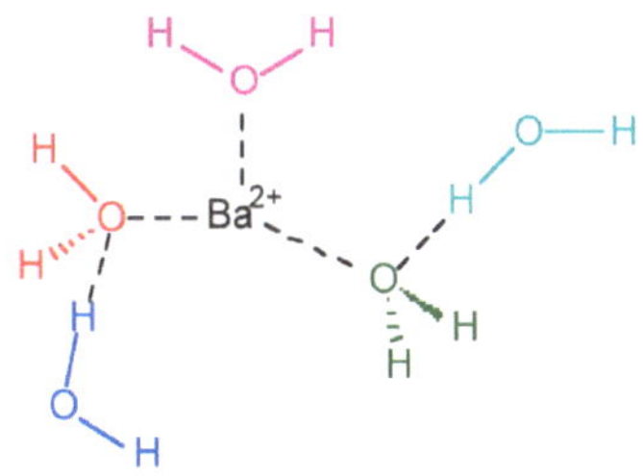


**Figure 5.4: Schematic representation of the extended water solvation sphere about $Ba^{2+}$ cation.**

## Electron transfer reactions

A consideration of the oxidation,

$$2\,H^+ + {}^1/_2\,O_2 + 2\,e^- \rightarrow H_2O \qquad E_0 = +0.82\ V$$

and reduction,

$$H^+ + e^- \rightarrow {}^1/_2\,H_2 \qquad E_0 = -0.41\ V$$

reactions that occur for pure water at neutral pH (where $[H^+] = 10^{-7}$) would suggest that water will not tolerate oxidants whose $E_0$ is greater than 0.82 V nor tolerate reductants whose $E_0$ is less than -0.41 V.

Thus, while water has a fair range to support redox reactions it is not very good at the extremes with strong reducing agents or strong oxidizing agents. Liquid ammonia is an excellent solvent for very strong reducing agents because of the stabilization of solvated electrons, i.e., $[e^-(NH_3)_6]$. In contrast, hydrochloric acid is a good solvent for reactions involving very strong oxidizing agents.

# Chapter 6: Periodic Trends

Within the main group (s- and p-block) elements of the Periodic Table (Figure 6.1) there are some general trends that we can observe for the elemental form, as well as the hydrides, oxides, and halides.

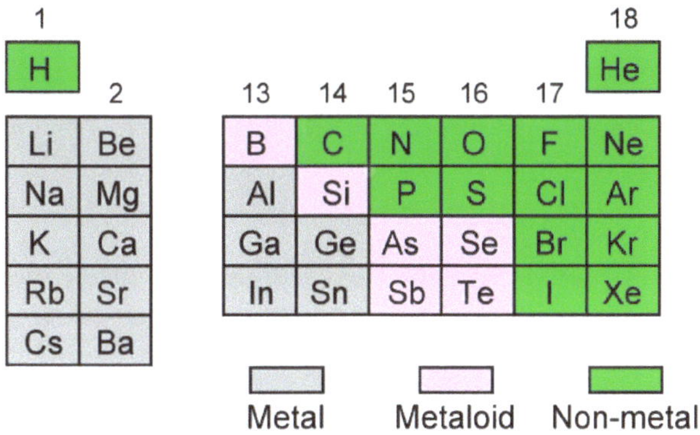


**Figure 6.1: The main group elements.**

## Periodic trends for the main group elements

Within the main group (s- and p-block) elements there are some general trends that we can observe.

- The further down a given Group the elements have increased metallic character, i.e., good conductors of both heat and electricity, and exhibit delocalized bonding.
- Moving from left to right across a Period the elements have greater non-metallic character, they are insulators with localized bonding.
- Within the p-block at the boundary between the metallic elements (Figure 6.1, grey elements) and nonmetal elements (Figure 6.1, green elements) there is positioned boron and silicon that are metalloid in character (Figure 6.1, pink elements), i.e., they have low electrical conductivity but it increases with temperature.

As an example of these changes Table 6.1 shows the trends across one Period.

| Element | Properties |
|---|---|
| Na | Electro-positive metal |
| Mg | Electro- positive metal |
| Al | Metal but forms covalent bonds |
| Si | Metalloid semiconductor metal/non-metal character |
| P | E-E bonding in elements |
| S | E-E bonding in elements |
| Cl | Simple molecule |
| Ar | Mono atomic gas |

**Table 6.1: Summary of trends for elements across the Periodic Table.**

## Periodic trends for the main group hydrides

The properties of main group hydrides are dependent on the difference in electronegativity between the element and hydrogen (Table 6.2). Elements on the left of the Periodic Table are highly electropositive and form ionic hydrides, while those of the center and right are covalent in character. However, of those with covalent E-H bonds, there is a change from polymeric hydrides to molecular compounds. For example, the Group 13 element hydrides (i.e., $BH_3$) form hydrogen-bridged oligomers (i.e., $B_2H_6$). In contrast, HCl is a diatomic molecule.

| Hydride | E electronegativity | H electronegativity | E-H polarity | Structure | Comments |
|---|---|---|---|---|---|
| NaH | 0.9 | 2.1 | $M^+ H^-$ | Ionic | Reacts with $H_2O$ to liberate $H_2$ |
| $BH_3$ | 2.0 | 2.1 | $B^{\delta+}$-$H^{\delta-}$ | Oligomeric and polymeric | Reacts slowly with $H_2O$ |
| $CH_4$ | 2.5 | 2.1 | $C^{\delta-}$-$H^{\delta+}$ | Molecular | Insoluble in $H_2O$ |
| HCl | 3.0 | 2.1 | $Cl^{\delta-}$-$H^{\delta+}$ | Molecular | Dissolves in $H_2O$ to form $H^+$ and $Cl^-$ |

**Table 6.2: Summary of properties of selected main group hydrides as a function of the relative electronegativities.**

## Periodic trends for the main group oxides

As with hydrides the properties of main group oxides are dependent on the difference in electronegativity between the element and oxygen. Highly electropositive metals for ionic oxides, while other elements for covalent bonds (albeit polar in character) with oxygen. In addition, the aggregation of covalent oxides decreased across the Period from left to right (Table 6.3). As may also be seen from Table 1.14, oxides of elements on the left of the Periodic Table dissolve in water to form basic solutions, while those on the right form acidic solutions. There is a class of oxides (especially those of Group 13 and 14) that can react as either an acid or a base. These are known as amphoteric substances. The word is from the Greek prefix "*ampho*" meaning "both".

| **Oxide** | **Bonding** | **Reactivity with $H_2O$** | **Description** |
|---|---|---|---|
| $Na_2O$ | Ionic | Dissolves to give a strong base | Basic |
| $Al_2O_3$ | Covalent polymeric | Dissolves in both acidic and basic solution | Amphoteric |
| $SiO_2$ | Covalent polymeric | Dissolves in both acidic and basic solution | Amphoteric |
| $CO_2$ | Covalent molecular | Dissolves to give a weak acid | Acidic |
| $SO_3$ | Covalent molecular | Dissolves to give a strong acid | Acidic |

**Table 6.3: Comparison of oxides across the Periodic Table.**

In summary, oxides of the main group elements show two trends.

1. From left to right across a Period, the oxides change from ionic to oligomeric/polymeric covalent to molecular covalent.
2. From left to right across a Period, the oxides change from basic to amphoteric to acidic.

## Periodic trends for the main group chlorides

The trend between ionic and non-ionic/covalent in moving across a Period is also true for the chlorides of the main group elements. Those on the left (i.e., Group 1 and 2) are ionic and soluble in water, while those to the right tend to give acidic solutions due to reactions with the water and the formation of hydrochloric acid, e.g.,

$$SiCl_4 + 2\,H_2O \rightarrow SiO_2 + 4\,HCl_{(aq)}$$

# Chapter 7: Hydrogen

## Discovery of Hydrogen

Hydrogen gas, $H_2$, was first artificially synthesized by Phillip von Hohenheim (known as Paracelsus, Figure 7.1) by mixing metals with strong acids. He was unaware that the flammable gas produced by this chemical reaction was a new chemical element.

**Figure 7.1: Renaissance physician, botanist, alchemist, and astrologer Paracelsus (1493 - 1541).**

In 1671, Robert Boyle (Figure 7.2) rediscovered the reaction between iron lings and dilute acids, which results in the production of hydrogen gas. He noted that these fumes were highly flammable and that the flame gave off a lot of heat but not much light. In 1766, Henry Cavendish (Figure 7.3) was the first to recognize hydrogen gas as a discrete substance, by identifying the gas from a metal-acid reaction as flammable air. One of the richest men in Britain at the time, he lived in London and spent his time in his private laboratory at his home. In 1781 he was the first person to find that the gas produces water when burned. This was a key experiment in disproving the Aristotelian theory of the four elements: earth, fire, air and water. As a consequence of his work he is usually given credit for its discovery as an element. However, it was Antoine Lavoisier (Figure 2.4) who in 1783 named the element hydrogen (from the Greek hydro meaning water and genes meaning creator) after he reproduced Cavendish's findings.

**Figure 7.2: Irish chemist, physicist, and inventor Robert Boyle (1627 - 1691).**

**Figure 7.3: British scientist Henry Cavendish (1731 - 1810).**

**Figure 7.4: French chemist and biologist Antoine-Laurent de Lavoisier (1743 - 1794).**

Using his invention, the vacuum flask, James Dewar (Figure 7.5) was the first to liquefy hydrogen in 1898. He produced solid hydrogen the next year.

**Figure 7.5: Scottish chemist and physicist Sir James Dewar FRS (1842 - 1923).**

## The Physical Properties of Hydrogen

| | |
|---|---|
| Chemical symbol | H |
| Atomic number | 1 |
| Electron configuration | $1s^1$ |
| Atomic weight | 1.00794 |
| Melting point | -259 °C |
| Boiling point | -253 °C |
| Density (gas) | 0.090 g/L |
| Density (liquid) | 0.070 g/L |
| Ionization potential | 13.598 eV |
| Pauling electronegativity | 2.1 |
| Ionic radius ($H^-$) | 1.54 Å |
| van der Waal radius | 1.2 Å |
| Explosive limit | 4 - 75% |
| Ignition temperature | 585 °C |

**Table 7.1: Physical properties of hydrogen.**

## Synthesis of Molecular Hydrogen

Although hydrogen is the most abundant element in the universe, its reactivity means that it exists as compounds with other elements. Thus, molecular hydrogen, $H_2$, must be prepared from other compounds. The following outlines a selection of synthetic methods.

## Steam reforming of carbon and hydrocarbons

Many reactions are available for the production of hydrogen from the reaction of steam with a carbon source. The choice of reaction is guided by the availability of raw materials and the desired purity of the hydrogen. The simplest reaction involves passing steam over coke at high temperatures (1000 °C).

$$C_{(s)} + H_2O_{(g)} \rightarrow H_{2(g)} + CO_{(g)}$$

Coke is a grey, hard, and porous carbonaceous material derived from destructive distillation of low-ash, low-sulfur bituminous coal. As an alternative to coke, methane may be used at a slightly higher temperature (1100 °C).

$$CH_{4(g)} + H_2O_{(g)} \rightarrow 3\,H_{2(g)} + CO_{(g)}$$

In each case the carbon monoxide formed in the reaction can react further with steam in the presence of a suitable catalyst to generate further hydrogen.

$$CO_{(g)} + H_2O_{(g)} \rightleftharpoons H_{2(g)} + CO_{2(g)}$$

This reaction is known as the water gas-shift reaction and was discovered by Italian physicist Felice Fontana (Figure 7.6) in 1780.

**Figure 7.6: Italian physicist Felice Fontana (1730 - 1805).**

The dominant industrial process for hydrogen production uses natural gas or oil refinery feedstock in the presence of a nickel catalyst at 900 °C.

$$C_3H_{8(g)} + 3\ H_2O_{(g)} \rightarrow 7\ H_{2(g)} + 3\ CO_{(g)}$$

## Electrolysis of water

Electrolysis of acidified water in with platinum electrodes is a simple (although energy intensive) route to hydrogen.

$$2\ H_2O_{(l)} \rightarrow 2\ H_{2(g)} + O_{2(g)}$$

On a larger scale hydrolysis of warm aqueous solutions of barium hydroxide can yield hydrogen of purity greater than 99.95%. Hydrogen is also formed as a side product in the production of chlorine from electrolysis of brine (NaCl) solutions in the presence of a mercury electrode.

$$2\ NaCl_{(aq)} + 2\ Hg_{(l)} \rightarrow Cl_{2(g)} + 2\ NaHg_{(l)}$$

The sodium mercury amalgam reacts with water to yield hydrogen.

$$2\ NaHg_{(l)} + 2\ H_2O_{(l)} \rightarrow H_{2(g)} + 2\ NaOH_{(aq)} + 2\ Hg_{(l)}$$

Thus, the overall reaction can be written as:

$$2\ NaCl_{(aq)} + H_2O_{(l)} \rightarrow H_{2(g)} + 2\ NaOH_{(aq)} + Cl_{2(g)}$$

However, this method is being phased out for environmental reasons.

## Reaction of metal with acid

Hydrogen is produced by the reaction of highly electropositive metals with water, and less reactive metals with acids, e.g.,

$$Fe_{(s)} + 2\ H_3O^+_{(aq)} \rightarrow H_{2(g)} + 2\ H_2O_{(aq)} + Fe^{2+}_{(aq)}$$

This method was originally used by Henry Cavendish (Figure 7.3) during his studies that led to the understanding of hydrogen as an element (Figure 7.7).

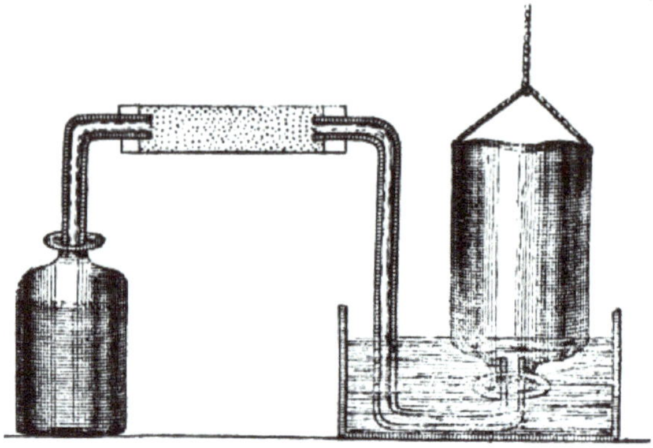

**Figure 7.7: Cavendish's apparatus for making hydrogen in the left-hand jar by the reaction of a strong acid with a metal and collecting the hydrogen gas above water in the right-hand inverted jar.**

The same method was employed by French inventor Jacques Charles (Figure 7.8) for the first flight of a hydrogen balloon on 27th August 1783. Unfortunately, terrified peasants destroyed his balloon when it landed outside of Paris.

**Figure 7.8: Jacques Alexandre César Charles (1746 - 1823).**

## Hydrolysis of metal hydrides

Reactive metal hydrides such as calcium hydride ($CaH_2$) undergo rapid hydrolysis to liberate hydrogen.

$$CaH_{2(s)} + 2\,H_2O_{(l)} \rightarrow 2\,H_{2(g)} + 2\,OH^-_{(aq)} + Ca^{2+}_{(aq)}$$

This reaction is sometimes used to inflate life rafts and weather balloons where a simple, compact means of generating $H_2$ is desired.

# Atomic hydrogen

Atomic hydrogen has the electron configuration of $1s^1$ and as such represents the simplest atomic configuration. However, as a consequence there is dispute as to its proper position within the Periodic Table. Its electron configuration is similar to the valence electron configuration of the alkali metals ($ns^1$) suggesting it be listed at the top of Group 1 (1A). However, its reaction chemistry is dissimilar to the alkali metals. Hydrogen is also one electron short of a Nobel gas configuration, and therefore it is possible to think of its relationship to the halogens.

## Vapor phase

Atomic hydrogen (H·) is highly reactive and consequently has a short lifetime due to its reaction chemistry. Consequently, in order to generate and observe the reactivity they must be generate at low pressures. Thermolysis of hydrogen compound (commonly halide) or photolysis at an energy above the bond dissociation energy results in the homoleptic cleavage of the H-X bond to generate the appropriate radical species,

$$\text{H-X} \xrightleftharpoons{\Delta \text{ or } h\nu} \text{H}\cdot + \text{X}\cdot$$

Alternatively, atomic hydrogen can be generated from elemental hydrogen,

$$\text{H-H} \xrightleftharpoons{\Delta \text{ or } h\nu} \text{H}\cdot + \text{H}\cdot$$

The reverse reaction (recombination of two hydrogen atoms) is highly exothermic (-434 kJ/mol) and forms the basis of the heat generated in arc welding.

## Solution

Atomic hydrogen may be generated in aqueous solution through the solvation of electrons,

$$e^{-}_{(aq)} + H_3O^{+} \rightleftharpoons H^{\cdot}_{(aq)} + H_2O$$

The formation equilibrium constant ($K_{eq}$) is very small resulting in very low concentrations being generated ($10^{-5}$ M). As expected, solvated atomic hydrogen is a strong reducing agent:

$$Ag^{+} + H^{\cdot} \rightarrow Ag + H^{+}$$

$$2I^{-} + 2\,H^{\cdot} \rightarrow 2\,H^{+} + I_2$$

## Solid state

Hydrogen atoms may be trapped in the solid-state lattice upon generation by photolysis of HX. Observation by electron spin resonance (ESR) of a signal split by s = ½ nucleus (i.e., $^{1}H$) results in a doublet with a coupling of 1428 MHz.

# The Proton

The proton, $H^{+}$, is the name given to hydrogen in the +1 oxidation state.

## Gas phase

The proton can be formed from the photolysis of atomic hydrogen in the vapor phase at low pressure.

$$H^{\cdot}_{(g)} + h\nu \rightarrow H^{+}_{(g)} + e^{-}_{(g)}$$

The proton is more reactive than the hydrogen atom because of its high charge density. In addition, the proton's small ionic radius, $1.5\times10^{-15}$ cm, means that it can get close to other atoms and hence form strong bonds.

$$H^{+}_{(g)} + NR_3 \rightarrow HNR_3{}^{+}_{(g)}$$

The strength of the bonding interaction is such that it is very hard to measure directly. Instead the relative bond strength between the proton and an appropriate base, $B_1$, is measured in the presence of a competing base, $B_2$.

$$B_1H^{+}_{(g)} + B_{2(g)} \rightleftharpoons B_2H^{+}_{(g)} + B_{1(g)}$$

In measuring the exchange reaction, the relative proton affinity of $B_1$ and $B_2$ is measured. This is also known as the gas phase acidity, and as such it is a measure of the inherent acidity of a species X-H because it obviates any solvent effects.

## Liquid and solution

The high reactivity of the proton means that it does not exist free in solution. There are however many $H^+$ containing species. These are generally classified as acids.

$$\underset{acid}{B_1H^{+}_{(sol)}} + \underset{base}{B_{2(sol)}} \rightleftharpoons \underset{acid}{B_2H^{+}_{(sol)}} + \underset{base}{B_{1(sol)}}$$

The reaction between the acid and the base is a proton transfer reaction. While the proton travels from $B_1$ to $B_2$, it is never free in solution. Instead a bridged transition state or intermediate is formed, $B_1\cdots H^{+}\cdots B_2$.

### *Acidity and pH*

The most common solvent for $H^+$ is water. The acid form is usually defined as the hydronium ion or $H_3O^+$,

$$H_2O + H^{+} \rightarrow H_3O^{+}$$

The terms oxonium, hydroxonium and oxidanium are all used for the $H_3O^+$ cation. Although we commonly use $H_3O^+$ it is known from spectroscopy that larger complexes are formed such as $H_9O_4^+$ (Figure 7.9).

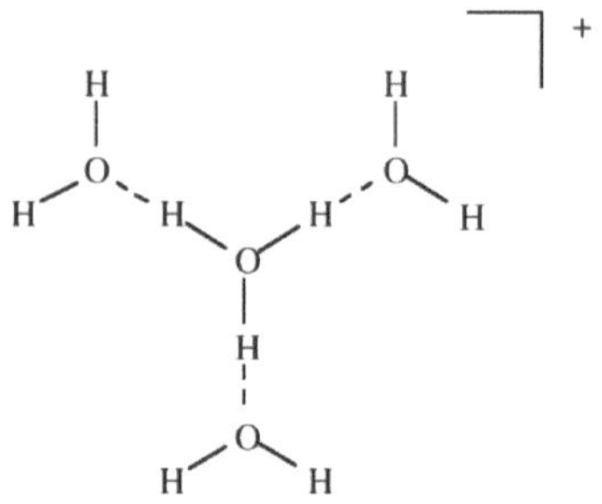

**Figure 7.9: Structure of $H_9O_4^+$.**

Acids and bases have been characterized in a number of different ways. In 1680 Robert Boyle (Figure 7.2) defined an acid as a compound that dissolved many other compounds, had a sour taste, and reacted with alkali (base).

Boyle's simple observational description was rationalized by Danish physical chemist Johannes Brønsted (Figure 7.10). Brønsted proposed that acids are proton donors, and bases are proton acceptors. An acid-base reaction is one in which a proton is transferred from a proton donor (acid) to a proton acceptor (base).

**Figure 7.10: Johannes Nicolaus Brønsted (1879 - 1947).**

Based upon Brønsted's proposal simple acids contain an ionizable proton. Examples of simple acids include neutral molecules (HCl, $H_2SO_4$), anions ($HSO_4^-$, $H_2PO_4^-$), and cations ($NH^+$). The most common Brønsted bases include metal hydroxides (MOH).

Brønsted noted that when an acid donates a proton it forms a conjugate base. The following are examples of an acid and its conjugate base.

$H_2O \rightarrow H^+ + OH^-$

*acid* *conjugate base*

$H_2SO_4 \rightarrow H^+ + HSO_4^-$

*acid* *conjugate base*

$NH_4^+ \rightarrow H^+ + NH_3$

*acid* *conjugate base*

The same occurs when a base accepts a proton it forms a conjugate acid. The following are examples of a base and its conjugate acid.

$H_2O + H^+ \rightarrow H_3O^+$

*base* *conjugate acid*

$HCO_3^- + H^+ \rightarrow H_2CO_3$

*base* *conjugate acid*

$F^- + H^+ \rightarrow HF$

*base* *conjugate acid*

Thus, the reaction between an acid and a base, results in the formation of the appropriate conjugate base and conjugate acid.

$$\text{acid}_1 + \text{base}_1 \rightleftharpoons \text{acid}_2 + \text{base}_2$$

A specific example is as follows:

$$HNO_3 + NH_3 \rightleftharpoons NH_4^+ + NO_3^-$$

*acid₁*  *base₁*  *acid₂*  *base₂*

In the equilibrium reactions shown above there is a competition between the two bases for the proton. As would be expected the strongest base wins. When a strong acid is added to (dissolved in) water it will react with the water as a base:

$$HCl + H_2O \rightarrow H_3O^+ + Cl^-$$

*acid*  *base*  *acid*  *base*

In contrast, when a strong base is added to (dissolved in) water it will react with the water as an acid:

$$H_2O + NH_3 \rightarrow NH_4^+ + OH^-$$

*acid*  *base*  *acid*  *base*

*pH a measure of acidity*

The acidity of a water (aqueous) solution depends on the concentration of the hydronium ion, i.e., $[H_3O^+]$. The acidity of a solution is therefore the ability of the solution to donate a proton to a base. The acidity or pH of a solution is defined as:

$$pH = -\log [H_3O^+_a]$$

It is important to note that the value is the activity of $H_3O^+$ and not the concentration. Activity is a measure of the effective concentration of a species in a mixture. The difference between activity and other measures of composition such as concentration arises because molecules in non-ideal gases or solutions interact with each other, either to attract or to repel each other.

The activity of the $H_3O^+$ ion can be measured by:

1. A glass electrode
2. Acid-base indicators.

## Proton transfer reactions

The proton transfer reaction is one of the simplest reactions in chemistry. It involves no electrons and low mass transfer/change, giving it a low energy of activation. For proton transfer between O-H or N-H groups and their associated bases the reaction is very fast. The proton transfer occurs across a hydrogen-bonded pathway during which the proton is never free.

# Hydrides

The combination of hydrogen with another element produces a hydride, $E_xH_y$. The formal charge or oxidation state of the hydrogen in these compounds is dependent on the relative electronegativity of the element in question.

## Ionic hydrides

Hydrogen compounds with highly electropositive metals, i.e., those in which the metal has an electronegativity of less than 1.2, are ionic with the hydrogen having a $s^2$ configuration ($H^-$). Typical ionic are those of the Group 1 (IA) metals and the heavier Group 2 (IIA) metals.

The ionic radius of the hydride ion is in between that of fluoride and chloride and (Table 7.2) as a consequence, in the solid state the hydride ion replicates that of a halide and similar solid-state structures are observed (Table 7.3).

| Ion | Ionic radius (Å) |
|---|---|
| $H^-$ | 1.40 |
| $F^-$ | 1.36 |
| $Cl^-$ | 1.81 |
| $O^{2-}$ | 1.40 |

**Table 7.2: Selected ionic radii.**

| Metal | Hydride | Fluoride | Chloride |
|---|---|---|---|
| Li | 4.085 | 4.0173 | 5.129 |
| Na | 4.880 | 4.620 | 5.640 |
| K | 5.700 | 5.347 | 6.292 |
| Rb | 6.037 | 5.640 | 6.581 |
| Cs | 6.376 | 6.008 | 7.020 |

**Table 7.3: Lattice parameters (*a*, Å) for hydrides, and halides of the Group 1 metal salts with cubic rock salt structures.**

Unlike the halide ions that are soluble in water, the hydride ion reacts with water,

$$H^- + H_2O \rightarrow OH^- + H_{2\,(g)}$$

and consequently, NaH and $CaH_2$ are commonly used as drying agents. The liberation of hydrogen was used as a commercial source of hydrogen for small-scale applications.

Group 1 and 2 metal hydrides can ignite in air, especially upon contact with water to release hydrogen, which is also flammable. Hydrolysis converts the hydride into the analogous hydroxide, which are caustic bases. In practice, most ionic hydrides are dispensed as a dispersion in oil, which can be safely handled in air.

## Covalent hydrides

The most common binary compounds of hydrogen are those in which hydrogen bonds have covalent bond character. The E-H bond is usually polar ranging from those in which the hydrogen is polarized positively (e.g., those with non-metals such as F, O, S, and C) to where it is negative (e.g., those with metals and metalloids such as B, Al, etc.). Magnesium hydride is intermediate between covalent and ionic since it has a polymeric solid similar to $AlH_3$ but reacts rapidly with water like ionic hydrides. Table 7.4 lists the important covalent hydrides of p-block elements. It should be noted that all organic hydrocarbons can be thought of as simply the hydrides of carbon!

| Group 13 | Group 14 | Group 15 | Group 16 | Group 17 |
|---|---|---|---|---|
| $B_2H_6$ | $C_nH_{2n+2}$, $C_nH_{2n}$, $C_nH_{2n-2}$ | $NH_3$, $N_2H_4$ | $H_2O$, $H_2O_2$ | HF |
| $(AlH_3)_n$ | $Si_nH_{2n+2}$ | $PH_3$, $P_2H_4$ | $H_2S$, $H_2S_2$ | HCl |
| $Ga_2H_6$ | $Ge_nH_{2n+2}$ | $AsH_3$ | $H_2Se$ | HBr |
| | $SnH_4$ | $SbH_3$ | $H_2Te$ | HI |

**Table 7.4: Important covalent hydrides of the p-block elements.**

The elements of Group 14 to 17 all form hydrides with normal covalent bonds in which the hydrogen is bonded by a single bond to the element in question. In contrast, the elements of Group 13 (as typified by boron) all exhibit a second type of covalent bond: the electron deficient hydrogen bridged bond. In this type of bond, the hydrogen nucleus is embedded in a molecular orbital that covers more than two atoms to create a multi-center two-electron bond. Diborane (Figure 7.11) represents the archetypal compound containing the hydrogen bridge bond. Hydrides are not limited to terminal (E-H) or those bridging two atoms (E-H-E) but are also known where the hydrogen bridges (or caps) more than two atoms, i.e., Figure 2.12.

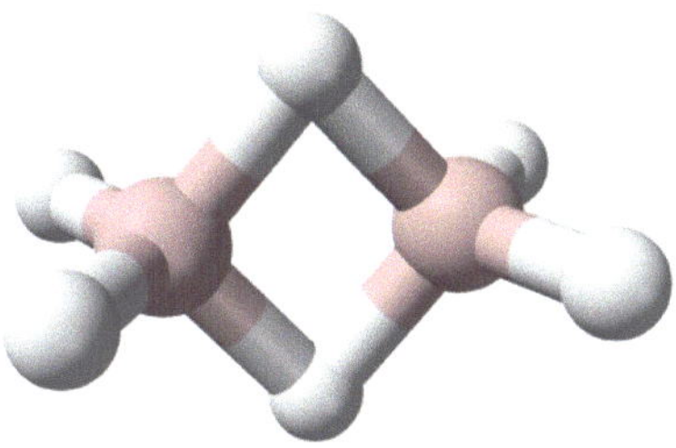

**Figure 7.11: Structure of diborane ($B_2H_6$): pink = boron; white = hydrogen.**

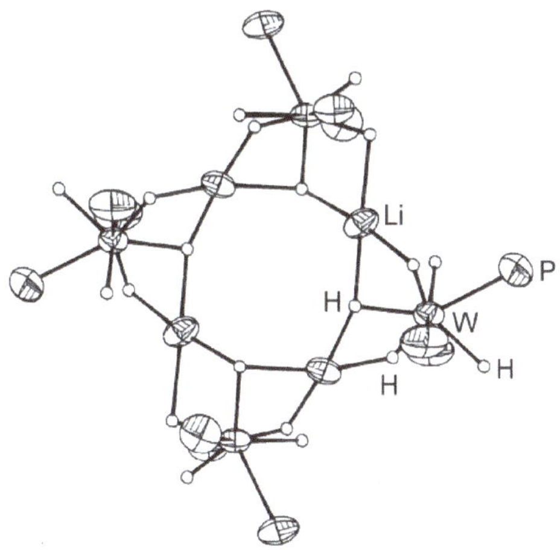


**Figure 7.12: The structure of $\{[(Me_3P)_3WH_5]Li\}_4$ showing the presence of terminal, bridging, and trifurcated hydrides. The methyl groups have been omitted for clarity. Adapted from A. R. Barron, M. B. Hursthouse, M. Motevalli, and G. Wilkinson, Transition metal polyhydride anions: a new synthetic route. X-ray crystal structure of $\{[(Me_3P)_3WH_5]^-Li^+\}_4$. *J. Chem. Soc., Chem. Commun.*, 1986, 81. Copyright: Royal Society of Chemistry (1986).**

*Synthesis of covalent hydrides*

Covalent hydrides can be made by a range of synthetic routes. The simplest is direct combination of the elements (in a similar manner to that used with ionic hydrides).

$$2\,H_2 + O_2 \rightarrow 2\,H_2O$$

The use of a hydride as a reagent to reduce a halide or oxide of the desired element.

$$SiCl_4 + LiAlH_4 \rightarrow SiH_4 + LiAlCl_4$$

Metal phosphides, carbides, silicides, and borides result in the formation of the hydride.

$$Ca_3P_2 + 6\,H_2O \rightarrow 2\,PH_3 + 3\,Ca(OH)_2$$

Hydride compounds can be interconverted in the presence of a catalyst, heat, or an electrical discharge. This is the basis of catalytic cracking of petroleum mixtures.

## Interstitial hydrides

Many transition metal, lanthanides and actinides absorb hydrogen to give a metallic hydride which retain the properties of a metal, although the presence of hydrogen does result in embrittlement of the metal. As such these hydrides are best considered as alloys since they do not have defined stoichiometries. For example, vanadium absorbs hydrogen to form an alloy with a maximum hydrogen content of $VH_{1.6}$. In a similar manner palladium forms $PdH_{0.6}$. The hydrogen atoms are present in interstitial sites in the metal's lattice; hence interstitial hydride. Interstitial hydrides show certain promise as a way for safe hydrogen storage.

## The hydrogen bond

Hydrogen bonds are formed between a species with a polar $X^{\delta-}$-$H^{\delta+}$ bond and a species with a lone pair ($Y^{\delta-}$), i.e., $X^{\delta-}$-$H^{\delta+}\cdots Y^{\delta-}$. The most common species for X is oxygen and nitrogen, and to a lesser extent carbon, fluorine, and sulfur; however, as long as the X-H bond is polar then hydrogen bonding is possible. Similarly, the most common Lewis bases that hydrogen bond involve oxygen, nitrogen, and fluorine as the donor atom. Again, there are many examples of other atoms, but as long as the atom has a lone pair that is chemically active, hydrogen bonding can occur.

The majority of hydrogen bonds are asymmetrical, that is the hydrogen is closer to one atom than the other (Figure 7.13), even when X and Y are the same element, i.e., O-H···O. While the typical hydrogen bond involves one Lewis base (lone pair donor), there are many examples where the hydrogen interacts with two Lewis base lone pairs (Figure 7.14).

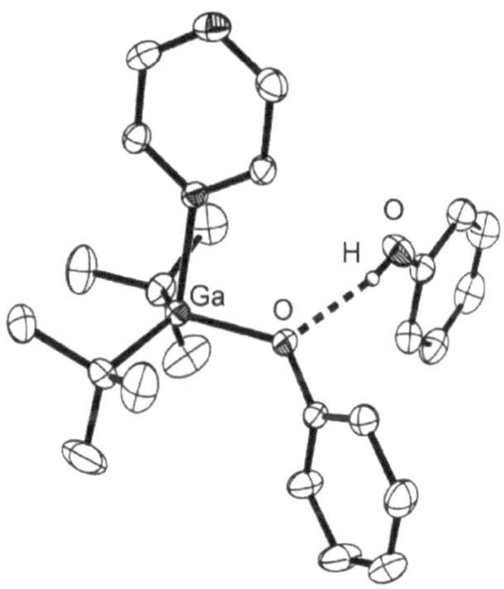


**Figure 7.13: Structure of $(^tBu)_2Ga(OPh)(pyz).(PhOH)$ from X-ray crystallographic data showing the presence of an asymmetrical hydrogen bond. Hydrogen atoms attached to carbon are omitted for clarity. Adapted from L. H. van Poppel, S. G. Bott and A. R. Barron, Molecular structures of $Ga(^tBu)_2(OPh)(pyz)\cdot PhOH$ and $[(^tBu)_2Ga(H_2O)(\mu\text{-}OH)Ga(^tBu)_2]_2(\mu\text{-}OC_6H_4O)\cdot 4(2\text{-}Mepy)$: intra- and inter-molecular hydrogen bonding to gallium aryloxides. *Polyhedron*, 2002, 21, 1877. Copyright: Elsevier (2002).**

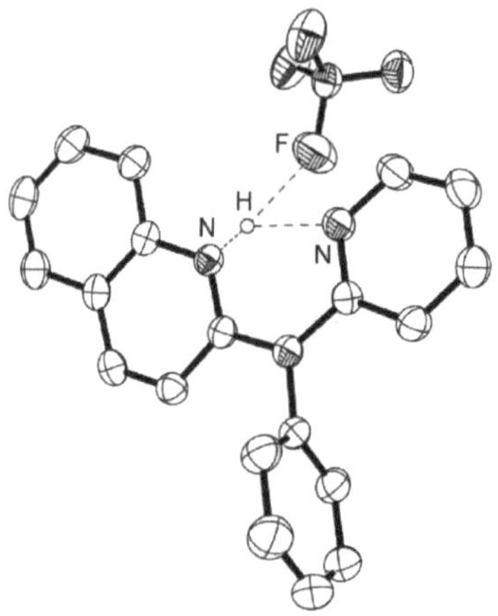


**Figure 7.14: Structure of [H{PhN(py)(quin)}]$BF_4$ from X-ray crystallographic data showing the presence of a trifurcated asymmetrical hydrogen bond. Hydrogen atoms attached to carbon are omitted for clarity. Adapted from J. J. Allen, C. E. Hamilton, and A. R. Barron, Molecular Structures of RN(H)Py (R = 2,4,6-$Me_3C_6H_2$, 2,6-$Et_2C_6H_3$, $Ph_3C$), and the Copper Complex [Cu{(2,4,6-$Me_3C_6H_2$)N(H)Py}$_2$]$BF_4$. *J. Chem. Cryst.*, 2009, 39, 573–580. Copyright: Springer Science (2009)**

## Classes of hydrogen bond

Although hydrogen bonds may be characterized with respect to the X and Y atom, it is more useful to classify them as either intramolecular or intermolecular hydrogen bonds. This is due to the difference in physical and chemical properties between these two classes.

### *Intramolecular*

Intramolecular hydrogen bonds (X-H···Y) arise where the X and Y atoms are in the same molecule (Figure 7.15).

### *Intermolecular*

If the hydrogen bond (X-H···Y) involves X and Y being from different molecules this is an intermolecular hydrogen bond. Within the range of intermolecular hydrogen bonded compounds there are two sub-categories: those involving discrete molecular species (oligomers) and those resulting in polymeric species.

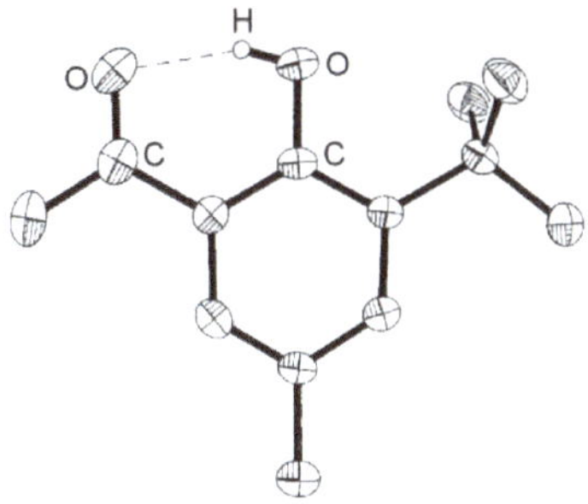


**Figure 7.15: Structure of 3-*tert*-butyl-2-hydroxy-5-methylacetophenone showing the presence of an intramolecular hydrogen bond. Hydrogen atoms attached to carbon are omitted for clarity. Adapted from M. B. Power, A. R. Barron, S. G. Bott, E. J. Bishop, K. D. Tierce and J. L. Atwood, Acylation and esterification of the aryloxide ligand in [AlMe(dbmp)$_2$]. Crystal structures of [AlMe(dbmp)(bhmap)], Hbhmap and O=C(dbmp)Bu$^t$(Hdbmp = 2,6-di-*tert*-butyl-4-methylphenol, Hbhmap = 3-*tert*-butyl-2-hydroxy-5-methylacetophenone). *J. Chem. Soc., Dalton Trans.*, 1991, 241. Copyright: Royal Society of Chemistry (1991).**

Carboxylic acids are a typical example of a discrete oligomeric species that are held together by inter- molecular hydrogen bonds (Figure 7.16a). A wide range of structurally analogous compounds also form head-to-tail hydrogen bonded dimers (e.g., Figure 7.17). In a polymeric hydrogen bonded species every molecule hydrogen bonds but in a random form. As an example, liquid primary alcohols form extended hydrogen networks (Figure 7.16b). Such an arrangement is labile and as such it is di cult to determine definitive speciation. Liquids that form this type of hydrogen-bonded network are known as associated liquids. In the solid state the networks generally adopt a more ordered structure. For example, as is seen in the structure of ice.

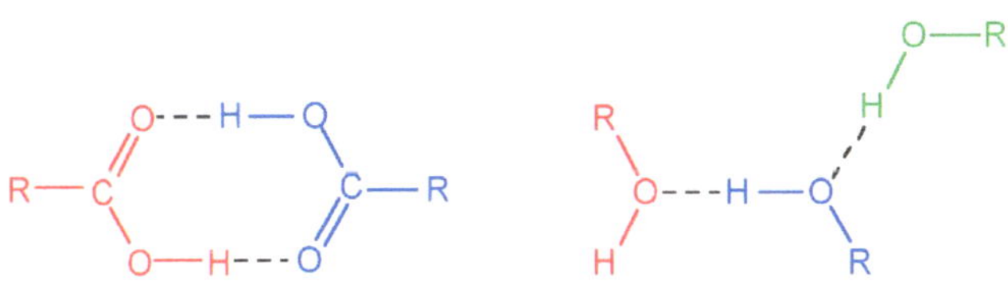


(a) (b)

**Figure 7.16: Structure of (a) the head-to-tail dimmer formed between two carboxylic acid molecules, and (b) a typical network of a primary alcohol in the liquid phase.**

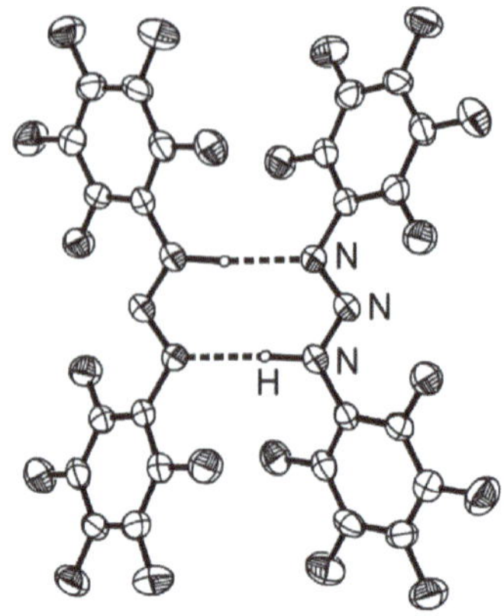


**Figure 7.17: Structure of the hydrogen bonded diner of $(C_6H_5)NNN(H)(C_6F_5)$. Adapted from J. T. Leman, J. Braddock-Wilking, A. J. Coolong, and A. R. Barron, 1,3-Diaryltriazenido compounds of aluminum. *Inorg. Chem.*, 1993, 32, 4324. Copyright: American Chemical Society (1993).**

## Methods of study

The study of the structure arising from hydrogen bonding and the properties exhibited due to the presence of hydrogen bonds is very important.

*Diffraction methods*

X-ray diffraction of single crystals is the most common structural method employed to determine the presence, effect, and strength of a hydrogen bond. Unfortunately, in order for the location of the hydrogen to be determined with some degree of accuracy, diffraction data of a high quality is needed and/or low temperature (e.g., -196 °C) data collection is required. Neutron scattering can be used where very accurate data is required because hydrogen atoms scatter neutrons better than they do X-rays. Figure 7.18 summarizes the key parameters that are obtained from X-ray (and neutron) diffraction experiments.

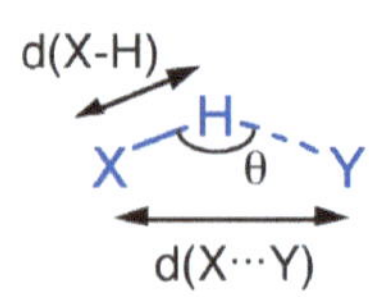


**Figure 7.18: Structural parameters obtained from diffraction methods.**

Given the electrostatic nature of a hydrogen bond between a polar X-H bond and a Lewis base it is reasonable that the X-H···Y angle (θ) is roughly linear (i.e., 180°). However, it is not always so and non-linear interactions are known where steric or conformational restrictions limit the orientation of the X-H bond with respect to Y.

The distance between X and Y, d(X···Y), is less than the sum of the van der Waal radii of X and Y (Table 7.5). This is in line with the relative strength of these interactions. As would be expected the shorter the X···Y distance the stronger the hydrogen bond.

| X | Y | Sum of van der Waal radii (Å) | Typical X···Y distance (Å) |
|---|---|---|---|
| O | O | 2.8 | 2.50 - 2.69 |
| O | N | 2.9 | 2.75 - 2.85 |
| N | N | 3.0 | 2.69 - 2.98 |

**Table 7.5: Comparison of the X···Y distance in hydrogen bonded species with the sum of the van der Waal radii.**

The bond distance to hydrogen, d(X-H), is often longer in hydrogen bonded species. For example, the O-H distance for an alcohol in the absence of hydrogen bonding is typically 0.97 Å. In contrast, the value typically seen for a hydrogen-bonded analog is 1.05 Å.

*Spectroscopy*

Spectroscopy is a simple method of comparing hydrogen-bonded systems in particular in the solution or liquid phase.

The X-H stretching frequency in the IR (and Raman) spectrum is dependent on the identity of X. For example, O-H = 3610 - 3640 $cm^{-1}$ and N-H = 3400 - 3500 $cm^{-1}$. However, the ν(X-H) is shifted to lower energy (lower frequency) as a consequence of hydrogen bonding. In addition, while non-hydrogen bonded X-H stretches are sharp, the presence of hydrogen bonding results in the peak being broadened. Figure 7.19 demonstrates both these effects. The O-H stretch for dilute $^{n}$BuOH in $CCl_4$ is a sharp peak at 3650 $cm^{-1}$ due to the lack of hydrogen bonding between the two components (Figure 7.19a), and the presence of hydrogen bonding between $^{n}$BuOH and $^{n}$BuOH is limited by the dilution. By contrast, a dilute solution of $^{n}$BuOH in $Et_2O$ results in a shift to lower frequency and a significant increase of peak width (Figure 7.21b) as a result of fairly strong O-H···O bonds. Finally, a dilute solution of $^{n}$BuOH in

$NMe_3$ results in a further shift to 3250 $cm^{-1}$ and a very broad peak (Figure 7.19c). The broadening of the peaks is due to the distribution of X-H distances within a X-H···Y hydrogen bond.

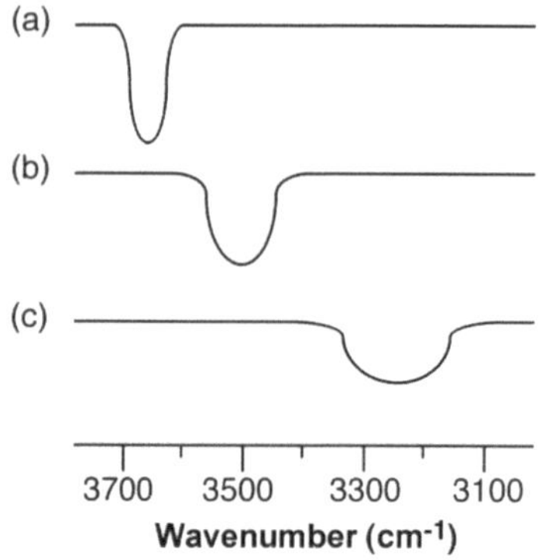


**Figure 7.19: Schematic representation of the O-H stretch region in the IR spectra of a dilute solution of $^{n}$BuOH in (a) $CCl_4$, (b) $Et_2O$, and (c) $NEt_3$.**

The presence of hydrogen bonding results the shift to higher ppm (lower frequency) of the $^{1}$H NMR resonance for the proton. This shift is due to the decrease in shielding of the proton. A dilute solution of $^{n}$BuOH in $CCl_4$ shows a resonance typical of a non-hydrogen bonded compound (Figure 2.20a), while that for $^{n}$BuOH in $NEt_3$ (Figure 2.20b) shows a significant low field shift. Very strong intra or intermolecular hydrogen bonded species show a very large $^{1}$H NMR shift (e.g., Figure 2.20c).

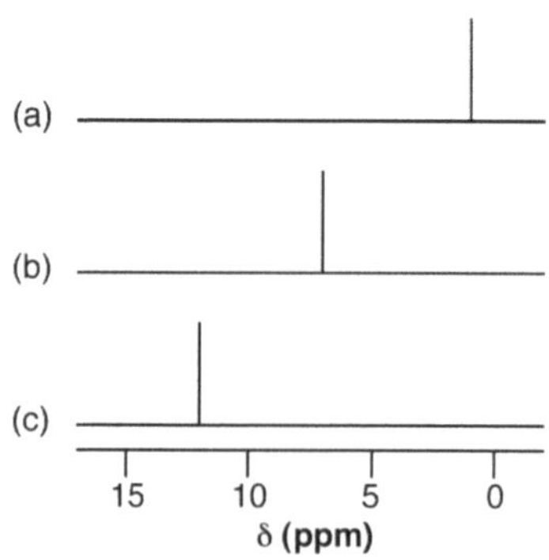


**Figure 7.20: Schematic representation of the relative shift of the OH resonance in the $^{1}$H NMR spectra for (a) $^{n}$BuOH in $CCl_4$, (b) $^{n}$BuOH in $NEt_3$, and (c) concentrated acetic acid.**

# Effects of hydrogen bonding

*Physical effects*

The presence of intermolecular hydrogen bonding provides additional attractive forces between molecules. Thus, properties that depend on intramolecular forces are affected.

Liquids with significant hydrogen bonding exhibit higher boiling points, higher viscosity, and higher heat of vaporization ($\Delta H_v$) as compared to analogous compounds without extensive hydrogen bonding. For solids the presence of hydrogen bonding results in an increase in the melting point of the solid and an increase in the associated heat of fusion ($\Delta H_f$).

The archetypal case for the effect of hydrogen bonding is the melting and boiling points of the hydrides of the Group 16 elements, i.e., $H_2E$. For a series of analogous compounds with the same molecular structure it would be expected that the boiling points would be related to the molecular mass. However, as can be seen from Table 7.6, the melting and boiling points of water are anomalously higher than those of its heavier analogs. In fact, from Figure 7.21 it is clear that just considering $H_2S$, $H_2Se$, and $H_2Te$, the expected trend is observed, and it is similar to that for the Group 14 hydrides ($CH_4$, $SiH_4$, etc.). Therefore, water must have additional intermolecular forces as compared to its heavier homologs. This observation is consistent with the strong hydrogen bonding in water, and the very weak if nonexistent hydrogen bonding in the sulfur, selenium, and tellurium analogs.

| Compound | Molecular weight (g/mol) | Melting point (°C) | Boiling point (°C) |
|---|---|---|---|
| $H_2O$ | 18.01 | 0 | 100 |
| $H_2S$ | 34.08 | -85.5 | -60.7 |
| $H_2Se$ | 80.98 | -60.4 | -41.5 |
| $H_2Te$ | 129.62 | -49 | -2 |

**Table 7.6: Summary of physical properties for the hydrides of the Group 16 elements.**

A similar but not as pronounced trend is observed for the Group 15 hydrides, where ammonia's higher values are associated with the presence of significant hydrogen bonding (Table 7.7).

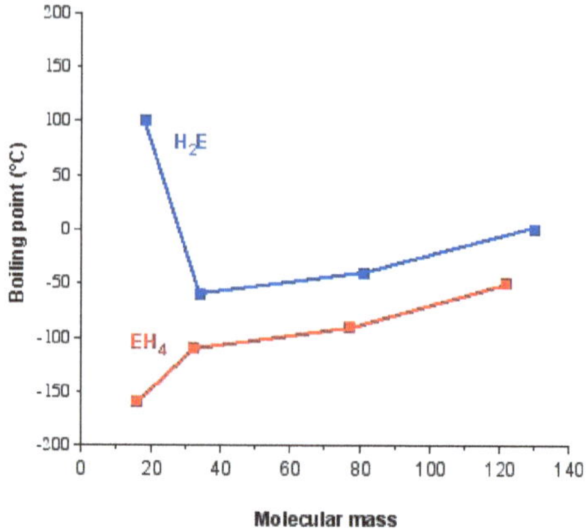


**Figure 7.21: Plot of boiling point as a function of molecular weight for the hydrides of the Group 14 ($EH_4$) and 16 ($H_2E$) elements.**

| Compound | Molecular weight (g/mol) | Melting point (°C) | Boiling point (°C) |
|---|---|---|---|
| $NH_3$ | 17.03 | -77.7 | -33.35 |
| $PH_3$ | 34.00 | -133.5 | -87.4 |
| $AsH_3$ | 77.95 | -113.5 | -55 |
| $SbH_3$ | 124.77 | -88.5 | -17 |

**Table 7.7: Summary of physical properties for the hydrides of the Group 15 elements.**

The types of hydrogen bond can also have a significant effect on the physical properties of a compound. For example, the *cis* isomer of hydroxybenzaldehyde melts at 1 °C, while the *trans* isomer has melting and boiling points of 112 °C. Both compounds exhibit strong hydrogen bonding in the solid state, however, as may be seen from Figure 7.22a, *cis*-hydroxybenzaldehyde (salicylaldehyde) has a configuration that allows strong intramolecular hydrogen bonding, which precludes any intermolecular hydrogen bonding. The melting point of *cis*-hydroxybenzaldehyde is going to be controlled by the van der Waal forces between adjacent molecules. In contrast, since intramolecular hydrogen bonding is precluded in the *trans* isomer (Figure 7.22b) it can form strong intermolecular hydrogen bonds in the solid state, and thus, it is these that de ne the melting point. The boiling points are controlled in a similar manner.

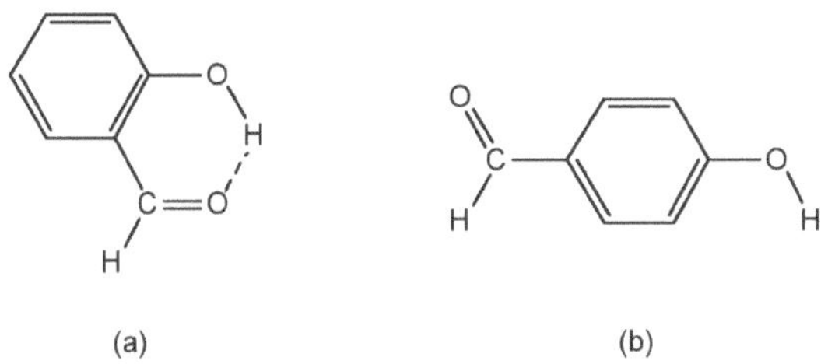

**Figure 7.22: Structures of (a) salicylaldehyde and (b) *trans*-hydroxybenzaldehyde.**

Melting and boiling are not the only physical properties that are affected by hydrogen bonding. Solubility can also be affected. Consider two isomers of $C_4H_{10}O$: $^nBuOH$ and $Et_2O$. The n-butanol is much more soluble in water than diethyl ether. The reason for this is that while both compounds can hydrogen bond to water, those between $^nBuOH$ and water are much stronger than those between $Et_2O$ and water, and thus, dissolution of $^nBuOH$ in water does not disrupt the very strong hydrogen bonding in water as much as $Et_2O$ does.

*Acid strength*

The acidity of a protic species can be affected by the presence of hydrogen bonding. For example, consider the di-carboxylic acid derivatives of ethylene (fumaric acid, $HO_2CCH{=}CHCO_2H$). Each of the carboxylic acid groups has sequential equilibria that may be defined by the pK values,

$$HO_2C\text{-}HC{=}CH\text{-}CO_2H \underset{}{\overset{pK_1}{\rightleftharpoons}} {}^-O_2C\text{-}HC{=}CH\text{-}CO_2H \overset{pK_2}{\rightleftharpoons} {}^-O_2C\text{-}HC{=}CH\text{-}CO_2^-$$

Table 7.8 lists the pK values for the *cis* and *trans* isomers. The acidity of the first and second carboxylic group for the *trans* isomer is similar, the difference being due to the increased charge on the molecule. In contrast, the second proton in the *cis* isomer is much less acidic than the first proton: why?

| Isomer | $pK_1$ | $pK_2$ | Ratio |
|---|---|---|---|
| *Trans* | 3 | 4.5 | 25:1 |
| *Cis* | 1.9 | 6.2 | 10,000:1 |

**Table 7.8: Acidity equilibrium constants for *cis* and *trans* isomers of fumaric acid ($HO_2CCH{=}CHCO_2H$).**

A consideration of the structure of the mono anion of the *cis* isomer (Figure 7.23) shows that a very strong intramolecular hydrogen bond is formed once the first proton is removed. This hydrogen bond makes the second acidic proton much more difficult to remove and thus lowers the acidity of the proton.

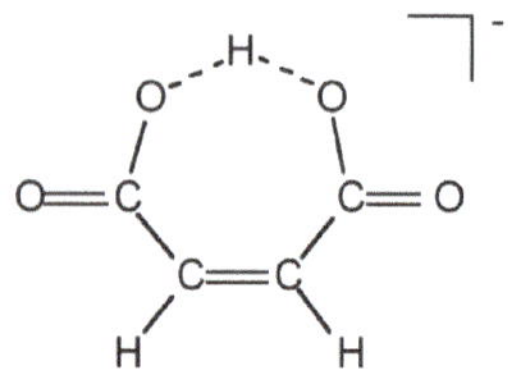


**Figure 2.23: Structure of *cis*-[$O_2C$-HC=CH-$CO_2H$]$^-$.**

*Solid state structure*

In the solid state, molecules that can form hydrogen bonds will tend to arrange themselves so as to maximize the formation of linear hydrogen bonds. The structure of ice is a typical example, where each water molecule's hydrogen bonds to four other molecules creating a diamond-like lattice (Figure 7.24).

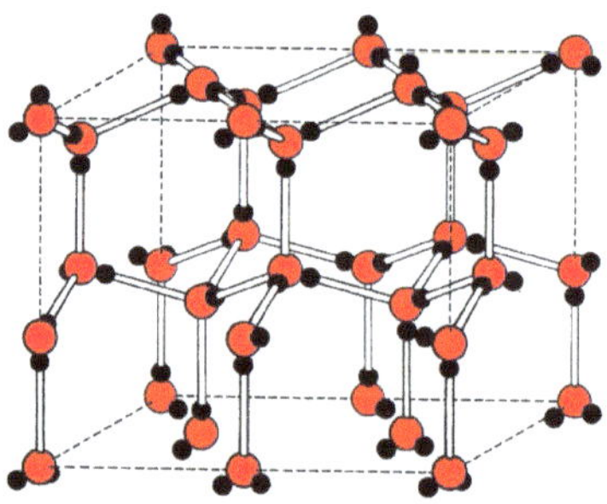

**Figure 7.24: The diamond-like structure formed in ice by hydrogen bonding.**

Ice isn't the only compound whose solid-state structure is defined by its hydrogen bonding. Cyanuric acid (1,3,5-triazine-2,4,6-triol) forms six strong intermolecular hydrogen bonds to three other molecules in the plane of the heterocyclic ring and thus creates a graphite-like structure (Figure 7.25).

The presence of hydrogen bonding can stabilize certain conformations over others that in the absence of hydrogen bonding would be more favored. For example, based on steric considerations $H_2NCH_2CH_2C(O)H$ would be expected to adopt a staggered conformation as is typical for compounds with free rotation about a C-C bond. However, due to the presence of a strong intramolecular hydrogen bond it adopts a sterically disfavored eclipsed conformation (Figure 7.26).

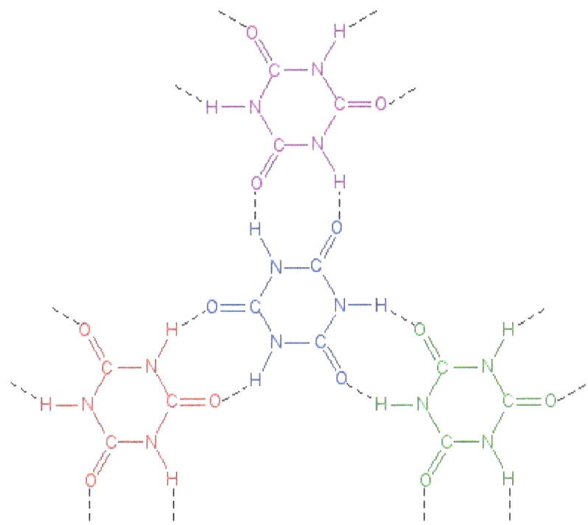

**Figure 7.25: The hydrogen bonding of cyanuric acid.**

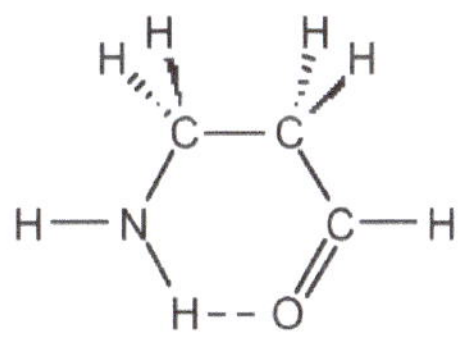

**Figure 7.26: Stable conformation of $H_2NCH_2CH_2C(O)H$.**

*Conformational stabilization*

This conformational stabilization is even more important in biopolymers such as peptides. If we assume that a peptide chain (Figure 7.27) made from a sequence of amino acids would not adopt any eclipsed conformations on steric grounds, and that the amide group (H-N-C=O) is near planar due to delocalization, then there will be $3^2$ possible conformations per nitrogen in the peptide chain, i.e., three each for the C-C and C-N bonds (Figure 7.27). Assuming a modest peptide has 50 amino acids, there will be $3^{2\times50}$ potential conformations of the polymer chain, i.e., over $5\times10^{47}$ conformations. However, in reality there are limits to the number of conformations observed since particular ones are stabilized by intramolecular hydrogen bonds. The most important of these are the α-helix (Figure 7.28) and β-sheet structures.

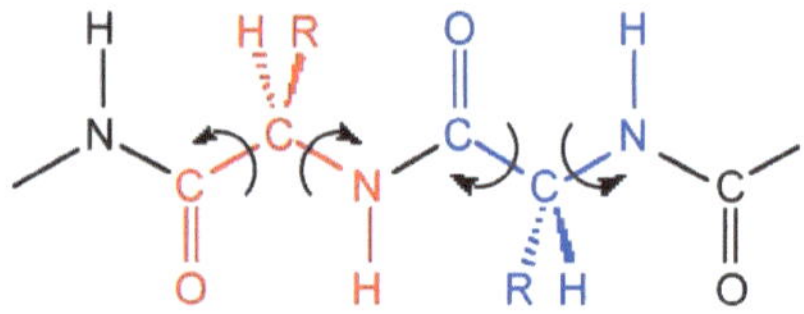


**Figure 7.27: A section of a peptide polymer chain showing the possible rotational freedom of the C-C and C-N bonds.**

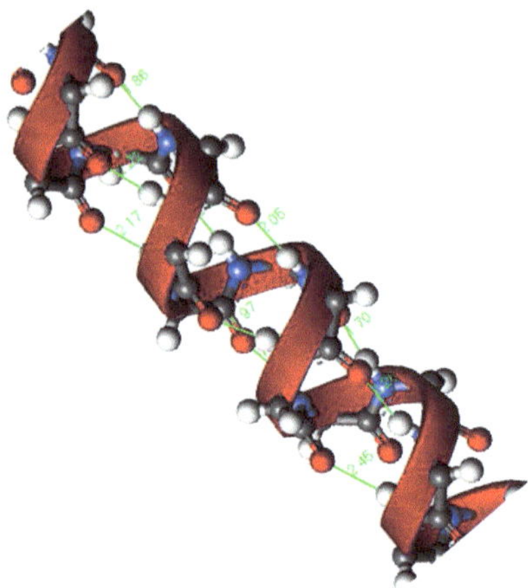

**Figure 7.28: A representation of a peptide α-helix structure showing the stabilizing hydrogen bonds.**

## Isotopes of hydrogen

Hydrogen has three isotopes (Table 7.9) and unlike other elements these are given special names to differentiate then from the most abundant isotope. Tritium is radioactive with a half-life of about 10 years.

| Isotope | Hydrogen-1 | Hydrogen-2 | Hydrogen-3 |
|---|---|---|---|
| Special name | Hydrogen | Deuterium | Tritium |
| Symbol | H | D | T |
| Atomic number | 1 | 1 | 1 |
| Number of neutrons | 0 | 1 | 2 |
| Mass number | 1 | 2 | 3 |
| Natural abundance | 99.9844% | 0.0156% | very small |

**Table 7.9: Summary of isotopes of hydrogen.**

# Synthesis of deuterium compounds

*Electrolysis of water*

The electrolysis of hydrogen-1 water ($H_2O$) in the presence of an alkali results in the formation of hydrogen and oxygen.

$$2\ H_2O_{(l)} \rightarrow 2\ H_{2(g)} + O_{2(g)}$$

In a similar manner the hydrolysis of deuterated water ($D_2O$) yields deuterium and oxygen.

$$2\ D_2O_{(l)} \rightarrow 2\ D_{2(g)} + O_{2(g)}$$

However, the rate of electrolysis of $D_2O$ is slightly slower than that of $H_2O$. Thus, the partial hydrolysis of water with a mixture of natural isotopes results in the slight enrichment of the water with $D_2O$. The level of enrichment in one step is less than 1%. In order to obtain high levels of $D_2O$ (e.g., ca. 30%) it is necessary to reduce the original volume of water by 1/100,000$^{th}$.

*Chemical equilibrium*

Proton exchange reactions can be used to enrich compounds in deuterium. For example, the reaction of HSD with water has a slight preference for the formation of $H_2S$, i.e., $K_{eq} = 1.012$.

$$H_2O_{(l)} + HSD_{(g)} \rightleftharpoons HOD_{(l)} + H_2S_{(g)}$$

Thus, bubbling HSD through water results in the enrichment of the water in HOD. However, about 30% enrichment is about the best that can be achieved by this method.

*Fractional distillation*

The boiling point of $H_2O$ is (by definition) 100 °C, in contrast the boiling point of $D_2O$ is 101.4 °C. Thus, it is possible to separate $H_2O$ from $D_2O$ by fractional distillation. This method provides the most suitable route to high isotopic enrichment and $D_2O$ of 99.8% can be produced this way. The term heavy water is used for $D_2O$ of greater than 99.8% enrichment.

## Uses of deuterium compounds

*Possible nuclear fusion*

The largest use of $D_2O$ is as a moderator and heat exchanger for fission nuclear reactors, however, the biggest potential application will be if nuclear fusion is realized as a commercial process.

The fusion of two deuterium atoms to form a helium atom and energy would be one source of energy,

$$^{2}_{1}H + ^{2}_{1}H \rightarrow ^{3}_{2}He + ^{1}_{0}n + \text{energy}$$

However, deuterium-tritium fusion is the most promising.

$$^{2}_{1}H + ^{3}_{1}H \rightarrow ^{4}_{2}He + ^{1}_{0}n + \text{energy}$$

The deuterium part of the fuel does not pose a great problem because about 1 part in 5000 of the hydrogen in seawater is deuterium. This amounts to an estimate that there is over $10^{15}$ tons of deuterium in the oceans. The tritium part of the fuel is more problematic since there is no significant natural source (Table 7.9), and the tritium would have to be obtained by breeding the tritium from lithium.

$$^{6}_{3}Li + ^{1}_{0}n \rightarrow ^{4}_{2}He + ^{3}_{1}H$$

Since a gallon of seawater could produce as much energy as 300 gallons of gasoline, there is clearly a large amount of energy that can potentially be realized through nuclear fusion. Unfortunately, this advantage is also a disadvantage since the temperatures attained are similar to the surface of the sun, which would vaporize any conventional container. Fusion experiments therefore use a magnetic field to contain the reaction. The shape of the field is like a bottle, hence the term magnetic bottle.

One demonstrated fusion process is the so-called hydrogen bomb or thermonuclear bomb in which a fission atom bomb is used to initiate a fusion reaction. The atomic bomb is surrounded by a layer of lithium deuteride. Neutrons from the atomic explosion (fission) cause the lithium to be converted into helium, tritium, and energy. The atomic explosion also supplies the 50,000,000 °C temperature needed for the subsequent fusion of deuterium with tritium. So, in-fact the hydrogen bomb is misnamed, and it should be called a deuterium bomb.

The original calculations to model the hydrogen bomb were performed using ENIAC (short for Electronic Numerical Integrator And Computer) that was originally designed to generate tables of trajectories of shells red from large artillery. The artillery ring tables were made by women mathematicians who were called 'calculators' hence the name used today. Built in 1946 ENIAC is often assumed to be the first programmable electronic computer; however, it was predated by the six Colossus machines that were used to successfully crack the German Enigma code as early as 1944. However, the existence of the Colossus machines was kept secret until 1975.

*Spectroscopy*

In the chemical laboratory deuterium compounds are commonly used in spectroscopy for:

- The assignment of resonances in IR, Raman, and NMR spectroscopy.
- As a non-proton containing solvent in $^{1}$H NMR spectroscopy.

A description of these applications is given below.

*Reaction mechanism and rate determination*

Given the larger mass of deuterium over hydrogen there is an associated difference in the rate of reactions (see below) and therefore investigations using hydrogen and deuterium analogs can provide information as to reaction mechanisms.

The spectroscopic differences between hydrogen and deuterium can also be used as a tracer to uniquely determine the source of particular substituents. For example, the magnesium (or Grignard) reduction of a ketone yields upon hydrolysis the secondary alcohol. If the reaction is carried out in a deuterated solvent and $H_2O$ used for hydrolysis then the secondary carbon is deuterated,

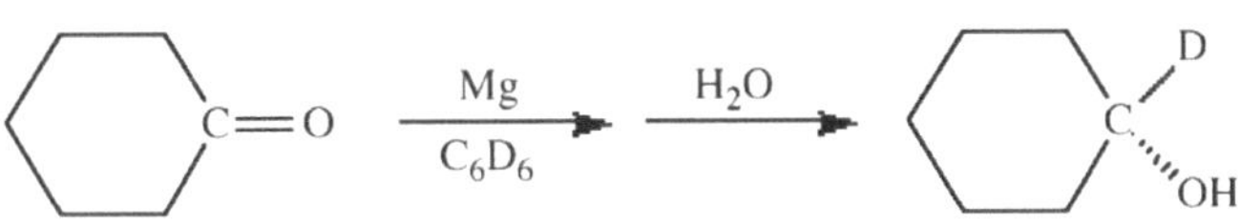

In contrast, if the reaction is carried out in a non-deuterated solvent and hydrolysis is accomplished with $D_2O$ then the deuterated alcohol is formed,

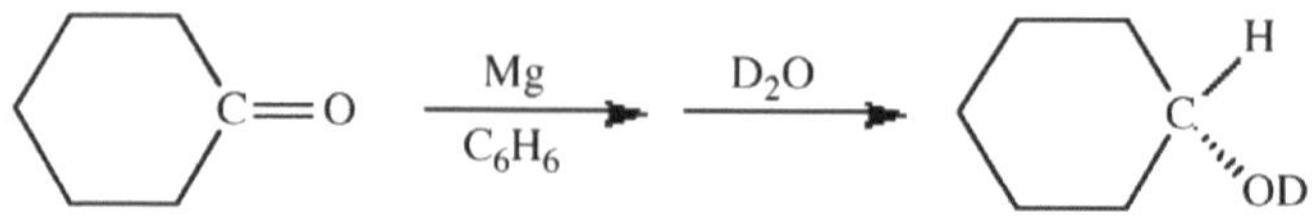


These experiments define that the initial reduction occurs at the ketone's α-carbon.

## Differences between hydrogen and deuterium

*Properties that depend on nuclei properties*

The nuclear magnetic moment of an atomic nucleus arises from the spins of the protons and neutrons within the nucleus. As a consequence, the magnetic moment for hydrogen and deuterium are very different and hence the conditions for detection by $^1$H NMR are very different. Thus, in observing the $^1$H NMR spectrum of a compound not only are the deuterium atoms not observed, but the coupling is now H-D rather than H-H (Figure 7.29).

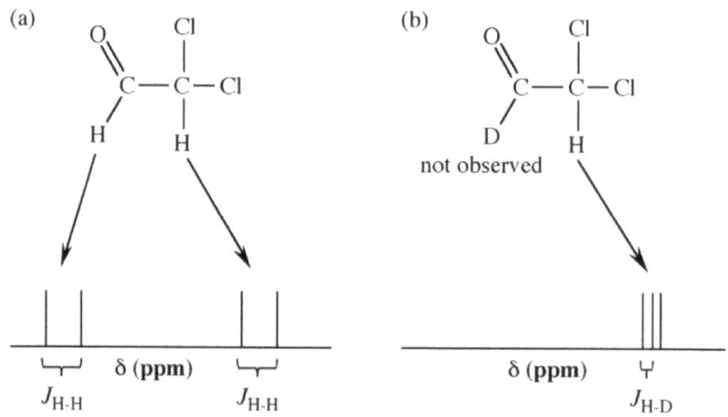


**Figure 7.29: An example of the differences in the $^1$H NMR spectrum upon deuterium substitution.**

Deuterium is better at scattering neutrons than hydrogen. The H and D cross sections are very distinct and different in sign, which allows contrast variation in such experiments. Hydrogen's low electron density makes it di cult to determine its position by X-ray diffraction methods, neutron diffraction methods allow for highly accurate structure determination. Hydrogen can be seen by neutron diffraction and scattering; however, it has a large incoherent neutron cross-section. This is nil for deuterium and thus delivers much clearer signals may be obtained for deuterated samples. Neutron scattering of deuterated samples is indispensable for many studies of macromolecules in biology.

*Properties that depend on mass*

The difference in mass between hydrogen and deuterium obviously results in a difference in molecular mass of their analogous compounds. This difference can be used for analysis by mass spectrometry, but it also results in different densities of compounds. For example, the density of $H_2O$ at 25 °C is 0.997 g/cm$^3$, while the density of $D_2O$ at 25 °C is 1.104 g/cm$^3$.

The vibrational frequency for a diatomic molecule, H-X, can be defined by the equation:

$$\nu_{H\text{-}X} = \frac{1}{2\pi}\sqrt{\frac{f_{H\text{-}X}}{\mu_{H\text{-}X}}}$$

where $f_{H\text{-}X}$ is the H-X bond force constant, and $\mu_{H\text{-}X}$ is the reduced mass.

$$\mu_{H\text{-}X} = \frac{m_H \cdot m_X}{m_H + m_X}$$

If substitute H for D the D-X force constant is the same as the H-X force constant, but the reduced mass is twice the value for the H-X bond. As a result, the ratio of the vibrational frequency of an H-X bond to that of the analogous D-X bond is given by the following equation.

$$\frac{\mu_{D\text{-}X}}{\mu_{H\text{-}X}} = \left(\frac{\mu_{H\text{-}X}}{\mu_{D\text{-}X}}\right)^{1/2} = \frac{1}{\sqrt{2}}$$

With the change in vibrational energy there is concomitant change in the bond strength.

$$E_{D\text{-}X} > E_{H\text{-}X}$$

Thus, the rate of reactions will be faster for hydrogen derivative than the deuterium analog. The ratio of the rate constants will be dependent on the involvement of H-X bond breaking or forming in the rate limiting step (the slowest reaction step within the overall reaction mechanism). When an H-X bond is made or broken in the rate limiting step, then the ratio of the rate constants upon deuterium substitution will be:

$$\frac{k_{H\text{-}X}}{k_{D\text{-}X}} \approx 7$$

This is known as the primary isotope effect. In this case where H-X bond breaking or forming is not part of the rate limiting step, then the isotope effect will be much smaller and is known as a secondary isotope effect.

The position of equilibrium reactions that involve hydrogen exchange,

$$\text{X-H} + \text{Y-D} \rightleftharpoons \text{X-D} + \text{Y-H}$$

will be affected by the presence of deuterium to favor the deuterium being concentrated in the more stable bond. This is the basis of the concentration of HOD from HSD and water.

## Nuclear fusion

The process of nuclear fission and radioactive decay are both associated with the conversion of an atom with a large nucleus to an atom (or atoms) with a smaller nucleus. In the process, mass is lost, and energy is produced. However, what happens if two atoms with small nuclei are combined to give a single atom with a larger nucleus? In such a process the nuclei would be fused together, and this process is called nuclear fusion.

One of the simplest fusion processes involves the fusing of two hydrogen-2 (deuterium) atoms:

$${}^{2}_{1}\text{H} + {}^{2}_{1}\text{H} \rightarrow {}^{4}_{2}\text{He}$$

The mass of each deuterium atom is 2.0140 amu, while the mass of the resulting helium is 4.0026 amu. The mass defect of the reaction is 0.0254 amu or 0.63% of the original mass. While this percentage of the original mass may not seem much, it should be noted that the mass defect for the conversion of uranium-238 to lead-206 is only 0.026%, and that for splitting uranium-235 is 0.056%. Based upon these comparisons it is clear that fusion of hydrogen produces 24× the energy kg/kg than natural radioactivity and 11× that of nuclear fission.

In addition to being, a plentiful source of energy, fusion is actually the most important process in the universe. Since the statement in 1847 of the Law of conservation of energy (the total amount of energy in an isolated system remains constant) scientists had wondered how the sun works. No source of energy was known in the 19$^{th}$ century that could explain the sun. Based upon the age of the earth it is known that the sun is 4,550,000,000 years old, and it was marveled at the continued source of energy over that span of time. By the 1920s nuclear energy was defined as being the most powerful source of energy, and British astrophysicist Arthur Eddington (Figure 7.30) suggested that the sun's energy arises from the fusion of hydrogen into helium.

**Figure 7.30: Astrophysicist Arthur Stanley Eddington (1882 - 1944).**

In 1929 American Henry Norris Russell (Figure 7.31) studied the spectrum of the sun. Based upon his studies he calculated the composition of the sun to be 90% hydrogen, 9% helium, and 1% of all other elements up to iron (but nothing of higher atomic number). Given the composition it became clear that the only reaction possible to account for the sun's energy was the fusion of hydrogen. In 1938 Hans Bethe (Figure 7.32) demonstrated a model for how the sun worked. The energy source is the fusion of hydrogen to give helium (Figure 7.33), while in massive starts the presence of heavier elements such as carbon, oxygen, nitrogen, neon, silicon, and iron are the result of the fusion of helium.

**Figure 2.31: Henry Norris Russell (1877 - 1957).**

**Figure 2.32: Hans Albrecht Bethe (1906 - 2005).**

Why is it that uranium, thorium and other radioactive elements undergo radioactive decay, but hydrogen does not undergo spontaneous fusion? The reason for this difference is that any change that uranium undergoes occurs inside a nucleus that is already formed, while fusion requires that two nuclei must come together. This process results in extremely large repulsive forces. So why does it happen in a star? The temperatures inside a star approach

15,000,000 °C, while the high-pressure results in a density of approximately 160 g/cm$^3$, which for comparison is 8× that of gold. Under these conditions the nuclei are free to move in a sea of electrons. Nuclear fusion takes place in the core of the sun where the density is at the highest, but an explosion does not result due to the extreme gravity of the sun: 333,000× that of Earth.

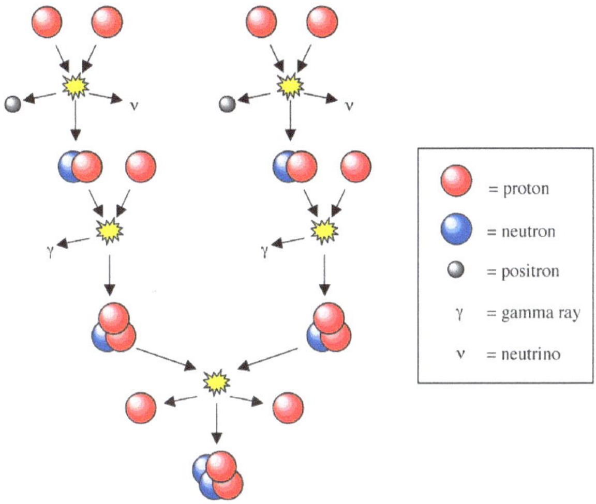


**Figure 7.33: The proton-proton chain dominates in stars the size of the Sun or smaller.**

Research into controlled fusion, with the aim of producing fusion power for the production of electricity, has been conducted for over 50 years. It has been accompanied by extreme scientific and technological difficulties but has resulted in progress. At present, break-even (self-sustaining) controlled fusion reactions have not been demonstrated, however, work continues since the fuel for such a fusion reaction is hydrogen (in its compounds including water) is the 3rd most abundant element on the Earth.

## Storage of hydrogen for use as a fuel

Dihydrogen is a colorless and odorless gas at room temperature which is highly flammable, releasing a large amount of energy when combusted. As compared with combustion of the current fuels which operate automobiles, for example petrol or diesel, the energy released when hydrogen is combusted is more than three times greater. The heat of combustion for hydrogen is 141.9

kJ/mol as compared to 47.0 kJ/mol and 45.0 kJ/mol for gasoline and diesel, respectively.

Furthermore, the combustion of hydrocarbons releases the greenhouse gas carbon dioxide ($CO_2$) into the atmosphere and is therefore not a "clean" fuel. When hydrogen is combusted in the presence of oxygen (from air) the only product is water,

$$2\,H_{2(g)} + O_{2(g)} \rightarrow 2\,H_2O_{(g)}$$

Both its clean reactivity and the large chemical energy make $H_2$ extremely appealing for use as a fuel in automobiles.

If hydrogen has such a potential as a fuel why has it not been widely implemented? Dihydrogen is a gas at room temperature. Gases, compared to the other states of matter (liquid and solid), occupy the most volume of space, for a given number of molecules. Octane and the other hydrocarbons found in gasoline are liquids at room temperature, demanding relatively small fuel tanks. Liquids are therefore easier to store than compressed gases.

Hydrogen has a high energy content per weight (more than three times as much as gasoline), but the energy density per volume is rather low at standard temperature and pressure. Volumetric energy density can be increased by storing the gaseous hydrogen under increased pressure or storing it at extremely low temperatures as a liquid. Hydrogen can also be adsorbed into metal hydrides and highly porous materials (Table 7.10). The current available methods of storing hydrogen include compressed hydrogen and liquefied hydrogen, however many promising methods exist, namely metal organic materials (MOMs), metal hydrides and carbon nanostructures.

| **Material** | **H-atoms per cm$^3$ (×10$^{22}$)** | **% of weight that is $H_2$** |
|---|---|---|
| $H_2$ gas, 200 bar (2850 psi) | 0.99 | 100 |
| $H_2$ liquid, 20 K (-253 °C) | 4.2 | 100 |
| $H_2$ solid, 4.2 K (-269 °C) | 5.3 | 100 |
| $MgH_2$ | 6.5 | 7.6 |
| $Mg_2NiH_4$ | 5.9 | 3.6 |
| $FeTiH_2$ | 6.0 | 1.89 |
| $LaNi_5H_6$ | 5.5 | 1.37 |

**Table 7.10: Comparison of hydrogen storage ability of metal hydrides.**

## Liquid hydrogen

Liquid hydrogen is made possible by cryogenically cooling it to below its boiling point, -253 °C. As a liquid, the same amount of gaseous hydrogen will require much less volume, and therefore is feasible to individual automobile use. A refrigeration system is required to keep the liquid cooled, for if the system temperature rises above hydrogen's critical point (-241 °C), the liquid will become a gas. There must exist a vacuum insulation between the inner and outer walls of liquid hydrogen tank system, for heat cannot travel through a vacuum. There is a tradeoff, however, because the tank must be an open system to prevent overpressure. This will lead directly to heat loss, though minimal.

The relative tank size has a broad range, with small tanks having a volume of 100 L, and large spherical tanks sizing all the way up to 2000 $m^3$. Refrigeration systems are not a likely feature for every automobile, and open systems may pose a hazard should an accident occur. Cooling hydrogen down to a liquid is a convenient method of storage, however, and its implementation most likely will be limited to large stationary tanks as well as mobile multi-axle trucks.

## Compressed hydrogen

Compressing gas is the process of applying an external force which minimizes the distance between gas particles, therefore forcing the system to occupy less volume. This is attractive since many particles can exist in a reasonably sized tank. At room temperature and atmospheric pressure, 4 kg of hydrogen occupies a volume of 45 $m^3$, which corresponds to a balloon with a diameter of 5 m. Clearly compression is required to store and transport the gas. When it comes to individual mobility however, these tanks are still far too large for the average sized automobile.

Compressed tanks are regularly filled to 200 atmospheres in most countries. Storing 4 kg of hydrogen still requires an internal volume of 225 L (about 60 gallons). This amount can be divided into 5 tanks with 45 L internal volume.

## Metal hydrides for storage

Metal hydrides are coordinated complexes and/or crystal systems which reversibly bind hydrogen. The hydrogen is favorably incorporated into the complex and may be released by applying heat to the system. A major method to determine a particular complex's effectiveness is to measure the amount of hydrogen that can be released from the complex, rather than the amount it can store (Table 7.10).

Some issues with metal hydrides are low hydrogen capacity, slow uptake and release kinetics, as well as cost. The rate at which the complex accepts the hydrogen is a factor, since the time to fuel a car should ideally be minimal. Even more importantly, at the current stage of research, the rate at which hydrogen is released from the complex is too slow for automobile requirements. This technology is still a very promising method, and further research allows for the possibility of highly binding and rapid reversal rates of hydrogen gas.

# Chapter 8: Group 1 The Alkali Metals

## The Elements

The Group 1 metals have a particular name: the alkali metals. This is due to the formation of alkali (basic) solutions upon their reaction with water. Table 8.1 lists the derivation of the names of the alkali metals.

| Element | Symbol | Name |
|---|---|---|
| Lithium | Li | Greek lithos meaning stone |
| Sodium | Na | Latin natrium or Arabic natrun meaning soda |
| Potassium | K | From the Latin kalium, and from Arabic al-qali meaning plant ashes |
| Rubidium | Rb | Latin rubidus meaning deepest red |
| Caesium | Cs | Latin caesius meaning blueish grey |
| Francium | Fr | Named after France |

**Table 8.1: Derivation of the names of each of the alkali metal elements.**

It should be noted that 'caesium' is the international spelling standardized by the IUPAC, but in the United States it is commonly spelled as 'cesium'.

## Discovery

*Lithium*

Petalite ($Li_2O.Al_2O_3.8SiO_2$) was first discovered in 1800 by José Bonifácio de Andrade e Silva (Figure 8.1), who discovered the mineral in a Swedish mine on the island of Utö. However, it was not until 1817 that Johan August Arfwedson (Figure 8.2) working in the laboratory of Jöns Jakob Berzelius (Figure 8.3), discovered the presence of a new element while analyzing petalite ore. Named from the Greek '*lithos*' meaning stone reflected its discovery in a mineral, as opposed to sodium and potassium, which had been discovered in plant tissue; its name was later standardized as lithium. The element was not isolated until 1821, when William Brande (Figure 8.4) isolated the element by performing electrolysis on lithium oxide, a process previously employed by Sir Humphry Davy to isolate potassium and sodium.

**Figure 8.1: Portuguese statesman and naturalist José Bonifácio de Andrada e Silva (1763 - 1838).**

**Figure 8.2: Swedish chemist Johan August Arfwedson (1792 - 1841).**

*Sodium*

Elemental sodium was first isolated by Sir Humphry Davy (Figure 8.5) in 1806 by passing an electric current through molten sodium hydroxide.

*Potassium*

The name *kalium* was taken from the word alkali, which came from Arabic *al qali* meaning the 'calcined ashes'. The name potassium was made from the English word potash, meaning an alkali extracted in a pot from the ash of burnt wood or tree leaves. Potassium metal was discovered in 1807 by Sir Humphry Davy (Figure 8.5), who derived it from caustic potash (KOH), by the use of electrolysis of the molten salt.

**Figure 8.3: Swedish chemist Jöns Jacob Berzelius (1779 - 1848).**

**Figure 8.4: English chemist William Thomas Brande FRS (1788 - 1866).**

**Figure 8.5: British chemist and inventor Sir Humphry Davy FRS (1778 - 1829).**

*Rubidium*

Rubidium was discovered using spectroscopy in 1861 by Robert Bunsen (Figure 8.6) and Gustav Kirchho (Figure 8.7) in the mineral lepidolite. The first rubidium metal was produced by Bunsen from the reaction of rubidium chloride (RbCl) with potassium.

**Figure 8.6: German chemist Robert Wilhelm Eberhard Bunsen (1811 - 1899).**

**Figure 8.7: German physicist Gustav Robert Kirchho (1824 - 1887).**

*Caesium*

Like Rubidium, caesium was discovered spectroscopically by Bunsen (Figure 8.6) and Kirchho (Figure 8.7) in 1860 in mineral water from Dürkheim, Germany. The residues of 44,000 liters of mineral water yielded several grams of a caesium salt. Its identification was based upon the bright blue lines in its spectrum and it was the first element discovered by spectral analysis. The first caesium metal was subsequently produced in 1882 by electrolysis of caesium chloride.

*Francium*

Originally known as eka-caesium, francium was discovered in 1939 by Marguerite Perey (Figure 8.8) of the Curie Institute in Paris, France when she purified a sample of actinium-227, which had been reported to have a decay energy of 220 keV. However, Perey noticed decay particles with an energy level below 80 keV. Perey thought this decay activity was caused by a previously unidentified element, which exhibited chemical properties of an alkali metal. This led Perey to believe that it was element 87, caused by the alpha decay of actinium-227. Perey named the new isotope actinium-K (now referred to as francium-223) and in 1946, she proposed the name *catium* for her newly discovered element, however, she subsequently suggested *francium*, after France.

**Figure 8.8: French physicist Marguerite Catherine Perey (1909 - 1975), a student of Marie Curie, and the first woman to be elected to the French Académie des Sciences.**

## Abundance

The abundance of the alkali metals is given in Table 8.2. Sodium's high abundance is due mainly to large underground deposits of rock salt (NaCl). However, sodium is also more abundant in seawater (10,800 ppm) as compared to potassium (380 ppm).

| Element | Terrestrial abundance (ppm) |
|---|---|
| Li | 20 (Earth's crust), 40 (soil), 0.17 (sea water) |
| Na | 23,000 (Earth's crust), 10,500 (sea water) |
| K | 21,000 (Earth's crust), 14 (soil), 380 (sea water) |
| Rb | 90 (Earth's crust), 30 - 250 (soil), 0.1 (sea water) |
| Cs | 3 (Earth's crust), 0.0001 (soil), 0.0003 (sea water) |
| Fr | Essentially nil |

**Table 8.2: Abundance of alkali metal elements.**

## Isotopes

The naturally abundant isotopes of the alkali metals are listed in Table 8.3. All of the isotopes of francium are radioactive. Lithium-7 and sodium-23 are both useful NMR nucleus having I = ½.

| Isotope | Natural abundance (%) |
|---|---|
| Lithium-6 | 7.5 |
| Lithium-7 | 92.5 |
| Sodium-23 | 100 |
| Potassium-39 | 93 |
| Potassium-40 | 0.0118 |
| Potassium-41 | 6.9 |
| Caesium-133 | 100 |

**Table 8.3: Abundance of the major isotopes of the alkali metals.**

Potassium has three isotopes (Table 3.3), of which potassium-40 is radioactive and provides the basis for the determination of the age of rocks between $10^5$ and $10^9$ years old, i.e., those formed in proterozoic and cenozoic periods of geological time. The decay of potassium-40 occurs with a half-life of $1.31 \times 10^9$ years, by two routes. That associated with a beta particle decay accounts for 89% of the decay:

$$^{40}_{19}K \rightarrow ^{97}_{20}Ca + ^{0}_{-1}e$$

While that associated with an electron capture and by positron emission decay accounts for 11% of the decay to give argon-40. Since many rocks contain potassium containing minerals the decay of potassium-40 after solidi cation of the rock will result in the formation of argon trapped in the rock. The argon-40 content is determined by mass spectrometry, while the potassium content is determined by flame spectrophotometry. The ratio of the two, will allow for the determination of the elapsed time since the rock solidified.

$$\frac{[^{40}_{18}Ar]}{[^{40}_{19}K]}$$

Caesium has at least 39 known isotopes (more than any other element except francium) ranging from caesium-112 caesium-151; however, caesium-133 is the only naturally occurring stable isotope. The other isotopes have half-lives from a few days to fractions of a second. The radiogenic isotope caesium-137 is produced from the detonation of nuclear weapons and is produced in nuclear power plants and was released to the atmosphere most notably from the 1986 Chernobyl accident.

## Physical properties

Many of the physical properties of the alkali metals (Table 8.4) are typical of metals, e.g., thermal and electrical conductivity. However, due to the relatively weak inter-atomic forces (weak M-M bonding) they are soft and readily cut with a knife.

| Element | Melting point (°C) | Boiling point (°C) | Density (g/cm$^3$) | Electrical resistivity (Ω·cm) |
|---|---|---|---|---|
| Li | 453 | 1615 | 0.534 | 12.17 @ 86 °C |
| Na | 370 | 1156 | 0.968 | 5.23 @ 29 °C |
| K | 336 | 1032 | 0.89 | 7.01 @ 22.8 °C |
| Rb | 312 | 961 | 1.532 | 12.53 @ 53 °C |
| Cs | 301 | 944 | 1.93 | 37.38 @ 28.1 °C |

**Table 8.4: Selected physical properties of the alkali metal elements.**

## Reactivity

All the alkali metals are highly reactive and are as a consequence of the stability of the $M^+$ ion, are strong reducing agents (Table 8.5). The metals react readily with hydrogen and oxygen.

$$M + {}^{1}/_{2}\,H_2 \rightarrow M^+H^-$$

$$M + O_2 \rightarrow M^+(O_2)^-$$

| Reduction | Reduction potential (V) |
|---|---|
| $Li^+ + e^- \rightarrow Li$ | -3.045 |
| $Na^+ + e^- \rightarrow Na$ | -2.7109 |
| $K^+ + e^- \rightarrow K$ | -2.924 |
| $Rb^+ + e^- \rightarrow Rb$ | -2.925 |
| $Cs^+ + e^- \rightarrow Cs$ | -2.923 |

**Table 8.5: Electrochemical reduction potential for alkali metals.**

All of the alkali metals react with water to liberate hydrogen.

$$M + H_2O \rightarrow M(OH) + {}^{1}/_{2}\,H_2$$

The reactions of alkali metals with water are exothermic and the heat generated is sufficient to ignite the hydrogen. In addition, the solutions formed are highly alkaline. Caution should be taken when handling alkali metals and storage should always be under mineral oil. A similar, but less violent, reaction is also observed with ammonia when catalyzed by transition metal ions.

$$M + NH_3 \rightarrow M(NH_2) + {}^{1}/_{2}\,H_2$$

In the absence of a catalyst, the Group 1 metals dissolve in liquid ammonia to form solutions with characteristic properties.

- Highly reducing.
- Blue color.
- ESR signal due to solvated electrons.

As an example, the dissolution of sodium in liquid ammonia results in the formation of solvated $Na^+$ cations and electrons.

$$Na_{(s)} \rightarrow Na_{(solv)} \rightleftharpoons Na^+_{(solv)} + e^-_{(solv)}$$

The solvated electrons are stable in liquid ammonia and form a complex: $[e^-(NH_3)_6]$. It is this solvated electron that gives the strong reducing properties of the solution as well as the characteristic signal in the ESR spectrum associated with a single unpaired electron. The blue color of the solution is often ascribed to these solvated electrons; however, their absorption is in the far infra-red region of the spectrum. A second species, $Na^-_{(solv)}$, is actually responsible for the blue color of the solution.

$$2\ Na_{(solv)} \rightleftharpoons Na^+_{(solv)} + Na^-_{(solv)}$$

The formation of the sodium anion is confirmed by complexation of the cation with a cryptan ligand ("C") such as a crown ether.

$$Na^+_{(solv)} + Na^-_{(solv)} + \text{"C"} \rightarrow [Na(C)]^+ + Na^-$$

The resulting complex is found to be isostructural to the iodide analog in the solid state.

$$Na^+_{(solv)} + I^-_{(solv)} + \text{"C"} \rightarrow [Na(C)]^+ + I^-$$

## Vapor phase

All the alkali metals form $M_2$ dimers in the vapor phase in an analogous manner to hydrogen. As with dihydrogen the bonding is associated with the molecular orbital combination of the two-valence s-orbitals (Figure 8.9).

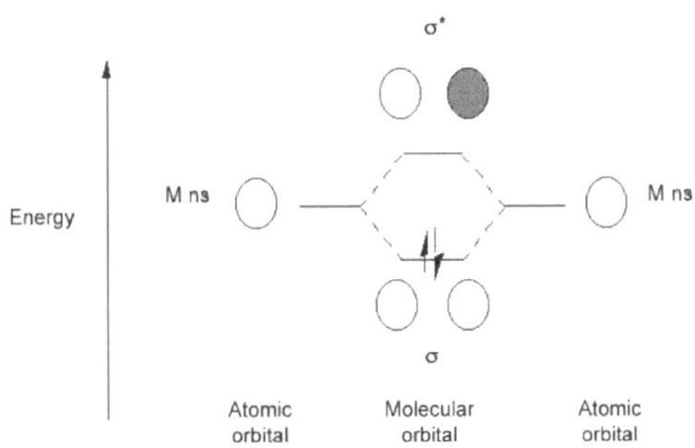


**Figure 8.9: Molecular orbital diagram for the formation of $M_2$.**

Sodium vapor is commonly used for lighting in a gas discharge lamp, which uses sodium in an excited state to produce light (Figure 8.10). There are two varieties of such lamps: low pressure and high pressure.

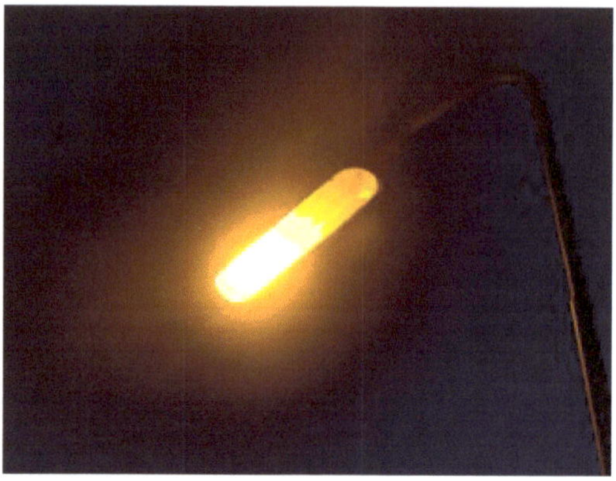

**Figure 8.10: A low pressure sodium streetlamp.**

## Compounds of the alkali metals

The chemistry of the alkali metals is dominated by the stability of the +1 oxidation state and the noble gas configuration of the $M^+$ cation. The alkali metals all have low first ionization energies (Table 8.6) but very high second ionization energies. As a consequence of the stability of $M^+$, the Group 1 metals have the least variation in chemistry of the any Group in the periodic table. The only exceptions are the subtle trends that exist for lithium due to its small size (Table 8.7). All of the metals are more electropositive than hydrogen (Table 3.8).

| Element | $1^{st}$ ionization energy (kJ/mol) |
|---|---|
| Li | 526 |
| Na | 502 |
| K | 425 |
| Rb | 409 |
| Cs | 382 |

**Table 8.6: First ionization potentials for the alkali metals.**

| Element | Atomic radius (Å) | Ionic radius (Å) | Covalent radius (Å) | Van der Waals radius (Å) |
|---|---|---|---|---|
| Li | 1.52 | 0.68 | 1.52 | 1.82 |
| Na | 1.86 | 0.97 | 1.53 | 2.27 |
| K | 2.31 | 1.33 | 1.90 | 2.75 |
| Rb | 2.44 | 1.47 | 2.47 | - |
| Cs | 2.62 | 1.67 | 2.65 | - |
| Fr | - | 1.80 | 2.70 | - |

**Table 8.7: Radii of alkali metals. Some values are unknown.**

| Element | Electronegativity |
|---|---|
| H | 2.20 |
| Li | 0.98 |
| Na | 0.93 |
| K | 0.82 |
| Rb | 0.82 |
| Cs | 0.79 |
| Fr | 0.70 |

**Table 8.8: Pauling electronegativities of Group 1 elements.**

## Solid state

In the solid state the compounds of the alkali metals generally form ionic lattices, e.g., $Na^{+}Cl^{-}$. These structures are essentially electrostatic in nature and the lattice energy is usually defined as the enthalpy of formation of the ionic compound from gaseous ions and as such is invariably exothermic.

$$M^{+}_{(vap)} + X^{-}_{(vap)} \rightarrow MX_{(s)}$$

In all cases the lattice energies are high and is found to be proportional to the ratio of the charges on the ions and the sum of the ionic radii (r).

$$U \propto \frac{z^{+}.z^{-}}{\Sigma(r)}$$

The ionic radii for alkali metal cations are given in Table 8.7; those for common anions are given in Table 8.9.

| Anion | Ionic radius (Å) |
|---|---|
| $F^-$ | 1.33 |
| $Cl^-$ | 1.81 |
| $Br^-$ | 1.96 |
| $I^-$ | 2.20 |
| $H^-$ | 1.54 |
| $O^{2-}$ | 1.32 |

**Table 8.9: Ionic radii of common anions.**

The ratio of the ionic radii ($r^+/r^-$) neatly defines the structural type observed for alkali metal salts (Table 8.10). The unit cells for ZnS (zinc blende), NaCl (rock salt), and CsCl are shown in Figures 3.10, 3.12, and 8.11, respectively.

| $r^+/r^-$ | Structural type | Metal coordination number |
|---|---|---|
| 0.225 - 0.414 | ZnS (zinc blende) | 4 |
| 0.414 - 0.732 | NaCl (rock salt) | 6 |
| 0.732 - | CsCl | 8 |

**Table 8.10: Defining MX structural types for alkali metal salts.**

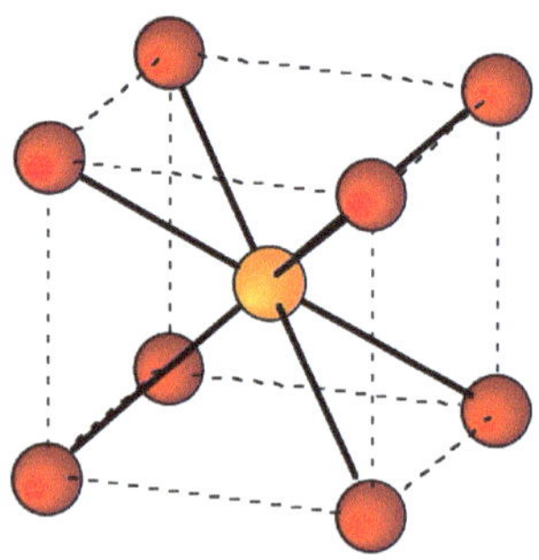

**Figure 8.11: Unit cell of CsCl.**

As an example, the structure of KBr can be predicted from the data in Table 8.7 and Table 8.9. The ionic radius of $K^+$ is 1.33 Å, while that for $Br^-$ is 1.96 Å. The ratio of the ionic radii ($r^+/r^-$) is 0.67. Hence, KBr has a NaCl (rock salt) structure.

# Complexes

The coordination complexes of the alkali metal cations ($M^+$) involve electrostatic, or ion-dipole, interactions (Figure 8.12) that have no preferred direction of interaction. Thus, the ionic radius of the cation (Table 8.7) controls the coordination numbers of the metal in its complexes (Table 8.11).

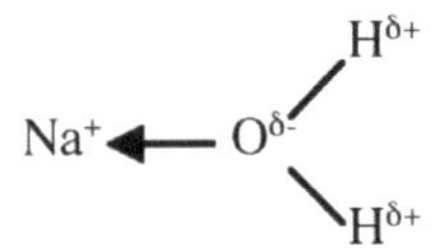


**Figure 8.12: A schematic representation of the ion-dipole or electrostatic interaction of Na with water.**

| Aquo ion | n |
|---|---|
| $[Li(H_2O)_n]^+$ | 4 |
| $[Na(H_2O)_n]^+$ | 6 |
| $[K(H_2O)_n]^+$ | 6 |
| $[Rb(H_2O)_n]^+$ | 6-8 |
| $[Cs(H_2O)_n]^+$ | 8 |

**Table 8.11: Coordination number (n) for alkali metal ions in aqueous complexes $[M(H_2O)_n]^+$.**

In general, the alkali metal ions form complexes with hard donor such as oxygen ($H_2O$, ROH, $RCO_2^-$, etc.) or nitrogen (e.g., $NH_3$, $NR_3$, etc.). The aquo complexes readily exchange the water for other ligands,

$$[M(H_2O)_n]^+ + L \rightleftharpoons [M(H_2O)_{n-1}L]^+ + H_2O$$

however, the equilibrium constants are small when the ligand is similar in size to water. As a consequence of the low equilibrium constants for monodentate ligands, the alkali metal cations, $M^+$, favor coordination to polydentate ligands such as ethylenediaminetetraacetic acid (EDTA, Figure 8.12), polyethers, and even natural polyesters or polypeptides. In each case the polydentate ligand wraps itself around the cation.

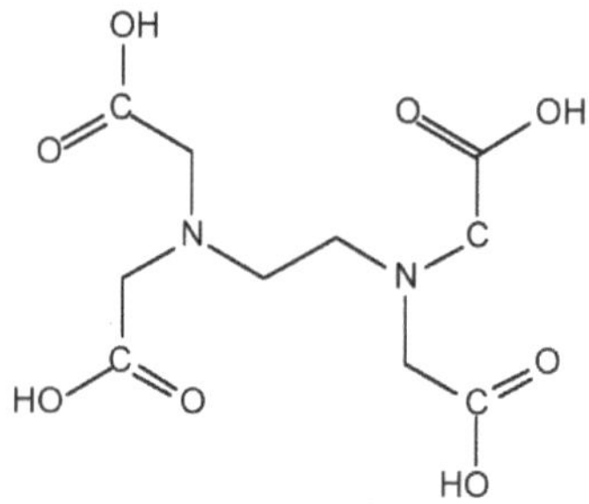

**Figure 8.12: Structure of ethylenediaminetetraacetic acid (EDTA).**

Macrocyclic ligands represent a special class of polydentate ligand. They are defined as being a cyclic compound with nine or more members including all the heteroatoms and with three or more donor atoms. and are given the special name of cryptands when they are synthetic bi- and poly-cyclic multidentate ligands. Crown ethers is the name applied to macrocyclic ligands that consist of a ring containing several ether groups. Figure 8.13 shows several common macrocyclic ligands.

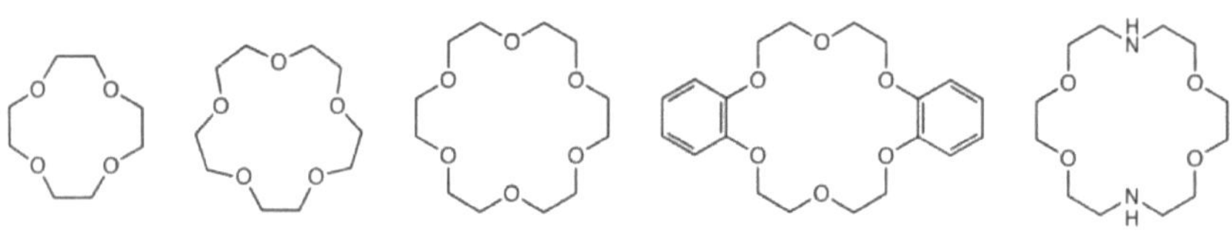

**Figure 8.13: The structures of some common macrocyclic ligands.**

Macrocyclic ligands are generally oxygen or nitrogen donor ligands, and they form highly stable 1:1 complexes with alkali metal ions in which all or most of the coordination sites on the metal are occupied, i.e., $[M(L)]^+$ rather than $[M(L)(H_2O)_n]^+$. Since the external surface of the macrocyclic ligands comprises of organic residue (e.g., $CH_2$ groups) the complexes are soluble in organic solvents. Thus, crown ethers are commonly used to solubilize salts (e.g., NaCl) in organic solvents. They have also been used to create solutions of nanoparticles, such as carbon nanotubes (Figure 8.14), that are ordinarily highly insoluble in common organic solvents (Table 8.12).

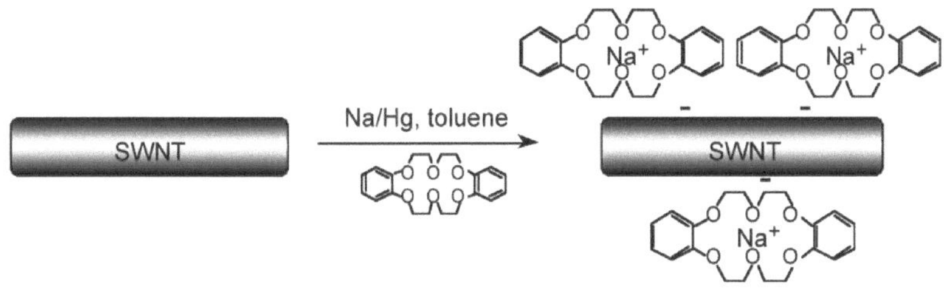


**Figure 8.14: Representation of the reaction between Na/Hg amalgam, dibenzo-18-crown-6, and purified single walled carbon nanotubes (SWNTs) in toluene and the formation of the [Na(dibenzo-18-crown- 6)]$_n$[SWNT] complex. Adapted from R. E. Anderson and A. R. Barron, Solubilization of single-wall carbon nanotubes in organic solvents without sidewall functionalization. *J. Nanosci. Nanotechnol.*, 2007, 7, 3436. Copyright: American Scientific Publishers (2007).**

| Solvent | SWNT concentration (mg/L) |
|---|---|
| $CH_2Cl_2$ | 14.05 |
| DMF | 11.42 |
| hexane | 4.65 |
| toluene | 3.79 |
| EtOH | 2.37 |
| MeOH | 2.37 |
| $CHCl_3$ | 0.30 |
| $H_2O$ | 0.10 |

**Table 8.12: The concentration of Na/dibenzo-18-crown-6 solubilized reduced single walled carbon nanotubes (SWNTs) in various solvents.**

The most important factor in the coordination of various macrocyclic ligands to alkali metal ions is the relative size of the ion and the ligand cavity. For example, 4,7,13,16,21,24-hexaoxa-1,10- diazabicyclo[8,8,8]hexacosane is a potentially octa-dentate ligand; however, the binding efficiency is very dependent on the identity of the $M^+$ ion (Figure 8.15). The low binding constant for lithium is probably as a consequence that the ligand would have to distort to coordinate to the small cation. Conversely, the lower equilibrium for caesium is due to its being too large to fit completely into the ligand cavity.

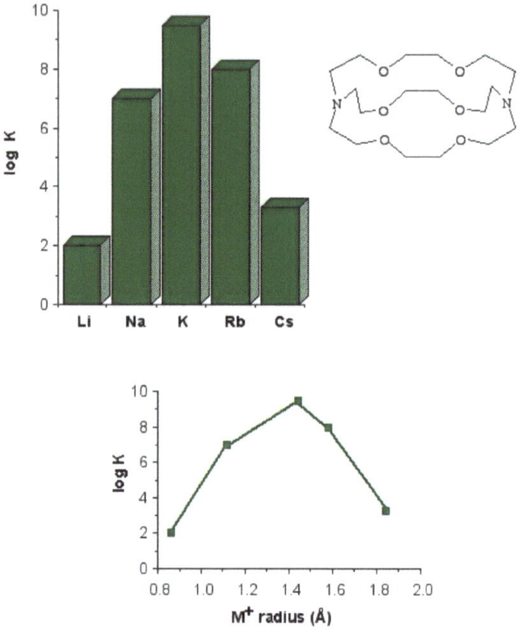


**Figure 8.15: Relative stability constants (log K) of alkali metal ions to 4,7,13,16,21,24-hexaoxa-1,10-diazabicyclo[8,8,8]hexacosane and relationship to the $M^+$ ionic radius.**

One application of the size effect for macrocyclic ligands is the ability to selectively bind different metals. For example, the 4,7,13,16,21-pentaoxa-1,10-diazabicyclo[8,8,5]tricosane and 4,7,13,16,21,24-hexaoxa-1,10-diazabicyclo[8,8,8]hexacosane ligands shown in Figure 8.16 have very different binding constants to $Na^+$ and $K^+$, as a consequence of the relative size of the cation and the ligand cavity (Table 8.13).

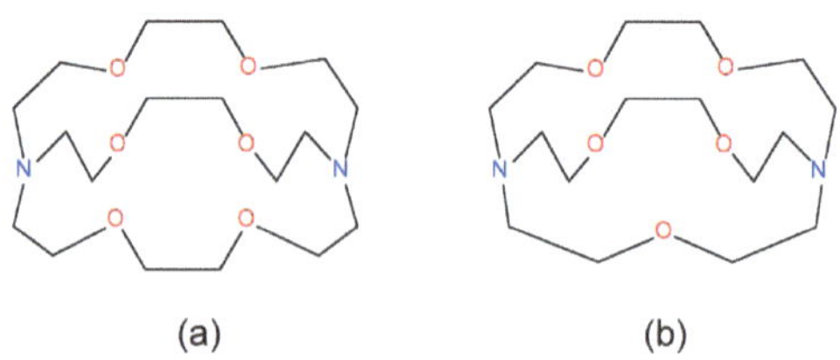


**Figure 8.16: Structure of (a) 4,7,13,16,21-pentaoxa-1,10-diazabicyclo[8,8,5]tricosane [2,2,1] and (b) 4,7,13,16,21,24-hexaoxa-1,10-diazabicyclo[8,8,8]hexacosane [2,2,2] .**

| Cation | [2,2,2] | [2,2,1] |
|---|---|---|
| $Na^+$ | 800 | 250,000 |
| $K^+$ | 250,000 | 700 |

**Table 8.13: Equilibrium constants for the $M^+$ complexes for macrocyclic ligands with different cavity sizes. The structures of the ligands are shown in Figure 8.16.**

# The anomalous chemistry of lithium

While lithium shows many properties that are clearly consistent with its position in Group 1, it also has key differences to the other alkali metals. In fact, in many ways it is more similar to its diagonal neighbor magnesium (Mg) than the other Group 1 metals.

## Charge/radius

The ionic radius for the +1 cation of lithium is very small in comparison with its next highest homolog, sodium (Table 8.14). This results in a correspondingly high value for the charge density ($z/r$). As may be seen from Table 8.14 the charge density for lithium is significantly higher than that of its Group 1 relations.

| Element | z | r (Å) | z/r (Å$^{-1}$) |
|---|---|---|---|
| Li | +1 | 0.68 | 1.47 |
| Na | +1 | 0.97 | 1.03 |
| K | +1 | 1.33 | 0.75 |
| Mg | +2 | 0.66 | 3.03 |

**Table 8.14: Comparison of charge densities for lithium, sodium, potassium, and magnesium.**

As a result of the high charge density, the $Li^+$ ion is a highly polarizing ion. One of the main consequences of this is that lithium tends to form polar covalent bonds rather than ionic interactions. For example, alkyl lithium compounds (RLi) contain covalent Li-C bonds in a similar manner to the Mg-C bonds in Grignards (RMgX, where X = Cl, Br).

## Lattice energy

Lithium compounds have high lattice energies as compared to the other Group 1 metals (Table 8.15). As a consequence, $Li_2O$, $Li_3N$, and LiF are all insoluble in water, whereas their sodium compounds are highly soluble.

| Compound | Lattice energy (kJ/mol) |
|---|---|
| LiF | -1046 |
| NaF | -923 |
| KF | -821 |
| $MgF_2$ | -2957 |

**Table 8.15: Comparison of lattice energies for compounds of lithium, sodium, potassium, and magnesium.**

## Coordination number

The small size of lithium results in a lower coordination number (4) for compounds and complexes than observed for the other Group 1 metals. However, lithium and magnesium complexes and organometallic compounds both have most commonly four-coordinate metal centers (in the absence of large steric constraints).

## Chemical reactivity

A review of some of the reactions of lithium, magnesium and the other Group 1 metals shows the anomalous behavior of lithium and its similarity to magnesium. Both lithium and magnesium react with carbon or nitrogen to form the corresponding carbide and nitride. Whereas sodium and the other Group 1 metals show no reaction under ambient conditions. The combustion of either lithium or magnesium in air results in the formation of the oxides, $Li_2O$ and MgO, respectively. In contrast, sodium forms the peroxide, $Na_2O_2$.

It is not only in the reactivity of the elements that this relationship between lithium and its diagonal neighbor exists. Many of the compounds of lithium have a similar reactivity to those of magnesium rather than sodium. For example, the carbonates of lithium and magnesium decompose under thermolysis to yield the oxides,

$$Li_2CO_3 \rightarrow Li_2O + CO_2$$

and,

$$MgCO_3 \rightarrow MgO + CO_2$$

in contrast, sodium carbonate ($Na_2CO_3$) is stable to thermolysis.

## Organolithium compounds

One of the major uses of lithium is in the synthesis of organolithium compounds, RLi. They have great importance and utility in industry and chemical research. Their reactivity resembles that of Grignard reagents, but they are generally more reactive.

### Synthesis

The best general method for RLi synthesis involves the reaction of an alkyl or aryl chloride with lithium metal in benzene or an aliphatic hydrocarbon (e.g., hexane),

$$RCl + 2\,Li \rightarrow RLi + LiCl$$

While it is possible to use diethyl ether ($Et_2O$), the solvent slowly attacks the resultant alkyl lithium compound,

$$Et_2O + {}^{n}BuLi \rightarrow EtOLi + H_2C{=}CH_2 + {}^{n}BuH$$

Metal-hydrogen exchange, metal-halogen exchange, and metal-metal exchange can also be used.

$$RH + R'Li \rightarrow R'H + RLi$$

$^{n}$BuLi + 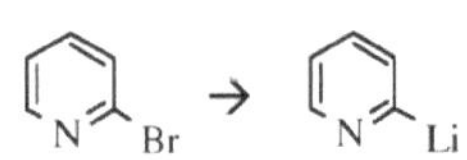 + $^{n}$BuBr

$$2\,Li + R_2Hg \rightarrow 2\,RLi + Hg$$

All organolithium compounds are produced as solutions and are hence used in synthetic protocols by volume of solution. It is therefore important to know the exact concentration of RLi in solution. The simplest approach to quantify the amount of organolithium is to react a known volume with water, and then titrate (with acid) the resultant base that is formed.

$$RLi + H_2O \rightarrow LiOH + RH$$

However, while the concentration of freshly prepared samples of organolithium reagents can the theoretically measured in this way, real samples always contain some amount of LiOH or other bases. A simple titration inevitably results in an over estimation of the organolithium reagent. To overcome this a double titration method is used.

*Gillman double titration method*

The careful addition of a known volume of an organolithium reagent solution (between 0.5 and 1.5 mL) to an excess of water yields a solution of LiOH that can be titrated with a standardized solution of hydrochloric acid, using phenolphthalein as the indicator. The presence of any LiOH in the original organolithium solution will be incorporated into this titration, and thus the result will be a measure of the total base content in the solution, i.e.,

$$\text{Total base content} = \text{LiOH formed from the reaction of RLi with } H_2O + \text{LiOH present as impurity in the RLi solution}$$

In order to determine the amount of LiOH present as impurity in the organolithium solution it is necessary to react the RLi without the formation of base, then titrate the resulting solution. To do this, an aliquot (the same amount as used before) of the organolithium is reacted slowly with 1,2-dibromoethane ($BrCH_2CH_2Br$) dissolved in dry diethyl ether ($Et_2O$). After 5 min of stirring, the solution is diluted with an excess of water and then titrated with a standardized solution of hydrochloric acid, again using phenolphthalein as the indicator. The difference of the two titrations gives the exact concentration of the organolithium.

For example, an aliquot of $^{n}$BuLi in hexanes (0.50 mL) was added to degassed water (20 mL). After any visible reaction had ceased, a few drops of a phenolphthalein solution in water/methanol are added resulting in a pink color indicative of a basic pH. The resulting mixture is titrated with standardized hydrochloric acid ([HCl] = 0.1034 N) until complete disappearance of the pink color (7.90 mL).

A second aliquot of $^{n}$BuLi in hexanes (0.50 mL) is added to 1,2-dibromoethane (0.20 mL, $Et_2O$). After 5 min of stirring, the mixture was diluted with water (20 mL) and after addition of the phenolphthalein indicator titrated (with vigorous stirring due to the biphasic nature of the system) with standardized hydrochloric acid ([HCl] = 0.1034 N) until complete disappearance of the pink color (0.25 mL).

The concentration of nBuLi is calculated as follows:

Step 1.

$$[\text{total base}] = \frac{\text{volume HCl} \times [\text{HCl}]}{\text{volume }^{n}\text{BuLi}} = \frac{7.90 \times 0.1034}{0.50} = 1.633$$

Step 2.

$$[\text{residual base}] = \frac{\text{volume HCl} \times [\text{HCl}]}{\text{volume }^{n}\text{BuLi}} = \frac{0.25 \times 0.1034}{0.50} = 0.013$$

Step 3.

$$[^{n}\text{BuLi}] = [\text{total base}] - [\text{residual base}] = 1.633 - 0.013 = 1.620\text{ M}$$

## Properties

Alkyl lithium compounds are either low melting solids or liquids, and often with high volatility (depending on the substituent) due to the covalent nature of the bonding. They are soluble in aliphatics, aromatics, and ethers; however, while the reaction with ethers is generally slow, alkyl lithium compounds can polymerize tetrahydrofuran (THF).

Organolithium compounds react rapidly with air and water (both vapor and liquid). The reaction with water is the basis of the Gillman double titration method for determining the concentration of organolithium reagents in solution.

## Structure

The structure of organolithium compounds is dominated by their highly oligomeric nature as a result of 3-center 2-electron bridging bonds. In all cases the extent of oligomerization is dependent on the identity of the alkyl (or aryl) group. The alkyl-bridged bond is similar to those found for beryllium and aluminum compounds.

In the vapor phase any particular organolithium derivative show a range of oligomeric structures. For example, the mass spectrum of EtLi shows ions associated with both tetramers (e.g., $[Et_3Li_4]^+$) and hexamers (e.g., $[Et_5Li_6]^+$). The structures of the different oligomers have been predicted by molecular orbital calculations (Figure 8.17).

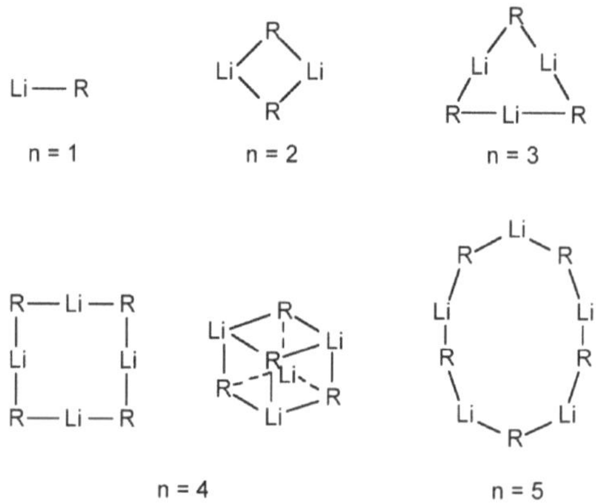


**Figure 8.17: Proposed vapor phase structures for various oligomers of RLi.**

Solution molecular weight measurements indicate the oligomerization is present (in the absence of a coordinating ligand such as $Et_2O$ or an amine). The extent of oligomerization depends on the steric bulk of the alkyl group (Table 8.16). Oligomerization and solution structures have also been investigated by $^7$Li and $^{13}$C NMR spectroscopy.

| R | $[RLi]_n$ | R | $[RLi]_n$ |
|---|---|---|---|
| Me | 4 | Et | 6 |
| $^n$Bu | 6 | $^t$Bu | 4 |

**Table 8.16: Extent of oligomerization (n) for organolithium compounds $[RLi]_n$ in benzene solution.**

There are a large number of X-ray crystallographically determined structures for organolithium derivatives. The archetypal example is MeLi, which exists as a tetramer in the solid state (Figure 8.21). The lithium atoms are arranged as a tetrahedron and the carbon atoms are positioned on the center of the facial planes, i.e., the carbon is equidistant from each of the lithium atoms. In contrast, EtLi has a similar tetrahedral structure, but the α-carbon of the ethyl groups are asymmetrically arranged such that it is closer to one lithium atom than the other two.

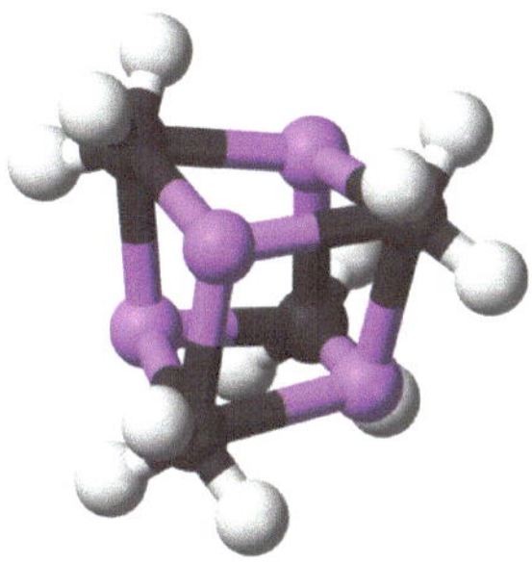

**Figure 8.19: Solid state structure of MeLi. Copyright: Ben Mills (2007).**

It is possible to prepare monomeric organolithium compounds by the addition of amines, especially chelate ligands such as ethylenediamine (en) and tetramethylethylenediamine (TMED). The reactivity of RLi is increased dramatically by the addition of such Lewis bases. For example, $PhCH_2Li$ shows an increased reactivity of $10^4$ with the addition of TMED.

The bonding in organolithium compounds is difficult to describe:

- Based upon the relative electronegativity of Li (0.98) and C (2.5) it would be expected to have significant ionic character.
- Organolithium compounds form electron deficient oligomers typical of covalent bonding and clearly not as a result of ionic aggregation.
- The spin-spin coupling, *J*(C-Li), observed by NMR spectroscopy (10-15 Hz) are consistent with a covalent bond.
- Organolithium compounds undergo radical reactions.

However, the overall description of the bonding in RLi is that of a covalent interaction with significant polar (ionic) character, i.e., $M^{\delta+}$-$C^{\delta-}$.

## Organolithium compounds in organic synthesis

Organolithium compounds perform many of the reactions commonly observed for Grignard reagents; however, lithium reagents are generally more reactive than their Grignard analogs.

*Hydrolysis*

Organolithium compounds react with water to give the hydrocarbon and lithium hydroxide,

$$CH_3Li + H_2O \rightarrow CH_4 + LiOH$$

Lithium alkyls also react with other hydroxylic compounds such as alcohols and carboxylic acids,

$$CH_3Li + HOR \rightarrow CH_4 + LiOR$$

One important use of the hydrolysis reaction is specifically deuteration,

$$CH_3CH_2(CH_3)_2CLi + D_2O \rightarrow CH_3CH_2(CH_3)_2CD + LiOD$$

*Reaction with carbonyls*

Organolithium compounds react with organic carbonyls (aldehydes, ketones, and esters) to yield the alcohol on hydrolysis,

$$MeLi + (^tBu)_2C{=}O \rightarrow (^tBu)_2MeCOLi \xrightarrow{H_2O} (^tBu)_2MeCOH + LiOH$$

This synthetic route is particularly useful since lithium reagents are far more reactive than the analogous Grignard, allowing reactions to be carried out at lower temperatures and minimizing enolization side reactions.

The high reactivity of alkyl lithium compounds means that they react with carboxylic acids to yield the ketone rather than the lithium carboxylate.

$$(C_6H_5)CO_2H \xrightarrow[H_2O]{2\ RLi} (C_6H_5)C(O)R$$

Organolithium compounds generally react with α,β-unsaturated ketones to give the 1,2-addition product,

$$C_6H_5(H)C{=}C(H)C(O)C_6H_5 \xrightarrow[H_2O]{2\ C_6H_5Li} C_6H_5(H)C{=}C(H)C(OH)(C_6H_5)_2$$

However, lithium dialkylcuprates, which are formed from the alkyl lithium and copper(I) iodide,

$$2\ RLi + CuI \rightarrow R_2CuLi + LiI$$

add exclusively by the 1,4-addition,

$$C_6H_5(H)C{=}C(H)C(O)C_6H_5 \xrightarrow[H_2O]{R_2CuLi} C_6H_5(H)(R)CH_2C(O)(C_6H_5)_2$$

*Transmetallation*

One of the most useful methods of preparing organometallic compounds is the exchange reaction of one organometallic compound with a salt of a different metal,

$$RM + M'X \rightleftharpoons RM' + MX$$

This is an equilibrium process whose equilibrium constant is defined by the reduction potential of both metals. In general, the reaction will proceed so that the more electropositive metal will form the more ionic salt (usually chloride).

Lithium reagents may be used to prepare a wide range of organometallic compounds.

$$4\ n\text{-}C_4H_9Li + SnCl_4 \rightarrow Sn(C_4H_9)_4 + 4\ LiCl$$

# Chapter 9: Group 2 Alkaline Earth Metals

## The Elements

The Group 2 metals have a particular name: the alkaline earth metals. The name is derived from the observation that they have such high melting points (Table 9.1) that they remain solids (earths) in a fire. Table 9.2 lists the derivation of the names of the alkali metals.

| Alkali metal | Melting point (°C) | Alkaline earth metal | Melting point (°C) |
|---|---|---|---|
| Li | 180.54 | Be | 1287 |
| Na | 97.72 | Mg | 650 |
| K | 63.38 | Ca | 842 |
| Rb | 39.31 | Sr | 777 |
| Cs | 28.44 | Ba | 727 |
| Fr | 27 (estimated) | Ra | 700 |

**Table 9.1: Melting points of the alkaline earth metals in comparison with the alkali metals.**

| Element | Symbol | Name |
|---|---|---|
| Beryllium | Be | From the Greek *berullos* meaning to become pale, in reference to the pale semiprecious gemstone beryl |
| Magnesium | Mg | From the Magnesia district in Greece |
| Calcium | Ca | From the Latin *calcis* meaning lime |
| Strontium | Sr | From the mineral strontianite named after the Scottish village of Strontian |
| Barium | Ba | From the Greek *bary*, meaning heavy |
| Radium | Ra | From the Latin *radius* meaning ray |

**Table 9.2: Derivation of the names of the alkaline earth metals.**

## Discovery

*Beryllium*

Beryllium was discovered by Louis-Nicolas Vauquelin (Figure 9.1) in 1798 as a component in beryls and emeralds; however, Friedrich Wöhler (Figure

9.2) and Antoine Bussy (Figure 9.3) independently isolated the metal in 1828 by reacting potassium with beryllium chloride,

$$BeCl_2 + 2\,K \rightarrow Be + 2\,KCl$$

**Figure 9.1: French pharmacist and chemist Louis Nicolas Vauquelin (1763 - 1829).**

**Figure 9.2: German chemist Friedrich Wöhler (1800 - 1882) also known for his synthesis of urea and, thus, the founding of the field of natural products synthesis.**

**Figure 9.3: French chemist Antoine Alexandre Brutus Bussy (1794 - 1882).**

*Magnesium*

Magnesium is found in large deposits of magnesite and dolomite, and in mineral waters where the $Mg^{2+}$ ion is soluble. In 1618 a farmer at Epsom in England attempted to give his cows water from a well. The farmer noticed that the water seemed to heal scratches and rashes. These Epsom salts were recognized to be hydrated magnesium sulfate, $MgSO_4$. The metal was first isolated in 1808 by Sir Humphry Davy (Figure 8.5) via the electrolysis of a mixture of magnesia and mercury oxide. Antoine Bussy (Figure 9.3) subsequently prepared magnesium by heating magnesium chloride and potassium in a glass tube,

$$MgCl_2 + 2\,K \rightarrow Mg + 2\,KCl$$

When the potassium chloride was washed out, small globules of magnesium remained.

*Calcium*

Calcium oxide or lime was known in ancient Rome, while even in 975 AD, Plaster of Paris (calcium sulphate) was reported to be useful for setting broken bones. The element itself was not isolated until 1808 when Sir Humphry Davy (Figure 8.5) electrolyzed a mixture of lime and mercuric oxide (HgO). His work was based upon prior work by Jöns Berzelius (Figure 8.3) who had prepared calcium amalgam (an alloy of calcium and mercury) by electrolyzing lime in mercury.

*Strontium*

Discovered in lead mines in 1787 the mineral *strontianite* was named after the Scottish village of Strontian. Although it was realized that this mineral was different from others that contained barium, it wasn't until 1798 that Thomas Hope (Figure 9.4) suggested the presence of a new element. As with calcium, metallic strontium was first isolated by Sir Humphry Davy (Figure 8.5) in 1808 using electrolysis of a mixture containing strontium chloride and mercuric oxide.

**Figure 9.4: Scottish chemist Thomas Charles Hope (1766 - 1844).**

*Barium*

Barium minerals were known by alchemists in the early Middle Ages. Stones of the mineral barite found in Bologna, Italy (also known as Bologna stones), were known to glow after exposure to light. Carl Scheele (Figure 9.5) identified barite in 1774 but did not isolate barium. Barium was first isolated as $Ba^{2+}$ in solution by Sir Humphry Davy (Figure 8.5) in 1808. The oxidized barium was at first called *barote*, by Guyton de Morveau (Figure 9.6), which was changed by Antoine Lavoisier (Figure 7.4) to *baryta*, from which barium was derived to describe the metal.

**Figure 4.7: Carl Wilhelm Scheele (1742 - 1786). Author Isaac Asimov has called him "hard-luck Scheele" because he made a number of chemical discoveries before others who are generally given the credit.**

**Figure 9.6: French chemist and politician Louis-Bernard Guyton de Morveau (1737-1816) changed his name to Guyton-Morveau after the French Revolution.**

*Radium*

Radium was discovered by Marie Curie (Figure 9.7) and her husband Pierre (Figure 9.8) in 1898 while studying pitchblende. After removing uranium, they found that the remaining material was still radioactive. They then separated out a radioactive mixture consisting mostly of barium and an element that gave crimson spectral lines that had never been documented before. In 1910, radium was isolated as a pure metal by Curie and André-Louis Debierne

(Figure 9.9) through the electrolysis of a radium chloride solution by using a mercury cathode and distilling in an atmosphere of hydrogen gas.

**Figure 9.7: Marie Curie (1867-1934). Copyright Academy of Achievement.**

**Figure 9.8: French physicist Pierre Curie (1859 - 1906).**

**Figure 9.9: French chemist André-Louis Debierne (1874 - 1949).**

## Abundance

The abundance of the alkaline earth elements is given in Table 9.3. Beryllium is rare but found in the mineral beryl ($Be_3Al_2Si_6O_{18}$). While magnesium is widespread within the Earth's crust, commercial sources tend to be from sea water as well as the carbonate minerals magnesite ($MgCO_3$) and dolomite [$(Ca,Mg)CO_3$]. Calcium is also commonly found as the carbonate; however, strontium and barium are present as the sulphates celestine ($SrSO_4$) and barites ($BaSO_4$), respectively.

| Element | Terrestrial abundance (ppm) |
|---|---|
| Be | 2.6 (Earth's crust), 6 (soil), $2\times10^{-7}$ (sea water) |
| Mg | 23,000 (Earth's crust), 10,000 (soil), 1,200 (sea water) |
| Ca | 41,000 (Earth's crust), 20,000 (soil), 400 (sea water) |
| Sr | 370 (Earth's crust), 200 (soil), 8 (sea water) |
| Ba | 500 (Earth's crust), 500 (soil), 0.001 (sea water) |
| Ra | $6\times10^{-7}$ (Earth's crust), $8\times10^{-7}$ (soil), $1\times10^{-10}$ (sea water) |

**Table 9.3: Abundance of alkaline earth elements.**

Calcium is a key element for living. Not only is it present as the skeletal material for shellfish and crabs ($CaCO_3$) its phosphate derivative, hydroxyapetite [$Ca_5(OH)(PO_4)_3$] is the structural material of bones and teeth. Calcium is also present in soft tissue at a level of ca. 22g/kg. Calcium is a vital metal for the following:

- Links large molecules together.
- Used in the activation of muscles

- Used in enzyme activation by stabilization of particular conformations of proteins to be acted upon by enzymes.

## Isotopes

The naturally abundant isotopes of the alkaline earth elements are listed in Table 9.4. All of the 25 isotopes of radium are radioactive, and while radium-223, radium-224, and radium-228 are found in nature as decay products of either uranium or thorium, they are only present in trace amounts.

| Isotope | Natural abundance (%) |
|---|---|
| Beryllium-9 | 100 |
| Magnesium-24 | 78.99 |
| Magnesium-25 | 10 |
| Magnesium-26 | 11.01 |
| Calcium-40 | 96.941 |
| Calcium-42 | 0.647 |
| Calcium-43 | 0.135 |
| Calcium-44 | 2.086 |
| Calcium-46 | 0.004 |
| Calcium-48 | 0.187 |
| Strontium-84 | 0.56 |
| Strontium-86 | 9.86 |
| Strontium-87 | 7.0 |
| Strontium-88 | 82.58 |
| Barium-130 | 0.106 |
| Barium-132 | 0.101 |
| Barium-134 | 2.417 |
| Barium-135 | 6.592 |
| Barium-136 | 7.854 |
| Barium-137 | 11.23 |
| Barium-138 | 71.7 |
| Radium-226 | 100 |

**Table 9.4: Abundance of the major isotopes of the alkaline earth elements.**

Although beryllium-7 and beryllium-10 are found as trace isotopes, they are so rare beryllium is considered mononuclidic element (a chemical element which is found essentially as a single nuclide, of only one atomic mass). Calcium has four stable isotopes plus two more isotopes (calcium-46 and calcium-48) that have such long half-lives ($2.8 \times 10^{15}$ and $4 \times 10^{19}$ years, respectively) that for all practical purposes they can be considered stable.

Measurement of the $^{87}Sr/^{86}Sr$ ratio allows for geological dating of minerals and rocks. Strontium-90 (half-life = 28.9 years) is a by-product of nuclear fission and found in nuclear fallout. For example, the 1986 Chernobyl nuclear accident contaminated released a large amount of strontium-90. Since strontium substitutes for calcium in bones it prevents excretion from the body and thus presents a significant health risk, however, strontium-89 is a short-lived artificial radioisotope that is used in the treatment of bone cancer.

Naturally occurring barium is a mix of seven stable isotopes (Table 9.4), but there are a total twenty-two isotopes known, most of which are highly radioactive and have half-lives in the several milliseconds to several day range. The only notable exception is barium-133 which has a half-life of 10.51 years.

## Industrial production

The industrial production of beryllium is usually from the reaction of beryl ($Be_3Al_2Si_6O_{18}$) with $Na_2(SiF_6)$ which yields the beryllium fluoride, $Na_2(BeF_4)$. Subsequent reactions with base give the hydroxide.

$$Na_2(BeF_4) + 2\ NaOH \rightarrow Be(OH)_2 + 4\ NaF$$

Reaction of the hydroxide with the ammonium salt of $HF_2^-$, followed by thermolysis gives beryllium fluoride ($BeF_2$). Finally, reduction of the fluoride with magnesium yields beryllium.

$$BeF_2 + Mg \rightarrow MgF_2 + Be$$

Although magnesium is an abundant metal in dozens of minerals the majority of commercial production comes from sea water, where it is present at about a level of 12% that of sodium. Calcium hydroxide is added to seawater to form magnesium hydroxide precipitate.

$$MgCl_2 + Ca(OH)_2 \rightarrow Mg(OH)_2 + CaCl_2$$

Subsequent reaction with hydrochloric acid yields concentrated magnesium chloride solution.

$$Mg(OH)_2 + 2\,HCl \rightarrow MgCl_2 + 2\,H_2O$$

Electrolysis of the magnesium chloride produces magnesium. At the cathode, the $Mg^{2+}$ ion is reduced to magnesium metal, while at the anode chlorine gas is formed,

$$Mg^{2+} + 2\,e^- \rightarrow Mg$$

$$2\,Cl^- \rightarrow Cl_{2(g)} + 2\,e^-$$

Strontium metal is produced in an analogous manner; however, it may also be prepared from strontium oxide by reduction with aluminum in vacuum at a temperature at which strontium distils off.

$$3\,SrO + 2\,Al \rightarrow 3\,Sr + Al_2O_3$$

## Reactivity and toxicity

The chemistry of the Group 2 elements is dominated by the +2 oxidation state and the noble gas configuration of the $M^{2+}$ cation.

Calcium, strontium, and barium react with water on contact to produce the hydroxide and hydrogen gas. Although the lighter alkaline earth metals do not react violently with water, they do burn in air.

Magnesium burns with a very bright white flame such that caution should be taken not to look at the flame directly. Magnesium is capable of reducing water,

$$Mg_{(s)} + 2\,H_2O \rightarrow Mg(OH)_{2(s)} + H_{2(g)}$$

and as a result, water cannot be used to extinguish magnesium fires; the hydrogen gas produced will only intensify the fire. In addition, magnesium also reacts with carbon dioxide,

$$2\,Mg_{(s)} + CO_2 \rightarrow 2\,MgO_{(s)} + C_{(s)}$$

precluding the use of carbon dioxide re extinguishers. Class D dry chemical re extinguisher or sand are used for magnesium fires.

Strontium and barium burn in air to produce both the oxide and the nitride, but since the metals do not react with nitrogen except at high temperatures, they only form the oxide spontaneously at room temperature.

Beryllium is a class 1 carcinogen, i.e., it is carcinogenic to both animals and humans. Beryllium is harmful if inhaled; if the concentration in air is high enough (greater than 100 $\mu g/m^3$) an acute condition can result, called acute beryllium disease, which resembles pneumonia. Acute beryllium disease was reported as being associated with the manufacture of fluorescent lighting tubes (a practice that ceased in 1949).

The human body absorbs strontium as if it were calcium, and while stable isotopes have posed no significant health threat, the uptake of radioactive strontium-90 can lead to various bone disorders and diseases, including bone cancer. All water or acid soluble barium compounds are extremely poisonous. At low doses, barium acts as a muscle stimulant, while higher doses affect the nervous system. Radium is highly radioactive and its decay product, radon gas, is also radioactive.

## Calcium the archetypal alkaline earth metal

### Solid state and solution chemistry

Calcium represents the typical Group 2 metal. The chemical properties of its compounds can be applied to all the heavier homologs.

*Solid state*

In the solid state the compounds of the alkaline metals generally form ionic lattices. In fact (except for beryllium and to a lesser extend magnesium) the lattice parameters calculated from the ionic radii of Group 2 metals are within 1% of the experimentally determined values, indicating the highly ionic character.

Due to the increased ionic charge the cations of the alkaline earths are less than that of their alkali metal neighbors. Thus, the ionic radius of $Ca^{2+}$ (0.99 Å) is less than that of $K^+$ (1.33 Å), but more similar to $Na^+$ (0.97 Å). This diagonal relationship is seen for the other metals in Group 2.

*Oxides*

Combustion of the Group 2 metals gives the monoxide, MO. In the case of SrO and BaO further reaction occurs by the absorption of oxygen under pressure to give the peroxides, $MO_2$. The peroxide and superoxides are not stable for the lighter homologs because the smaller $M^{2+}$ ions are more polarizing and cause the peroxide and superoxides to decompose to the monoxide. Calcium peroxide can be prepared by the reaction of the hydroxide with hydrogen peroxide.

$$Ca(OH)_2 + H_2O_2 + 6\,H_2O \rightarrow CaO_2.8H_2O$$

As typical for the Group 2 oxides, calcium monoxide is basic and reacts with water to give the hydroxide,

$$CaO + H_2O \rightarrow Ca(OH)_2$$

In fact, thin films of the Group 2 oxides of calcium, barium and strontium will readily absorb water and form the hydroxides.

*Carbonate*

Calcium carbonate ($CaCO_3$) is a common substance found in rock in all parts of the world and is the main component of shells of marine organisms, snails, pearls, and eggshells. Calcium carbonate is usually the principal cause of hard water. Most calcium carbonate used is extracted by quarrying or mining. However, pure calcium carbonate (e.g., for pharmaceutical use) can be produced from a pure quarried source (usually marble) or manufactured by the sequential reaction involving the thermal decomposition of the carbonate to the monoxide,

$$CaCO_3 \rightarrow CaO + CO_2$$

followed by the reaction with water to give the hydroxide, and finally, the reaction with carbon dioxide to reform the carbonate,

$$Ca(OH)_2 + CO_2 \rightarrow CaCO_3 + H_2O$$

Calcium carbonate crystallizes as a variety of mineral forms.

- Aragonite
- Calcite (Figure 9.10)
- Vaterite
- Chalk
- Travertine
- Limestone
- Marble

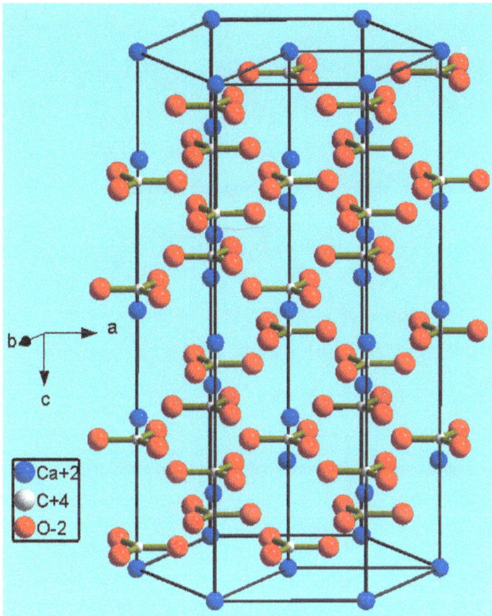


**Figure 9.10: Unit cell of calcite a common form of calcium carbonate ($CaCO_3$).**

Calcium carbonate is one of the most widely used mineral materials, the following represents a list of some of the main applications.

- Construction industry as a building material (marble) or as an ingredient of cement.
- Purification of iron from iron ore in a blast furnace.
- A drilling-fluids in the oil industry.
- Filler material for latex gloves.
- Filler (extender) in paints.
- Filler in plastics.
- Babies' diapers.
- DIY adhesives, sealants, and decorating fillers.

- Whiting in ceramics/glazing application.
- The filler in glossy paper.
- The production of blackboard chalk ($CaSO_4$).
- An abrasive in household cleaning products.
- Dietary calcium supplement.
- Inert filler for tablets and pharmaceuticals.
- Toothpaste.

*Halides*

Calcium chloride, bromide, and iodide are all ionic, water-soluble salts. In contrast, due to its high lattice energy for the small fluoride ion, $CaF_2$ is only slightly soluble (Table 9.5). The fluorite structure typified for $CaF_2$ (Figure 9.11) is found for most of the $MX_2$ ionic solids.

| Compound | Solubility @ 20 °C (g/100 mL) | Solubility @ 100 °C (g/100 mL) |
|---|---|---|
| $CaF_2$ | 0.0016 | 0.0017 |
| $CaCl_2$ | 74.5 | 159 |
| $CaBr_2$ | 142 | 312 |
| $CaI_2$ | 209 | 426 |

**Table 9.5: Solubility of calcium halides in water.**

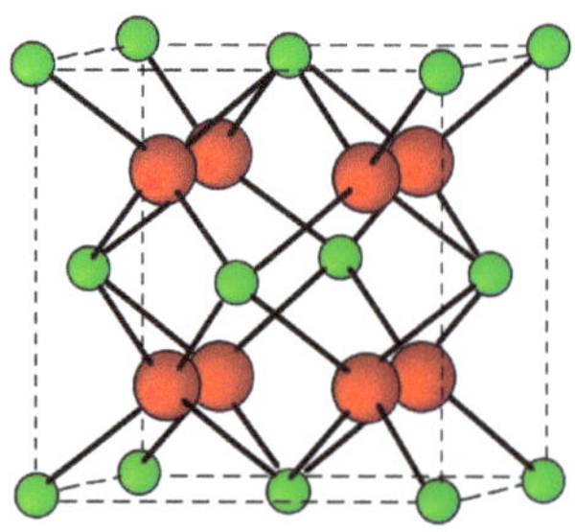

**Figure 9.11: Unit cell for the fluorite ($CaF_2$) structure. Calcium atoms are shown in green (small), fluorine atoms are shown in red (large), and the dashed lines show the unit cell.**

*Complexes*

The coordination complexes of the alkaline earth metal cations ($M^{2+}$) involve electrostatic, or ion-dipole, interactions that have no preferred direction of interaction. Calcium will tolerate a wide range of coordination numbers; however, 6 and 8 are the most common. Complexes generally form with oxygen donors in a wide range of ligands, including: ROH, $R_2O$, $R_2C=O$, $RCO_2^-$, and $PO_4^{2-}$. While complexes with nitrogen donor ligands are known they are usually present with oxygen donors as well. As with the Group 1 metals, the aquo complex readily exchanges the water for other ligands, however, since the bond energy is larger for $Ca^{2+}$ than for $Na^+$, the equilibrium constants are larger for analogous complexes.

## Calcium carbide: from gaslight to fertilizer

Calcium carbide has an interesting role in the societal and commercial changes that took place in the late 19$^{th}$ and early 20$^{th}$ centuries. However, in order to understand the effects of calcium carbide it is important to realize the state of the art of lighting in the late 18$^{th}$ century.

It was in 1792 that William Murdoch (Figure 9.12) first began experimenting with the use of gas, derived from the heating of coal and other materials, for lighting. Murdoch produced this coal gas, or manufactured gas and conveyed it through metal pipes, lighting his cottage and offices in Redruth, Cornwall (Figure 9.13).

**Figure 9.12: Scottish engineer, inventor, and pioneer of gas lighting William Murdoch (1754 - 1839).**

In 1802 as part of the public celebrations of the Peace of Amiens (between England and France) Murdoch made a public exhibition of his lighting by illuminating the exterior of the Soho Foundry in Birmingham, England. Then in 1807 an entrepreneur, Fredrick Winsor (Figure 9.14) displayed gaslights along the top of the wall between Carlton House and the Mall in London. This demonstration for city use was a revelation.

**Figure 9.13: Murdoch's house at Redruth (Cornwall, UK) was the first domestic residence to be lit by gas.**

**Figure 9.14: German inventor one of the pioneers of gas lighting in UK and France, Frederick Albert Winsor, originally Friedrich Albrecht Winzer (1763 - 1803).**

By 1823 Britain had 300 miles of gas pipe and by 1850 it was 2000 miles. Gaslight had a profound impact on society. Walking the streets at night was safer, and it allowed for longer working hours. It also made evening activities easier. As a consequence, reading and evening schools became popular pastimes. Unfortunately, gaslight was rather dull orange in color (Figure 9.15), but it was due to another area of chemical research that a brighter alternative was discovered.

In 1895 the Frenchman Henry Moissan (Figure 9.16) was trying to make diamonds by the reaction of carbon (graphite) with almost anything he could lay his hands on. Although highly unsuccessful, one of his experiments did prove useful. By reacting carbon with lime, the common name for calcium oxide

(CaO) at 2000 °C (in an electric arc furnace that he had helped develop) he produced calcium carbide ($CaC_2$).

**Figure 9.15: A gas light still in use on the streets of London.**

**Figure 9.16: French chemist Ferdinand Frederick Henri Moissan (1852 - 1907).**

Pure calcium carbide is colorless, but most samples have a color ranging from black to grayish-white, depending on the grade. As an ionic salt it has a high melting point (2160 °C). While the structure of calcium carbide (Figure 9.17) has a tetragonal lattice, it is related to that of a cubic rock salt structure, but where the anion is the linear $C_2^{2-}$ moiety.

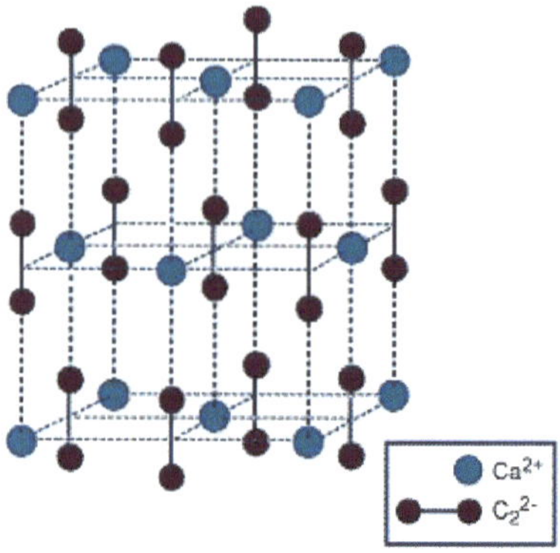


**Figure 9.17: The unit cell of calcium carbide ($CaC_2$). Adapted from N. N. Greenwood and A. Earnshaw, Chemistry of the Elements, Butterworth-Heinemann (1984).**

Although now used extensively, at the time of its discovery calcium carbide itself did not prove very interesting, its reaction with water had a profound effect on illumination. The reaction of calcium carbide with water yields acetylene,

$$Ca_2C + H_2O \rightarrow Ca(OH)_2 + HC{\equiv}CH$$

Unlike coal gas, acetylene burns with a very bright white flame. Although electricity was starting to become more commonly used it was very expensive and acetylene offered a cheaper alternative for domestic lighting. Thus, by 1899 there were over 250,000 acetylene gas jets in Germany alone. The reaction of calcium carbide to form acetylene was used in a variety of portable lamps. These so-called carbide lamps were used in slate, copper and tin mines, and were also used extensively as headlights in early automobiles and bikes (Figure 9.18).

Unfortunately for acetylene there was one discovery and one economic change that brought the use of acetylene as a light source to an end. In 1893 Auer von Welsbach (Figure 9.19) invented the gas mantel (Figure 9.20). By the impregnation of silk or cotton with a mixture of thorium dioxide and cerium(IV) oxide (99:1), and using this in combination with gas he was able to produce a very white flame.

**Figure 9.18: The carbide lamp on a 1911 Renault.**

**Figure 9.19: Austrian scientist and inventor Carl Auer Freiherr von Welsbach (1858 - 1929).**

**Figure 9.20: Gas mantles. Copyright Science Museum, London.**

The other impact on acetylene, was that by 1905 the cost of electricity was significantly lower, and as a consequence the price of $CaC_2$ dropped to 30%. There were stockpiles of calcium carbide all over Europe and America. This may have been the end of calcium carbide's usefulness, however, in 1895 Heinrich Caro (Figure 9.21) and Adolf Frank (Figure 9.22), at the German chemical giant Badische Anilin und Soda-Fabrik (BASF), were trying to make hydrogen cyanide (HCN) to use in its color dye business. In 1898, one of their colleagues demonstrated that what was actually produced during the reaction at temperatures exceeding 1000 °C was not cyanide, as they had hoped. It turned out that what Caro and Frank had found was that that when calcium carbide is reacted with nitrogen at 1000 °C it forms calcium cyanamide ($CaCN_2$),

$$CaC_2 + N_2 \rightarrow CaCN_2 + C$$

**Figure 9.21: German chemist Heinrich Caro (1834 - 1910).**

**Figure 9.22: German chemist, engineer, and businessman Adolph Frank (1834 - 1916).**

The cyanamide anion has the structure $[N=C=N]^{2-}$.

In contact with water calcium cyanamide decomposes and liberates ammonia:

$$CaCN_2 + 3\ H_2O \rightarrow 2\ NH_3 + CaCO_3$$

As such, $CaCN_2$ is an excellent solid fertilizer that is readily plowed into the soil. By 1908 calcium cyanamide was also found to be a plant protection agent, which, at a time when all weed control was performed mechanically, represented a great step forward. Consequently, the output of calcium cyanamide grew enormously. In 1910, 30,000 tons were produced, but in 1928, global production had reached 1.2 million tons. After a temporary decline, demand has again risen in recent years owing to the ban on several pesticides due to the environmental damage they cause. Even after 100 years of use, no harmful long-term effects to the earth or environment have been observed, nor have weeds or pests developed a resistance to calcium cyanamide.

## Differences for beryllium and magnesium

While the chemistry of strontium, barium (and radium) is similar to that of calcium, magnesium and beryllium show marked differences. In both cases these differences are due to the small size of the ions.

### Beryllium

Beryllium can be thought of as being even more covalent than magnesium. The small size (ca. 0.3 Å) results in a very high charge density of $Be^{2+}$. In addition, the ionization energy for beryllium is a large positive value (1$^{st}$ ionization energy = 899.5 kJ/mol, 2$^{nd}$ ionization energy = 14,848.7 kJ/mol). Both of these factors mean that the free ion does not exist. Instead, beryllium forms covalent compounds in a similar manner to its diagonal analog aluminum. Both beryllium and aluminum form covalent compounds or strongly solvated cations, and both form polymeric hydrides, chlorides, and alkyls.

Beryllium chloride is not a lattice structure with a concomitantly high melting and boiling point as observed for the other Group 2 metals (Table 9.6). Instead $BeCl_2$ is a polymer in the solid state (Figure 9.23a), and an equilibrium between a monomer (Figure 9.23b) and dimer (Figure 9.23c) in the vapor phase.

| M | Structure |
|---|---|
| Be | Polymer (4-coordinate Be) |
| Mg | Cadmium chloride structure (6-coordinate Mg) |
| Ca | Deformed rutile structure (6-coordinate Ca) |
| Sr | Deformed rutile structure (6-coordinate Sr) |
| Ba | $PbCl_2$ structure (9-coordinate Ba) or fluorite structure (8-coordinate Ba) |

**Table 9.6: Summary of structures for alkaline earth chlorides ($MCl_2$).**

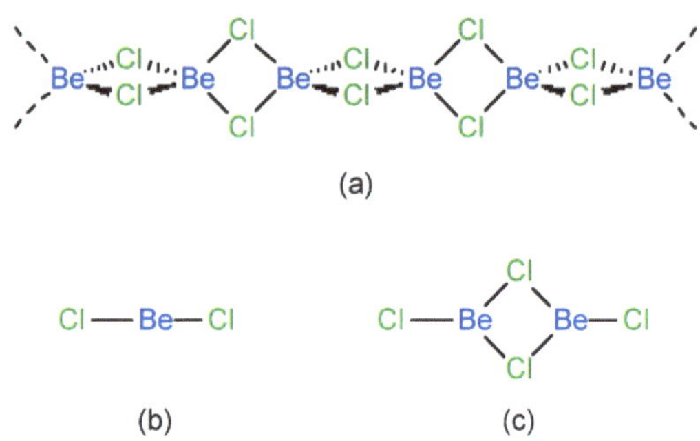

**Figure 9.23: Structures of $BeCl_2$.**

# Magnesium

The ionic radius for the +2 cation of magnesium is fairly small (0.65 Å). As a consequence, the charge density (*z/r*) is high, which results in a high polarizing power of the $Mg^{2+}$ ion. Thus, magnesium tends to form polar covalent bonds rather than ionic complexes. As with lithium there is a wide range of organometallic derivatives of magnesium, especially the Grignards (RMgX, where X = Cl, Br).

A further consequence of the covalent character of the bonding is that magnesium tends to form either 4-coordinate (tetrahedral) or 6-coordinate (octahedral) complexes with well-defined geometries.

## Organometallic compounds of magnesium

While beryllium makes a range of organometallic compounds, their hazardous nature has limited their study. In contrast, the ionic nature of calcium, strontium, and barium compounds limits the number of stable organometallic derivatives. However, the organometallic chemistry of magnesium is rich and extensive. The importance of Grignards (RMgX, where X = halide) and dialkyl magnesium compounds ($R_2Mg$) is due to their use in organic synthesis and as synthons for a range of organometallic compounds.

## Grignard reagents

Grignard reagents (and the Grignard reaction using these compounds) are named after Victor Grignard (Figure 9.24). After studying mathematics at Lyon, he transferred to chemistry, becoming a professor at the University of Nancy in 1910. During World War I, he was involved in the field of chemical warfare; however, it is for his major contribution to organic chemistry he is remembered.

**Figure 9.24: French chemist and Nobel Prize winner François Auguste Victor Grignard (1871 - 1935).**

*Preparation*

The general the synthesis of a Grignard reagent involves the reaction of an alkyl halide (RX, where X = Cl, Br, I) with magnesium metal in a suitable ether solvent,

$$RX + Mg \rightarrow RMgX$$

While diethyl ether ($Et_2O$) and tetrahydrofuran (THF) are commonly used as solvents, other polar non- protic solvents are suitable, including, triethylamine ($NEt_3$), dimethylsulphide ($Me_2S$), dimethylselenide ($Me_2Se$), and dimethyl-telluride ($Me_2Te$).

In general, the alkyl halide is added to an excess of magnesium suspended in the solvent. In most cases it is necessary to activate the magnesium, by the addition of iodine ($I_2$), 1,2-dibromoethane, or sonication. If the halide is very inert reaction can be promoted by the co-condensation of magnesium and THF under vacuum.

There is often an induction period after the initial addition of alkyl halide. However, since the reaction, is highly exothermic care should be taken to ensure that the reaction does not run-away. For this reason, it is normal to initially add a small quantity of the alkyl halide to ensure the reaction initiates. Once reaction is initiated, the addition of alkyl halide is maintained at a suitable rate to ensure the reaction is maintained until all the alkyl halide is consumed. The excess reaction magnesium is removed from the reaction mixture by filtration.

It is not always necessary to use a liquid or solid halide dissolved in the solvent. Bubbling methyl chloride (MeCl) through an $Et_2O$ suspension of magnesium yields MeMgCl. The advantage of a gaseous alkyl halide is that the reaction is very clean as all the magnesium is consumed and the excess alkyl halide is bubbled away.

The purity of the magnesium is very important. For his original experiments Grignard used magnesium of a purity of 99.2%. However, it is now more typical to use 99.8% pure magnesium. It is important that the magnesium not be too pure since it is thought that the transition metal impurities catalyze the reaction.

The relative order of reactivity of the alkyl halide follows the trend:

$$I > Br > Cl > F$$

In fact, alkyl fluorides are sufficiently inert that highly coordinating polar solvents such as THF or dimethyl- formamide (DMF) must be used. If the reaction is allowed to get too hot, then several possible side reactions can occur. In THF reaction with the solvent occurs,

$$RMgX + THF \rightarrow RH + H_2C{=}CH_2 + H_2C{=}C(H)MgX$$

Alternatively, a transition metal catalyzed radical coupling between the Grignard and unreacted alkyl halide is observed irrespective of the identity of the solvent,

$$RMgX + RX \rightarrow R\text{-}R + MgX_2$$

The mechanism for Grignard formation is thought to be radical in nature; however, a study of the surface of the magnesium during the reaction has shown the presence of corrosion pits. It is generally agreed that initiation occurs at surface dislocations, but the major reaction occurs at a polished surface.

The kinetics of the reaction is 1[st] order with respect to the alkyl halide concentration, but it has also been claimed to be 1[st] order with respect to the solvent concentration. It has therefore been concluded that the rate-determining step involves the metal solvent interface.

The reaction of magnesium with aryl bromides has been studied and is proposed to occur by two reactions. The first involves electron transfer between the aryl halide and the metal, while the second involves aryl radical formation.

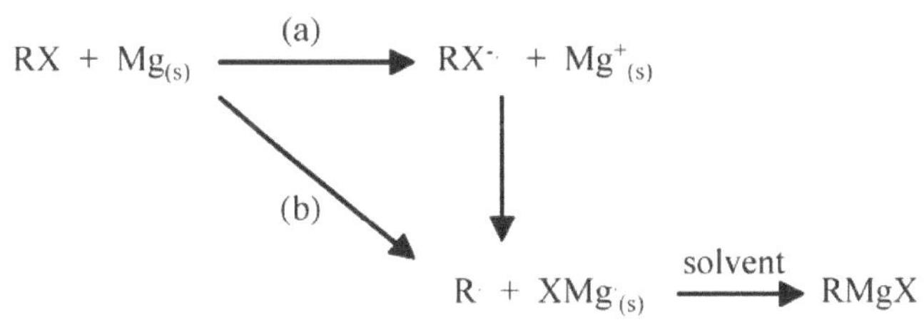


A number of alternative synthetic routes are used with polyhalogenated hydrocarbons, and, and where the alkyl radical is unstable,

$$X_3CH + {}^iPrMgX \rightarrow (X_3C)MgX + {}^iPrH$$

$$C_6Br_6 + EtMgX \rightarrow (C_6Br_5)MgX + EtBr$$

$$RX + R'MgX' \rightarrow RMgX' + R'X$$

*Structure*

The solid-state structure of Grignard reagents is controlled by the presence and identity of the solvent used in the synthesis. In this regard the size and the basicity of the solvent is important. For example, the structure of EtMgBr crystallized from diethyl ether exists as a 4-ccordinate monomer (Figure 9.25a), while the use of the sterically less demanding THF results in a 5-co-ordinate monomeric structure (Figure 9.25b). In contrast, the use of triethylamine yields a dimeric bromide bridged structure (Figure 9.25c), and the use of a chelate bidentate amine gives a structure (Figure 9.25d) similar to that observed with diethyl ether (Figure 9.25a).

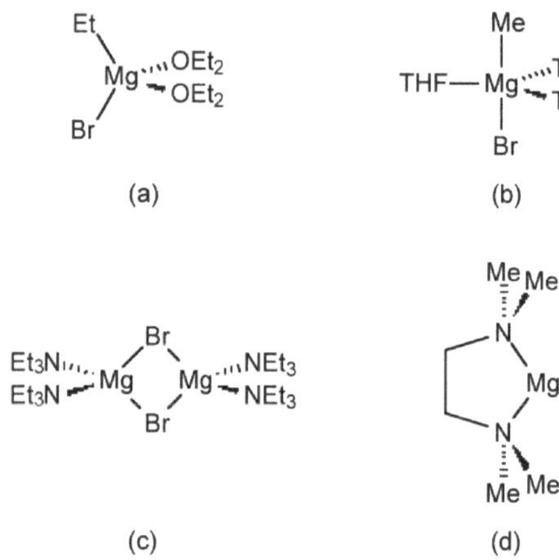


**Figure 9.25: Molecular structure of EtMgBr in (a) diethyl ether, (b) THF, (c) triethyl amine, and (d) tetramethyletheylenediamine (TMED).**

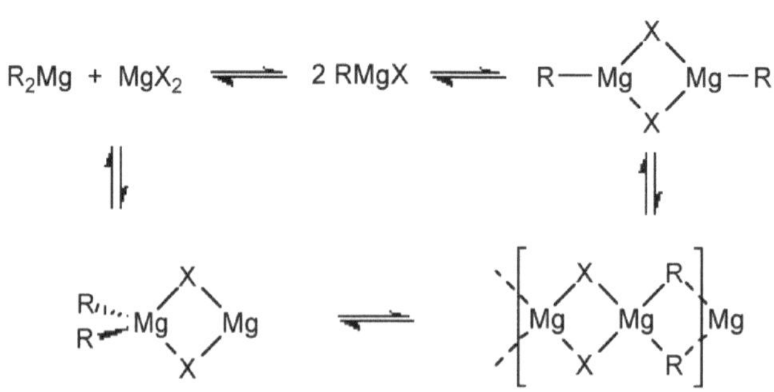


**Figure 9.26: Schematic representation of the extended Schlenk equilibrium observed for Grignard compounds in solution.**

In solution, Grignards are fluxional such that no single defined structure is present. The series of exchange reactions are known as an extended Schlenk equilibrium (Figure 9.26).

It is observed that Grignard solutions are also slightly conducting, and magnesium is deposited at both the anode and cathode suggesting the formation of $RMg^+$ and $[RMgX_2]^-$. The alkyl/halide exchange is thought to occur through a bridging intermediate (Figure 9.27).

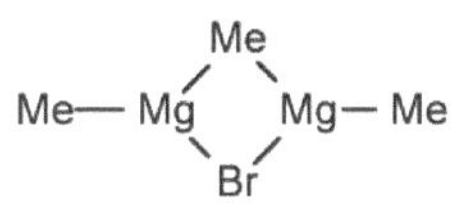


**Figure 9.27: Proposed structure for the alkyl/halide exchange bridging intermediate.**

## Dialkyl magnesium ($R_2Mg$)

Dialkyl magnesium compounds are involatile white solids. They generally have similar reactivity to their Grignard analogs.

*Synthesis*

The most common synthesis of $R_2Mg$ is by the reaction of a Grignard with dioxane ($C_4H_8O_2$), where the precipitation of the dihalide is the reaction driving force.

This method is useful for the synthesis of cyclic compounds,

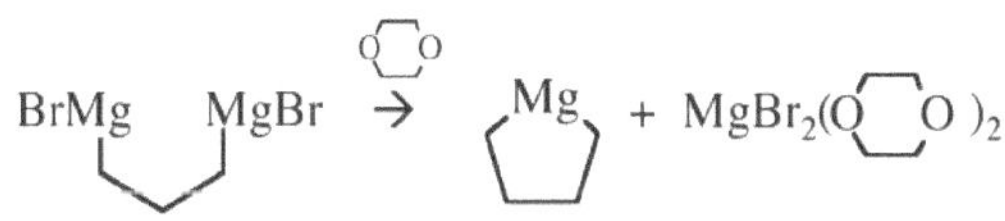


An alternative synthesis that does not require dioxane involves the metal exchange reaction between magnesium metal and a dialkyl mercury compound.

$$R_2Hg + Mg \rightarrow R_2Mg + Hg$$

Finally, in selected cases, magnesium will react with acidic hydrocarbons such as cyclopentadienyl at high temperatures (600 °C).

*Structure*

In the vapor phase dialkyl magnesium compounds are generally monomeric linear compounds. In solution, in the absence of coordinating solvents $R_2Mg$ form a variety of oligomers (Figure 9.28a-c) in solution as deter- mined by molecular weight measurements. In the presence of coordinating solvents 4-coordinate monomers predominate (Figure 9.28d).

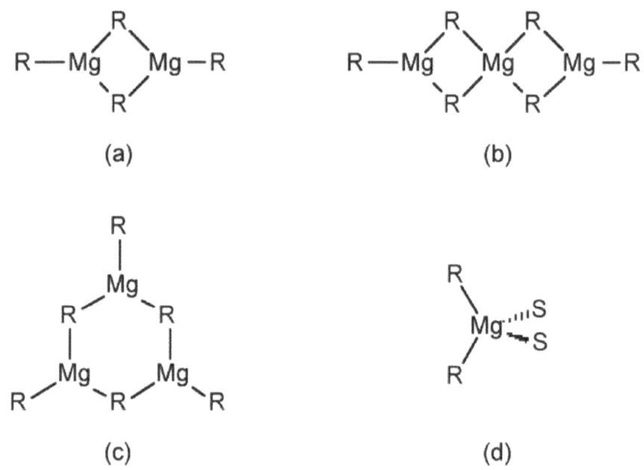

**Figure 9.28: Solution structure of $R_2Mg$ (R = Me, Et) in (a - c) non-coordinating solvents, and (d) diethyl ether.**

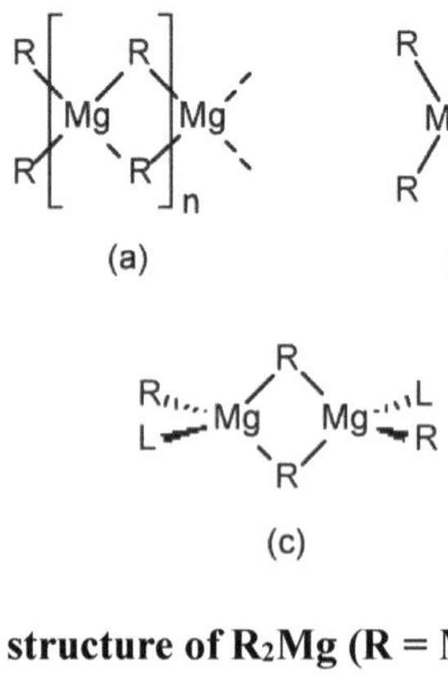

**Figure 9.29: Solid state structure of $R_2Mg$ (R = Me, Et) crystallized in (a) the absence and (b and c) the presence of a coordinating solvents.**

As similar trend is observed in the solid state, where polymers have been characterized in the absence of coordinating solvents (Figure 9.29a), while monomers or dimmers are generally observed when crystallized from a coordinating solvent (Figure 9.29b and c).

## Organomagnesium compounds in organic synthesis

*Hydrolysis and related reactions*

Grignard compounds react with water to give the hydrocarbon,

$$CH_3MgBr + H_2O \rightarrow CH_4 + BrMgOH$$

they also react with other hydroxylic compounds such as alcohols and carboxylic acids. One important use of the hydrolysis reaction is specifically deuteration,

$$CH_3CH_2(CH_3)_2CMgBr + D_2O \rightarrow CH_3CH_2(CH_3)_2CD + BrMgOD$$

The hydrogen atom on a terminal alkyne is sufficiently acidic that the reaction with Grignards occurs in an analogous manner to that of hydrolysis.

$$C_6H_5C{\equiv}CH + C_2H_5MgBr \rightarrow C_6H_5C{\equiv}CMgBr + C_2H_6$$

Once formed the alkynyl Grignard undergoes the same hydrolysis reaction.

$$C_6H_5C{\equiv}CMgBr + D_2O \rightarrow C_6H_5C{\equiv}CD + BrMgOD$$

*Reaction with $CO_2$*

Grignards react readily with carbon dioxide to form the carboxylate, which yields the associated carboxylic acid upon hydrolysis,

$$RMgX + CO_2 \rightarrow RCO_2MgX \xrightarrow{H_2O} RCO_2H + HOMgX$$

*Reaction with carbonyls*

Organomagnesium compounds react with organic carbonyls (aldehydes, ketones, and esters) to yield the alcohol on hydrolysis,

$$RMgX + R'_2C{=}O \rightarrow R'_2(R)COMgX \xrightarrow{H_2O} R'_2(R)COH + HOMgX$$

This synthetic route is useful for the formation of primary, secondary and terminal alcohols. Unfortunately, for some carbonyls there is a competing side reaction of enolization, where the starting ketone is reformed upon hydrolysis.

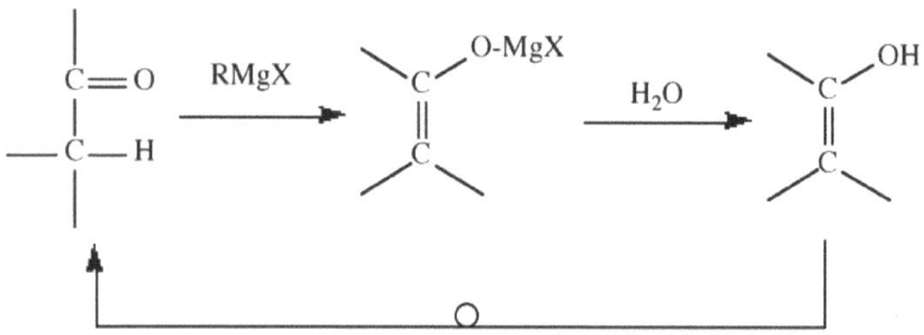


When the Grignard reagent has a β-hydrogen another side reaction occurs in which the carbonyl group is reduced, and an alkene is formed.

$$R_2C{=}O + (CH_3CH_2)MgX \rightarrow R_2(H)COH + H_2C{=}CH_2$$

Both the enolization and reduction occur via similar 6-membered cyclic transition states (Figure 9.30).

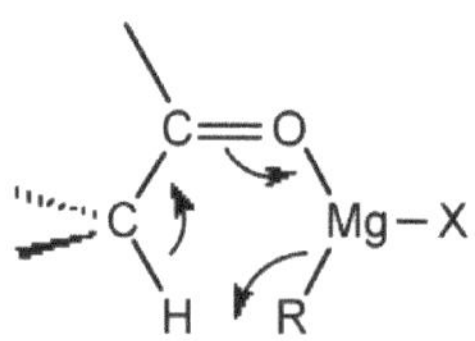


**Figure 9.30: Representation of the 6-membered transition state reaction for enolization of a ketone.**

Grignards react with α,β-unsaturated ketones to give either the 1,2-addition product or the 1,4-addition product, or both,

$$Ph(H)C{=}C(H)\text{-}C(O)Me + EtMgBr \xrightarrow{H_2O} Ph(H)C{=}C(H)\text{-}C(OH)EtMe + Ph(H)EtC\text{-}CH_2\text{-}C(O)Me$$

*Reaction with acyl halides*

Acyl halides react with Grignards to give ketones,

$$CH_3C(O)Cl + RMgX \rightarrow CH_3C(O)R + XMgCl$$

Best results are obtained if the reaction is carried out at low temperature and in the presence of a Lewis acid catalysts (e.g., $FeCl_3$).

*Reaction with epoxides*

Oxirane (epoxide) rings are opened by Grignards,

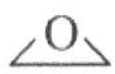

$$RMgBr \xrightarrow[H_2O]{} RCH_2CH_2OH$$

in a useful reaction that extends the carbon chain of the Grignard by two carbon atoms. This reaction is best performed with ethylene oxide since the magnesium halide formed is a Lewis acid catalyst for further reactions in the case of substituted oxiranes.

*Reaction with salts*

One of the most useful methods of preparing organometallic compounds is the exchange reaction of one organometallic compound with a salt of a different metal,

$$RM + M'X \rightleftharpoons RM' + MX$$

This is an equilibrium process, whose equilibrium constant is defined by the reduction potential of both metals. In general, the reaction will proceed so that the more electropositive metal will form the more ionic salt (usually chloride). Grignard reagents are particularly useful in this regard and may be used to prepare a wide range of organometallic compounds. For example:

$$2\,CH_3CH_2MgCl + CdCl_2 \rightarrow Cd(CH_2CH_3)_2 + 2\,MgCl_2$$

$$4\,CH_3MgCl + SiCl_4 \rightarrow Si(CH_3)_4 + 4\,MgCl_2$$

The reaction with a Grignard is milder than the analogous reaction with lithium reagents and leads to a lower incident of side-products.

# Chapter 10: Group 12

## The Elements

Although the Group 12 metals (Table 10.1) are formally part of the d-block elements from their position in the Periodic Table, their electronic configuration in both their elemental form ($d^{10}s^2$) and the vast majority of their compounds ($d^{10}$) is that of the main group elements. The common oxidation state for all the Group 12 elements is +2, and the chemistry of zinc and cadmium compounds in particular is very similar to the analogous magnesium derivatives.

| Element | Symbol | Name |
|---|---|---|
| Zinc | Zn | From German *zinke*, meaning tooth-like, pointed or jagged (metallic zinc crystals have a needle-like appearance), or meaning tin-like because of its relation to German word *zinn* meaning tin, or from Persian *seng* meaning stone. |
| Cadmium | Cd | From the Latin *cadmia*, meaning calamine |
| Mercury | Hg | From the Latin *hydrargyrum*, meaning watery or liquid silver |

**Table 10.1: Derivation of the names of the Group 12 metals.**

The IUPAC (International Union of Pure and Applied Chemistry) definition of a transition metal (or transition element) states that a transition metal is "an element whose atom has an incomplete d sub-shell, or which can give rise to cations with an incomplete d sub-shell." Thus, Group 12 elements are not transition metals.

## Discovery

### *Zinc*

Artifacts with a high zinc content (as much as 90%) have been found to be over 2500 years old, and possibly older. As such it is clear that several cultures had the knowledge of working with zinc alloys, in particular brass (a zinc/copper alloy). Zinc mines at Zawar, near Udaipur in India, have been active since the late 1$^{st}$ millennium BC. However, the smelting of metallic zinc appears to have begun around the 12$^{th}$ Century AD.

The isolation of purified metallic zinc was reported concurrently by several people. The extraction of zinc from its oxide (ZnO) was reported as early as 1668, while John Lane is supposed to have smelted zinc in 1726. The first Patent for zinc smelting was granted to English metallurgist William Champion in 1738; however, the credit for discovering pure metallic zinc is often given to Andreas Marggraf (Figure 10.1) in 1746.

**Figure 10.1: Engraving of German chemist Andreas Sigismund Marggraf (1709 - 1782).**

*Cadmium*

Cadmium was discovered in 1817 by Friedrich Stromeyer (Figure 5.2) as an impurity in calamine (zinc carbonate, $ZnCO_3$). Stromeyer observed that impure samples of calamine changed color when heated but pure calamine did not. Eventually he was able to isolate cadmium metal by roasting and reduction of the sulfide.

*Mercury*

Mercury was known to the ancient Chinese and was found in Egyptian tombs that date from 1500 BC. In China and Tibet, mercury use was thought to prolong life, heal fractures, and maintain generally good health. The ancient Greeks used mercury in ointments; the ancient Egyptians and the Romans used it in cosmetics that sometimes deformed the face.

**Figure 10.2: German chemist Friedrich Stromeyer (1776 - 1835).**

## Abundance

The Group 12 elements mainly occur in sulfide ores, however, as with their Group 2 analogs, carbonate is known, but not as economically viable. The major zinc containing ore is zinc blende (also known as sphalerite), which is zinc sulfide (ZnS). Other important ores include, wurtzite (ZnS), smithsonite (zinc carbonate, $ZnCO_3$), and hemimorphite (calamine, $Zn_2SiO_4$). The basic form of zinc carbonate (hydrozincite, $Zn_5(CO_3)_2(OH)_6$) is also mined where economically viable. The main source of cadmium is as an impurity in zinc blende; however, there are several other ores known, e.g., cadmoselite (cadmium selenide, CdSe) and otavite ($CdCO_3$). Mercury sulfide (cinnabar, HgS) is the major source of mercury, and in fact metallic liquid mercury droplets are often found in the ore. The terrestrial abundance of the Group 12 elements is given in Table 10.2.

| Element | Terrestrial abundance (ppm) |
|---|---|
| Zn | 75 (Earth's crust), 64 (soil), $30\times10^{-6}$ (sea water) |
| Cd | 0.1 (Earth's crust), 1 (soil), $1\times10^{-6}$ (sea water) |
| Hg | $50\times10^{-6}$ (Earth's crust), $2\times10^{-8}$ (soil), $40\times10^{-12}$ (sea water) |

**Table 10.2: Abundance of Group 12 elements.**

# Isotopes

The naturally abundant isotopes of the Group 12 metals are listed in Table 10.3.

| Isotope | Natural abundance (%) |
|---|---|
| Zinc-64 | 48.6 |
| Zinc-66 | 27.9 |
| Zinc-67 | 4.1 |
| Zinc-68 | 18.8 |
| Zinc-70 | 0.6 |
| Cadmium-106 * | 1.25 |
| Cadmium-108 * | 0.89 |
| Cadmium-110 | 12.49 |
| Cadmium-111 | 12.8 |
| Cadmium-112 | 24.13 |
| Cadmium-113 * | 12.22 |
| Cadmium-114 * | 28.73 |
| Cadmium-116 * | 7.49 |
| Mercury-196 | 0.15 |
| Mercury-198 | 9.97 |
| Mercury-199 | 16.87 |
| Mercury-200 | 23.1 |
| Mercury-201 | 13.18 |
| Mercury-202 | 29.86 |
| Mercury-204 | 6.87 |

**Table 10.3: Abundance of the major (non-synthetic) isotopes of the Group 12 metals. Isotopes labeled with * are radioactive.**

Many radioisotopes of zinc have been characterized. Zinc-65 that has a half-life of 244 days, is the most long-lived isotope, followed by $^{72}Zn$ with a half-life of 46.5 hours. The most common decay mode of an isotope of zinc with a mass number lower than 64 is electron capture, producing an isotope of copper,

$$^{30}_{n}Zn + e^{-} \rightarrow {}^{29}_{n}Cu$$

The most common decay mode of an isotope of zinc with mass number higher than 64 is *beta* decay (β-), which produces an isotope of gallium,

$$^{30}_{n}Zn \rightarrow ^{31}_{n}Ga + e^- + \nu_e$$

Naturally occurring cadmium is composed of 8 isotopes. For two of them, natural radioactivity was observed, and three others are predicted to be radioactive, but their decay is not observed, due to extremely long half-life times. The two natural radioactive isotopes are $^{113}$Cd (half-life = $7.7\times10^{15}$ years) and $^{116}$Cd (half-life = $2.9\times10^{19}$ years).

There are seven stable isotopes of mercury with the longest-lived radioisotopes being $^{194}$Hg (half-life = 444 years) and $^{203}$Hg (half-life = 47 days). $^{199}$Hg and $^{201}$Hg are the most often studied NMR-active nuclei, having spins of ½ and $^3/_2$ respectively.

## Properties

A summary of the physical properties of the Group 12 metals is given in Table 10.4. Because of the ns electron in the Group 12 metals are tightly bound, and hence relatively unavailable for metallic bonding, the metals are volatile with low boiling points, as compared to the Group 2 metals.

| Element | Melting point (°C) | Boiling point (°C) | Density (g/cm$^3$) |
|---|---|---|---|
| Zn | 419.53 | 907 | 7.14 |
| Cd | 321.07 | 767 | 8.65 |
| Hg | -38.83 | 356.73 | 13.534 (liquid) |

**Table 10.4: Selected physical properties of the Group 12 metals.**

The most notable anomaly in the Group 12 metals is the low melting point of mercury compared to zinc and cadmium. In order to completely understand the reasons for mercury's low melting point quantum physics is required; however, the key point is that mercury has a unique electronic configuration, i.e., [Xe] $5d^{10}6s^2$. The stability of the 6s shell is due to the presence of a filled 4f shell, because an f shell poorly screens the nuclear charge that increases the attractive coulomb interaction of the 6s shell and the nucleus. Such a configuration strongly resists removal of an electron and as such mercury behaves similarly to noble gas elements, which form weakly bonded and thus easily melting solids (Figure 10.3).

**Figure 10.3: Liquid mercury.**

## Industrial production

The vast majority (95%) of zinc is mined from of the zinc sulfide ores. The zinc is most often mixed with copper, lead, and iron. Zinc metal is produced by extraction, in which the ore is ground and then the minerals are separated from the gangue (commercially worthless mineral matter) by froth flotation (a process for selectively separating hydrophobic materials from hydrophilic). Roasting converts the zinc sulfide concentrate produced to zinc oxide.

$$2\,ZnS + 3\,O_2 \rightarrow 2\,ZnO + 2\,SO_2$$

Reduction of the zinc oxide with carbon, or carbon monoxide, at 950 °C into the metal is followed by distillation of the metal.

$$2\,ZnO + C \rightarrow 2\,Zn + CO_2$$

$$2\,ZnO + 2\,CO \rightarrow 2\,Zn + 2\,CO$$

Since cadmium is a common impurity in zinc ores, it is most often isolated during the production of zinc. Cadmium is isolated from the zinc metal by vacuum distillation if the zinc is smelted, or cadmium sulfate is precipitated out of the electrolysis solution.

Mercury is extracted by heating cinnabar (HgS) in a current of air and condensing the vapor.

$HgS + O_2 \rightarrow Hg + SO_2$

## Cadmium chalcogenides

The most studied non-oxide semiconductors are cadmium chalcogenides (CdE, with E = sulfide, selenide and telluride). CdE nanocrystals were probably the first material used to demonstrate quantum size effects corresponding to a change in the electronic structure with size, i.e., the increase of the band gap energy with the decrease in size of particles (Figure 10.4). These semiconductors nanocrystals are commonly synthesized by thermal decomposition of an organometallic precursor dissolved in an anhydrous solvent containing the source of chalcogenide and a stabilizing material (polymer or capping ligand). Stabilizing molecules bound to the surface of particles control their growth and prevent particle aggregation.

**Figure 10.4: Picture of cadmium selenide (CdSe) quantum dots, dissolved in toluene, fluorescing brightly, as they are exposed to an ultraviolet lamp, in three noticeable different colors (blue ~481 nm, green ~520 nm, and orange ~612 nm) due to the quantum dots' bandgap (and thus the wavelength of emitted light) depends strongly on the particle size; the smaller the dot, the shorter the emitted wavelength of light. The "blue" quantum dots have the smallest particle size, the "green" dots are slightly larger, and the "orange" dots are the largest.**

Although cadmium chalcogenides are the most studies semiconducting nanoparticles, the methodology for the formation of semiconducting nanoparticles was first demonstrated independently for InP and GaAs, e.g.,

$InCl_3 + P(SiMe_3)_3 \rightarrow InP + 3\,Me_3SiCl$

This method has been adapted for a range of semiconductor nanoparticles. In the case of CdE, dimethylcadmium $Cd(CH_2)_2$ is used as a cadmium source and bis(trimethylsilyl)sulfide, $(Me_3Si)_2S$, trioctylphosphine selenide or telluride (TOPSe, TOPTe) serve as sources of selenide in tri-octylphosphine oxide (TOPO) used as solvent and capping molecule. The mixture is heated at 230-260 °C, over a few hours while modulating the temperature in response to changes in the size distribution as estimated from the absorption spectra of aliquots removed at regular intervals. These particles, capped with TOP/TOPO molecules, are non-aggregated (Figure 10.5) and easily dispersible in organic solvents forming optically clear dispersions. When similar syntheses are performed in the presence of surfactant, strongly anisotropic nanoparticles are obtained, e.g., rod-shaped CdSe nanoparticles can be obtained.

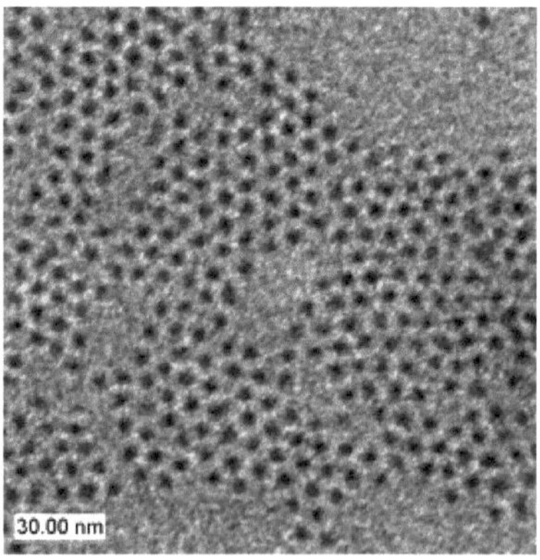


**Figure 10.5: TEM image of CdSe nanoparticles.**

Because $Cd(CH_3)_2$ is extremely toxic, pyrophoric and explosive at elevated temperature, other Cd sources have been used. CdO appears to be an interesting precursor. CdO powder dissolves in TOPO and HPA or TDPA (tetradecylphosphonic acid) at about 300 °C giving a colorless homogeneous solution. By introducing selenium or tellurium dissolved in TOP, nanocrystals grow to the desired size.

Nanorods of CdSe or CdTe can also be produced by using a greater initial concentration of cadmium as compared to reactions for nanoparticles. This approach has been successfully applied for synthesis of numerous other metal chalcogenides including ZnS, ZnSe, and $Zn_{1-x}Cd_xS$. Similar procedures enable the formation of MnS, PdS, NiS, $Cu_2S$ nanoparticles, nano rods, and nano disks.

# Organometallic chemistry

## Organometallic chemistry of zinc

The first dialkyl zinc derivatives, $Me_2Zn$ and $Et_2Zn$, were prepared in 1848 by Edward Franklin (Figure 10.6). He also prepared the monoalkyl derivatives, RZnX.

**Figure 10.6: British chemist Sir Edward Franklin FRS (1825 - 1899).**

Initially alkyl zinc compounds were used in organic synthesis, however, their use diminished significantly once Grignard reagents had been discovered. Little further was investigated of their chemistry until their use in the growth of electronic materials developed in the 1980's.

*RZnX*

The monoalkyl derivatives are not widely used but were historically the first to be prepared.

$$RX + Zn \rightarrow RZnX$$

While the iodide derivatives can be isolated as un-solvated derivatives, the chloride and bromides need to be prepared in the presence of dimethylformamide (DMF) or dimethyl sulfoxide (DMSO). Alternative methods of synthesis involve the reaction with a Grignard reagent or electrochemical synthesis with a zinc electrode.

$$ZnX_2 + RMgX \rightarrow RZnX + MgX_2$$

In the solid state RZnI exists as cage oligomers or polymeric chain structures. These structures are broken by the addition of strong Lewis bases to form Lewis acid-base complexes. In solution there exists a Schlenk equilibrium, whose presence has been determined by IR and Raman spectroscopy (Table 10.5).

$$2\ RZnI \rightleftharpoons R_2Zn + ZnI_2$$

| Stretch | IR ($cm^{-1}$) | Raman ($cm^{-1}$) |
|---|---|---|
| symmetric C-Zn-C | Not observed | 477 |
| asymmetric C-Zn-C | 551 | Not observed |
| C-Zn-I | 510 | 510 |

**Table 10.5: IR and Raman spectroscopic characterization of the components of the Schlenk equilibrium.**

*$R_2Zn$*

Dialkyl zinc compounds are prepared via the monoalkyl derivatives,

$$2\ Zn + 2\ RX \rightarrow 2\ RZnX \overset{\Delta}{\rightleftharpoons} R_2Zn\uparrow + ZnX_2$$

The Schlenk equilibrium is shifted at high temperatures by the distillation of the volatile $R_2Zn$ derivative. The zinc is usually alloyed with copper (10%) to improve the reaction rate. Alternative preparation methods include the reaction of $ZnX_2$ with a Grignard or by metal-metal exchange,

$$2\ RMgX + ZnX_2 \rightarrow R_2Zn + 2\ MgX_2$$

$$R_2Hg + Zn \rightarrow Hg + R_2Zn$$

# Organomercury compounds

*Synthesis*

The most common routes to organomercury compounds involve the direct reaction of mercury with an alkyl iodide to form the mercury analog of a Grignard reagent.

$$Hg + RX \rightarrow RHgX$$

The subsequent reaction of RHgI with potassium cyanide yields the appropriate dialkyl mercury derivative.

$$2\ RHgI + 2\ KCN \rightarrow R_2Hg + Hg + 2\ KI + (NC)_2$$

*Solvomercuration*

The general process of solvomercuration involves the addition of $HgX_2$ across an alkene double (or alkyne triple) bond in the presence of a solvent, Solvomercuration applies where HY are part of the solvent system, e.g., in water the process can be described as hydroxymercuration.

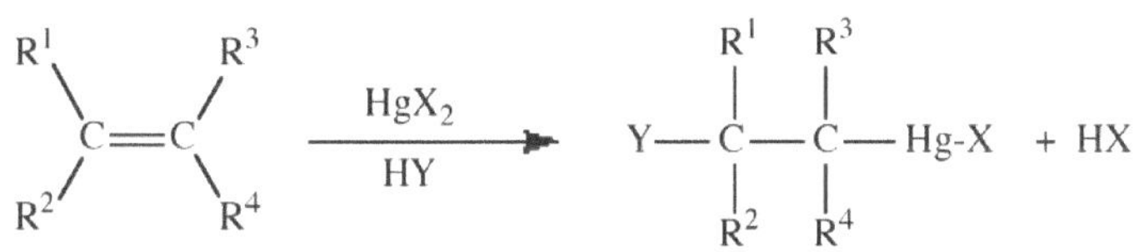

The addition to the alkene occurs with Markovniko addition, i.e., through the formation of the most stable carbonium ion resulting with the mercury adding to the less substituted carbon. The Hg-C bond can be cleaved by the addition of $NaBH_4$ to yield the C-H bond. It is common to employ mercury acetate, $Hg(OAc)_2$ ($OAc = O_2CCH_3$), with subsequent reaction with NaCl or NaI to form the halide, rather than the mercury halide directly. However, the acetate can also act as a neucleophile resulting in a mixture of products. The order of reactivity for alkenes follows the trend:

$$R_2C{=}CH_2 > R(H)C{=}CH_2 > cis\ R(H)C{=}C(H)R > trans\ R(H)C{=}C(H)R > R_2C{=}C(H)R > R_2C{=}CR_2$$

This is due to a combination of steric and electronic effects. The addition occurs in a *trans/anti* fashion,

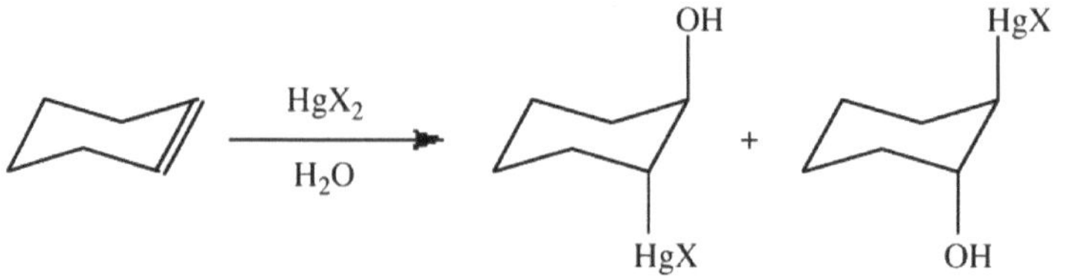


Mercury(II) halides dissociate in polar solvents and this species is commonly associated with the initial reaction step.

$$HgX_2 \rightleftharpoons HgX^+ + X^-$$

The mechanism of solvomercuration is best described by the reaction shown in Figure 10.7.

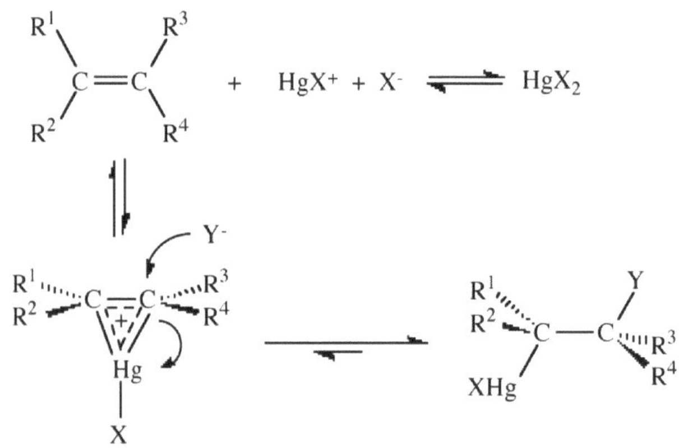


**Figure 10.7: Mechanism of solvomercuration.**

Evidence for the mechanism is twofold:

1. The addition is exclusively *trans*.
2. No rearrangement is observed even for $^{t}$Bu(H)C=C(H)$^{t}$Bu whose carbonium ion is known to undergo rearrangement.

Isotope studies indicate that the C-O bond formation is present in the transition state (Figure 10.8). The mercury bridge in the transition state may not be symmetrical, but the structure is similar to the addition of $Br^+$ and $AuX^+$ to alkenes.

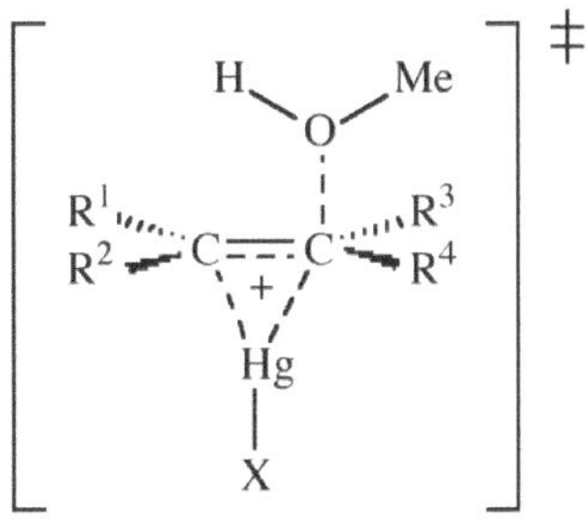

**Figure 10.8: Proposed transition state for the solvomercuration reaction where the solvent is methanol.**

The actual reaction was originally carried out with $Hg(OAc)_2$ in benzene at 110 °C for several hours in acetic acid solution. It was found that the reaction was catalyzed by the presence of $HClO_4$, $H_2SO_4$, and $HNO_3$, which were found to replace the acetate ion. The reaction rate is also increased by 690,000 times by the use of $Hg(O_2CCF_3)_2$ in $HO_2CCF_3$.

The solvomercuration of alkynes gives alkenylmercury compounds, but the reaction is more sluggish than for the reaction with alkenes, and the product is always the trans isomer.

*Mercuration of aromatic compounds*

Mercuration is an electrophilic aromatic substitution reaction that is possible for most 2n+2 π-electron species,

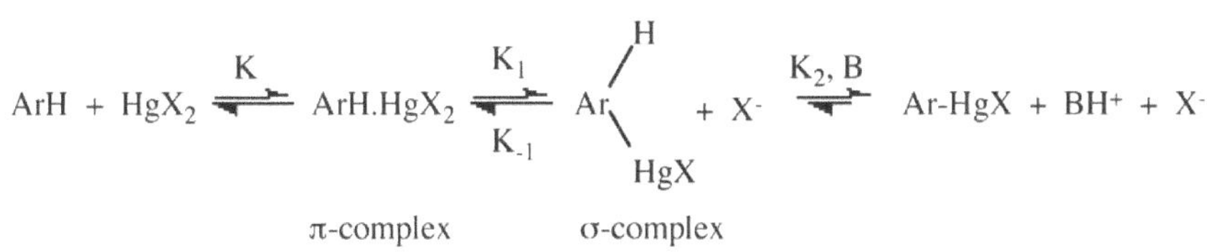

Evidence for the π-complex intermediate is indicated by UV spectroscopy, which shows an increase in the region 280-320 nm for the reaction of aromatic compounds with $Hg(O_2CCF_3)_2$ in $HO_2CCF_3$. The σ-complex has detected in liquid $SO_2$. As a preparative method mercuration suffers from a lack of selectivity, including an isomerization from para substitution at low temperature to meta substitution at high temperatures.

## Structure

Dialkyl mercury compounds, $R_2Hg$, are generally air stable and show little Lewis acid behavior. They are monomeric colorless liquids or low melting solids, e.g., boiling point = 92.5 °C for $Me_2Hg$. No solubility is observed in water, except for $(F_3C)_2Hg$.

$$(F_3C)_2Hg + H_2O \rightleftharpoons (F_3C)_2Hg(OH)^- + H^+$$

The hybridization at mercury involves the 6s and 6p orbitals; however, the 5d may be involved. X-ray crystal structures for both $R_2Hg$ and RHgX show linear structures. In general, $R_2Hg$ compounds are very weak Lewis acids, but adduct are formed if the alkyl group is sufficiently electron withdrawing, e.g., $(C_6F_5)_2Hg$. The geometry of the Lewis acid-base complex is not triangular as predicted from VSEPR theory, but a T-shape. The distortion from a linear C-Hg-C unit is minor. For example, in $(C_6F_5)_2Hg$ the C-Hg-C angle is 176.2°, while in $[(C_6F_5)_2Hg]_2$(diars), where diars = $Ph_2AsCH_2CH_2AsPh_2$ (Figure 10.9), the same angle is 173°. Furthermore, the donor atom-Hg bond distance (e.g., As-Hg = 3.4 Å) is only slightly shorter than the sum of the van der Waal radii (e.g., 3.5 Å).

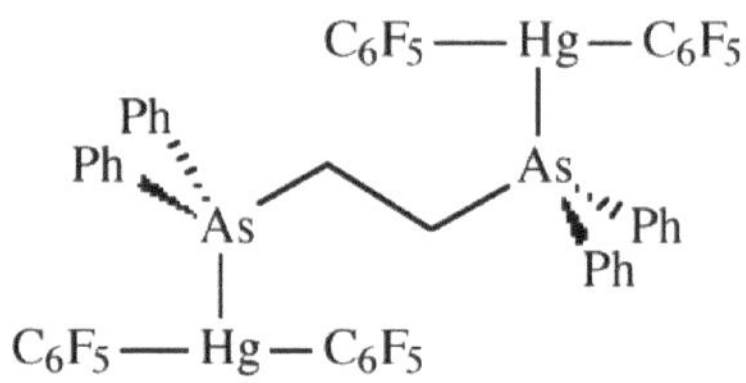


**Figure 10.9: The structure of $[(C_6F_5)_2Hg]_2$(diars).**

## The myth, reality, and history of mercury toxicity

Mercury has a reputation for being a dangerous element, but is its reputation deserved? Given the large-scale use of mercury today it is important to understand the risks and issues related to mercury exposure. Nowhere is this now important than with the use of mercury for small low energy fluorescent lights (Figure 10.10) that are being advocated by everyone from the electricity companies to Greenpeace.

When considering the issue of mercury toxicity, it is important to separate the effects of mercury metal (as a liquid or vapor) from the compounds of mercury.

**Figure 10.10: An example of a modern low energy mercury vapor fluorescent light.**

## Mercury metal

It was found very early on that people who worked with mercury, in mining for example, had very bad health. Other jobs that exposed people to mercury were mirror makers and hatters (people who manufactured hats). The problems in this latter occupation will forever live on with one of the central characters in Lewis Carroll's Alice's Adventures in Wonderland; the Mad Hatter.

Hats were made from felt, which is a non-woven textile of animal hair. Wool interlocks naturally due to the surface texture of the individual hairs, but rabbit and beaver have to be artificially roughened. This process was usually accomplished with nitric acid ($HNO_3$). It was found that if mercury was added to the nitric acid, a better quality of felt was produced. Unfortunately, when the felt was dried a fine dust was formed containing mercury. The hatters who shaped the felt inhaled large quantities of this dust were found to suffer from excessive salivation, erethism (presenting with excessive shyness, timidity and social phobia), and shaking of the limbs, which became known as hatter's shakes. The madness that was observed is the derivation of the phrase mad as a hatter. It is interesting that while Carroll's Mad Hatter is mad, he does not show the classic symptoms of mercury poisoning. In particular he can be in no way described as shy!

Hatters were not the only people that mercury caused a problem for. Chemists doing research using large quantities of mercury were also affected. They were given to violent headaches, tremors of the hands, socially troublesome inflammation of the bladder, loss of memory, and slow mental processes. In 1926 Alfred Stock (Figure 10.11) and his research group all suffered from symptoms. However, when the lab was cleaned of mercury the symptoms went away.

**Figure 10.11: German chemist Alfred Stock (1876 - 1946).**

Many other notable scientists have also suffered from mercury poisoning. Faraday (Figure 10.12), Pascal (Figure 10.13), and most probably Sir Isaac Newton (Figure 10.14) were affected. As part of his research studies, Newton boiled several pounds of mercury a day just before his period of insanity between 1692 and 1693. It is likely that the mercury vapor was the cause of his malady. However, in each case, the symptoms (and insanity) abated once the source of mercury was removed.

It is important to remember that in all the cases described above it is the inhalation of the mercury vapor that was the cause of the trouble. Solid alloys of mercury such as those found in dental fillings have never been shown to cause any medical issues directly. Despite this the US banned the use of Cu/Hg dental amalgams until 1850. More recently, it has been suggested that dentists are exposed at higher levels during the placing and removal of fillings. Dentists as a group have higher mercury levels than those associated with people with amalgam restorations, but experience no increase in disease or death rates, and in fact tend to be healthier than the general population.

**Figure 10.12: An engraving by John Cochran of English chemist and physicist Michael Faraday, FRS (1791 - 1867).**

**Figure 10.13: French mathematician, physicist, and religious philosopher Blaise Pascal (1623 - 1662).**

Although elemental mercury was clearly toxic, this did not stop its use in pharmacy for hundreds of years. In the 1500's mercury was used in the treatment (albeit ineffective) of syphilis. Syphilis was a new disease in Europe; it had been brought back from America by Columbus' sailors and was promptly spread through Europe by the French army, amongst others. Syphilis was much more fatal and had more dramatic symptoms than today.

**Figure 10.14: English physicist, mathematician, astronomer, natural philosopher, alchemist, and theologian Sir Isaac Newton FRS (1643 - 1727). Portrait by Godfrey Kneller.**

Initially mercury was used as an ointment, but the patients often got worse. Then there was the tub, which was a mercury vapor bath, and even calomel ($Hg_2Cl_2$) was used, but with little effect. These treatments were used for over four centuries, but none provided a cure, despite claims at the time. For example, John Hunter, a doctor who gave himself syphilis by mistake claimed he had been cured, but he actually died of a heart attack during an argument, so it is unlikely the mercury worked. Despite this it became known that a night with Venus results in a lifetime with Mercury.

The reasons that mercury was thought erroneously to cure syphilis are two-fold:

1. Until 1906 it was difficult to diagnose syphilis. It was often confused with gonorrhea, and therefore it is likely that some people did not have the far more deadly syphilis.
2. Syphilis occurs in three phases, each with remission between the phases. The period of remission between secondary and tertiary phases can be two to three years, and therefore it may appear that a cure is found. Especially as may patients (like John Hunter) died other deaths during this remission phase.

The prevalent use of mercury and its presence in many cadavers, led some doctors to assume that mercury was a natural part of the body. It was not just humans that were treated with mercury, cattle were also treated, and one druggist sold 25 tons of mercury to a single farmer in one year.

The density of mercury and its liquid state at room temperature led to another unusual application that was somewhat more successful, although equally dangerous: constipation. In medical texts of the time it was noted that mercury is given in the disease called Miserere, unto two or three pounds, and is voided again by siege to the same weight; it is better to take a great deal of it that a little, because a small quantity might be apt to stop in the circumvolutions of the guts, and if some acid humors should happen to join with it, a sublimate corrosive would be made; but when a large quantity of it is taken, there's no need to fearing this accident, because it passes through by its own weight.

It is interesting that the mention of the corrosive sublimate; this is in fact mercury(II) dichloride ($HgCl_2$) which unlike mercury(I) chloride ($Hg_2Cl_2$), is a very violent poison. Death is caused by renal failure. So, while there is no evidence for elemental mercury itself causing fatalities, its compounds are another matter to be considered.

## Organomercury compounds

In 1953 it was noted that the fishing village of Minamata in Japan had an epidemic in which a large number of people died. Initial thoughts of either an infectious disease or malnutrition were discounted; then it was found that the fish eaten by the villagers was highly contaminated by mercury.

It was found that the mercury came from the Chisso Corporation chemical plant that made acetaldehyde from acetylene using a mercury catalyst. The plant was losing 1 kg of mercury metal for every ton of acetaldehyde being produced. As a consequence, it was originally assumed that the poisoning of the village was due to inorganic mercury. Based upon prior incidents, an obvious response was to ban consumption of all fish and shellfish. As a consequence, no new cases were reported, however, people already effected continued to die. This was unlike any previous mercury poisoning.

Further analysis showed small quantities of water-soluble methyl mercury ($MeIIg^+$) was present and sequestered by the shellfish to give MeIIgSMe. While the lethal effects of organomercury compounds were known, the source of the methyl mercury was a mystery. A group of Swedish researchers showed that the bacterial action in river sediment or rotting fish converted inorganic mercury to either volatile $Me_2Hg$ or water soluble $MeHg^+$. With this discovery, it was understood how the anaerobic mud of the estuary near Minamata could perform this methylation.

Of course, Minamata was not the first report or an organomercury compound, but it was the first time that it was shown that mercury metal could be converted to a highly toxic organometallic derivative in the environment. The hazardous nature of organomercurials was found almost as soon as the first compounds were reported!

While working in Bunsen's research group in Marburg, Edward Franklin (Figure 10.6) discovered the synthesis of the zinc analog of a Grignard reagent.

$$Zn + MeI \rightarrow MeZnI$$

Subsequently, in 1851 Franklin moved to Owens College in Manchester where he extended his work to mercury.

$$Hg + RX \rightarrow RHgX$$

In his publication he noted that these organomercury compounds had a nauseous taste but didn't realize they were toxic.

In 1858 George Buckton (Figure 10.15) working at the then Royal College of Chemistry (now Imperial College) reported the synthesis of dimethyl mercury, as a volatile liquid.

$$2\,MeHgI + 2\,KCN \rightarrow Me_2Hg + Hg + 2\,KI + (NC)_2$$

**Figure 10.15: British chemist George Buckton (1818 - 1905).**

When Frankland moved his research to St Bartholomew's Hospital ('Barts') London he started looking into the chemistry of $R_2Hg$ with an assistant called Bill Odling in collaboration with Dr Carl Ulrich.

Ulrich died in 1865 as a consequence of exposure to $Me_2Hg$. In his own statement, he had inhaled a large quantity of the volatile compound without having taken the proper precautions. The following day his countenance had attained a dull, anxious, and confused expression and he was admitted to the hospital in a weak condition on 3rd February. On the 9th he became noisy and had to be put under mechanical restraint. The next day his breath and body began to smell offensively, and he was in a coma. He would rise from the coma periodically to utter incoherent howls. He died on the 14th of February.

A technician from the same research group (who is only identified as `T. C.') was also admitted to the hospital on 28th March of the same year. His symptoms were initially milder than Ulrich's, but soon developed. By that summer he was completely demented, with no control over his body functions. He stayed in this state for many months, only dying on 7th April 1866. Records indicate that a third assistant was also taken ill, but there was no further mention of him, so it is unknown if he died.

## Summary and the green future

Metallic mercury causes severe symptoms, but all records show that if the patient is removed from the source they recover. Thus, short term exposure to metallic mercury, while dangerous, is not fatal if proper precautions are taken. In contrast, mercury compounds offer different risks. As a general rule, inorganic mercury(I) compounds are far less toxic than their mercury(II) analogs; however, all should be treated with care.

Where mercury compounds offer the greatest risk of fatality is their organometallic derivatives. There is no known cure for exposure to even modest doses of organomercury compounds. Furthermore, the ability of elemental mercury to be transformed into water-soluble organomercury compounds such as $MeHg^+$, offers a future threat to public health.

The new generation of low energy consumption light bulbs contain mercury vapor. While they last longer than a traditional tungsten lament light bulb, they do have a lifetime. The presence of mercury means that they should be disposed-off separately from household waste to ensure that when the glass is broken the mercury is not released; however, this is unlikely. Most will be disposed of along with household waste which may be subsequently land filled. The lesson from Minamata should be that the bacterial action under anaerobic condition allows for the formation of water soluble $MeHg^+$, that can di use into the water table. Although the amount of mercury in each bulb is very small, the highly lethal nature (low $LD_{50}$) of orgnaomercury compounds should be considered in efforts to conserve energy by the use of the low energy light bulbs. At the very minimum protocols for their efficient disposal and recycling should be in place.

# Chapter 11: Group 13

## The Elements

Table 11.1 lists the derivation of the names of the Group 13 (IIIA) elements.

| Element | Symbol | Name |
|---|---|---|
| Boron | B | From the Arabic word *buraq* or the Persian word *burah* for the mineral borax |
| Aluminium (Aluminum) | Al | From alum |
| Gallium | Ga | From Named after the Latin word for France (*Gaul*) *Gallia* |
| Indium | In | Latin *rubidus* meaning *deepest red* |
| Thalium | Tl | From the Latin *thallus* meaning *sprouting green twig* |

**Table 11.1: Derivation of the names of each of the alkali metal elements.**

Aluminium is the international spelling standardized by the IUPAC, but in the United States it is more commonly spelled as aluminum.

## Discovery

*Boron*

Borax (a mixture of $Na_2B_4O_7.4H_2O$ and $Na_2B_4O_7.10H_2O$) was known for thousands of years. In Tibet it was known by the Sanskrit name of *tincal*. Borax glazes were used in China in 300 AD and the writings of the Arabic alchemist Geber (Figure 11.1) appear to mention it in 700 AD. However, it is known that Marco Polo brought some borax glazes back to Italy in the 13th century. In 1600 Agricola (Figure 11.2) in his treatise De Re Metallica reported the use of borax as a flux in metallurgy.

**Figure 11.1: A drawing of the father of chemistry Abū Mūsā Jābir ibn Hayyān known by Geber; the Latinized form of Jabir (721 - 815). Geber was a chemist and alchemist, astronomer and astrologer, engineer, geologist, philosopher, physicist, and pharmacist and physician.**

**Figure 11.2: A painting of German author Georg Bauer (1494 - 1555), whose penname was the Latinized Georgius Agricola, was most probably the first person to be environmentally conscious.**

Boron was not recognized as an element until its isolation by Sir Humphry Davy (Figure 8.5), Joseph Louis Gay-Lussac (Figure 11.3) and Louis Jacques Thénard (Figure 11.4) in 1808 through the reaction of boric acid and potassium. Davy called the element *boracium*. Jöns Jakob Berzelius (Figure 8.3) identified boron as an element in 1824.

**Figure 11.3: French chemist and physicist Joseph Louis Gay-Lussac (1778 - 1850).**

**Figure 11.4: French chemist Louis Jacques Thénard (1777 - 1857).**

*Aluminum*

Ancient Greeks and Romans used aluminum salts as dyeing mordants and as astringents for dressing wounds; alum ($KAl(SO_4)_2.12H_2O$) is still used as a styptic (an anti-haemorrhagic agent). In 1808, Sir Humphry Davy (Figure 8.5) identified the existence of a metal base of alum, which he at first termed *alumium* and later *aluminum*.

The metal was first produced in 1825 (in an impure form) by Hans Christian Ørsted (Figure 11.5) by the reaction of anhydrous aluminum chloride with potassium amalgam. Friedrich Wöhler (Figure 9.2) repeated the experiments of Ørsted but suggested that Ørsted had only isolated potassium. By the use of potassium, he is credited with isolating aluminum in 1827. While Wöhler is generally credited with isolating aluminum, Ørsted should also be given credit.

$$AlCl_3 + 3\,K \rightarrow Al + 3\,KCl$$

**Figure 11.5: Danish physicist and chemist Hans Christian Ørsted (1777 - 1851).**

In 1846 Henri Deville (Figure 11.6) improved Wöhler's method, and described his improvements in particular the use of sodium in place of the expensive potassium

$$AlCl_3 + 3\,Na \rightarrow Al + 3\,NaCl$$

**Figure 11.9: French chemist Henri Etienne Sainte-Claire Deville (1818 - 1881).**

*Gallium*

The element gallium was predicted, as *eka-aluminum*, by Mendeleev (Figure 11.7) in 1870, and subsequently discovered by Lecoq de Boisbaudran (Figure 11.8) in 1875; in fact, de Boisbaudran had been searching for the missing element for some years, based on his own independent theory.

**Figure 11.7: Russian chemist and inventor Dmitri Ivanovich Mendeleev (1834 - 1907).**

**Figure 11.8: French chemist Paul Émile (François) Lecoq de Boisbaudran (1838 - 1912).**

The first experimental indication of gallium came with the observation of two new violet lines in the spark spectrum of a sample deposited on zinc. Within a month of these initial results de Boisbaudran had isolated 1 g of the metal starting from several hundred kilograms of crude zinc blende ore. The new element was named in honor of France (Latin Gallia), and the striking similarity of its physical and chemical properties to those predicted by Mendeleev did much to establish the general acceptance of the periodic Law; indeed, when de Boisbaudran first stated that the density of Ga was 4.7 g/cm3 rather than the predicted 5.9 g/cm$^3$, Mendeleev wrote to him suggesting that he re-determine the value (the correct value is 5.904 g/cm$^3$).

*Indium*

While testing ores from the mines around Freiberg, Saxony, Ferdinand Reich (Figure 11.9) and Hieronymous Theodor Richter (Figure 11.10) when they dissolved the minerals pyrite, arsenopyrite, galena and sphalerite in hydrochloric acid, and since it was known that ores from that region contained thallium they searched for the green emission lines by spectroscopy. Although the green lines were absent, a blue line was present in the spectrum. As no element was known with a bright-blue emission they concluded that a new element was present in the minerals. They named the element with the blue spectral line indium. Richter went on to isolate the metal in 1864.

**Figure 11.9: German chemist Ferdinand Reich (1799 - 1882).**

**Figure 11.10: German chemist Hieronymus Theodor Richter (1824 - 1898).**

*Thallium*

After the publication of their improved method of flame spectroscopy by Robert Bunsen (Figure 8.6) and Gustav Kirchho (Figure 8.7) this method became an accepted method to determine the composition of minerals and chemical products. Two chemists, William Crookes (Figure 11.11) and Claude-Auguste Lamy, both started to use the new method and independently employed it in their discovery of thallium.

**Figure 11.11: English chemist and physicist Sir William Crookes, FRS (1832 - 1919) attended the Royal College of Chemistry (Imperial College) in London.**

Crookes was making spectroscopic determinations on selenium compounds deposited in the lead chamber of a sulfuric acid production plant near Tilkerode in the Harz mountains. Using a similar spectrometer to Crookes', Lamy was determining the composition of a selenium-containing substance that was deposited during the production of sulfuric acid from pyrite. Using spectroscopy both researchers both observed a new green line the atomic absorption spectrum and assigned it to a new element. Both set out to isolate the new element. Fortunately for Lamy, he had received his material in larger quantities and thus he was able to isolate sufficient quantities of thallium to determine the properties of several compounds and prepare a small ingot of metallic thallium. At the same time Crookes was able to isolate small quantities of elemental thallium and determine the properties of a few compounds. The claim by both scientists resulted in significant controversy during 1862 and 1863; interestingly this ended when Crookes was elected Fellow of the Royal Society in June 1863.

## Abundance

The abundance of the Group 13 elements is given in Table 11.2. Aluminum is the most abundant metal in the Earth's crust and is found in a wide range of minerals. While boron is not as common it is also found in a range of borate minerals. In contrast, gallium, indium, and thallium are found as impurities in other minerals. In particular indium and thallium are found in sulfide or selenide mineral rather than oxides, while gallium is found in both sulfides (ZnS) and oxides (bauxite). Although indium and thallium minerals are known, they are rare: indite ($FeIn_2S_4$), lorandite ($TlAsS_2$), crookesite ($Cu_7TlSe_4$).

| Element | Terrestrial abundance (ppm) |
|---|---|
| B | 10 (Earth's crust), 20 (soil), 4 (sea water) |
| Al | 82,000 (Earth's crust), 100,000 (soil), $5\times10^{-4}$(sea water) |
| Ga | 18 (Earth's crust), 28 (soil), $30\times10^{-6}$ (sea water) |
| In | 0.1 (Earth's crust), 0.01 (soil), $0.1\times10^{-6}$ (sea water) |
| Tl | 0.6 (Earth's crust), 0.2 (soil), $10\times10^{-6}$ (sea water) |

**Table 11.2: Abundance of Group 13 elements.**

## Isotopes

The naturally abundant isotopes of the Group 13 elements are listed in Table 11.3. Thallium has 25 isotopes that have atomic masses that range from 184 to 210. Thallium-204 is the most stable radioisotope, with a half-life of 3.78 years.

| Isotope | Natural abundance (%) |
|---|---|
| Boron-10 | 19.9 |
| Boron-11 | 80.1 |
| Aluminum-27 | 100 |
| Gallium-69 | 60.11 |
| Gallium-71 | 39.89 |
| Indium-113 | 4.3 |
| Indium-115 | 95.7 |
| Thallium-203 | 29.52 |
| Thallium-205 | 70.48 |

**Table 11.3: Abundance of the major isotopes of the Group 13 elements.**

The Group 13 elements offer potential as NMR nuclei (Table 11.4). In particular $^{11}B$ and $^{27}Al$ show promise for characterization in both solution and solid state.

| Isotope | Spin | Natural abundance (%) | Quadrupole moment ($10^{-30}$ $m^2$) | NMR frequency (MHz) at a field of 2.3488 T | Reference |
|---|---|---|---|---|---|
| Boron-10 | 3 | 19.58 | 8.459 | -10.746 | $BF_3.E_2O$ |
| Boron-11 | $^3/_2$ | 80.42 | 4.059 | -32.084 | $BF_3.E_2O$ |
| Aluminum-27 | $^5/_2$ | 100 | 14.66 | -26.057 | $Al(NO_3)_3$ |
| Gallium-69 | $^3/_2$ | 60.4 | 17.1 | -24.003 | $Ga(NO_3)_3$ |
| Gallium-71 | $^3/_2$ | 39.6 | 10.7 | -30.495 | $Ga(NO_3)_3$ |
| Indium-113 | $^9/_2$ | 4.28 | 79.9 | -21.866 | $In(NO_3)_3$ |
| Indium-115 | $^9/_2$ | 95.72 | 81.0 | -21.914 | $In(NO_3)_3$ |

**Table 11.4: Isotopes of Group 13 elements for NMR spectroscopy.**

## Industrial production

Borax is mined as a mixture of $Na_2B_4O_7.4H_2O$ and $Na_2B_4O_7.10H_2O$. Acidic cation gives boric acid, $B(OH)_3$, which can be reduced with sodium amalgam (Na/Hg) to give amorphous boron. Pure boron can be prepared by reducing boron halides (e.g., $BF_3$ and $BCl_3$) with hydrogen at high temperatures. Ultrapure boron, for the use in semiconductor industry, is produced by the decomposition of diborane ($B_2H_6$) and then further purified with the zone melting or Czochralski processes (Figure 11.12).

**Figure 11.12. Polish chemist Jan Czochralski (1885 - 1953).**

Gallium is very much less abundant than aluminum and tends to occur at low concentrations in sulfide minerals rather than as oxides, although gallium is also found associated with aluminum in bauxite. The main source of gallium is as a by-product of aluminum refining. At 19 ppm of the earth's crust, gallium is about as abundant as nitrogen, lithium and lead; it is twice as abundant as boron (9 ppm) but is more difficult to extract due to the lack of any major gallium-containing ore. Gallium always occurs in association either with zinc or germanium, its neighbors in the periodic table, or with aluminum in the same group. Thus, the highest concentrations (0.1 - 1%) are in the rare mineral germanite (a complex sulfide of Zn, Cu, Ge, and As); concentrations in sphalerite (ZnS), bauxite, or coal, are a hundred-fold less.

The lack of indium mineral deposits and the fact that indium is enriched in sulfides of lead, tin, copper, iron and zinc, makes the zinc production the main source for indium. The indium is leached from slag and dust of zinc production. Up until 1924, there was only about a gram of isolated indium on the planet, however, today worldwide production is currently greater 476 tons per year from mining and a 650 tons per year from recycling. This massive increase in demand is due to applications in LCD displays and solar cell applications.

*Aluminum*

Due to aluminum's position as the most abundant metallic element in the Earth's crust (7.5 - 8.1%) and its enormous industrial importance warrants detailed discussion of its industrial production. Aluminum only appears in its elemental form in nature in oxygen-deficient environments such as volcanic mud. Ordinarily, it is found in a variety of oxide minerals.

In comparison to other metals aluminum is difficult to extract from its ores. Unlike iron, aluminum oxides cannot be reduced by carbon, and so purification is only possible on an economic scale using electrolysis. Prior to electrolysis purified aluminum oxide is obtained by refining bauxite in the process of developed by Karl Bayer (Figure 11.13).

**Figure 11.13: Austrian chemist Karl Josef Bayer (1847 - 1904) whose father, Friedrich, founded the Bayer chemical and pharmaceutical company.**

Bauxite, the most important ore of aluminum, contains only 30-50% alumina, $Al_2O_3$, the rest being a mixture of silica, iron oxide, and titanium dioxide. Thus, the alumina must be purified before it can be used as the oxide or refined into aluminum metal. In the Bayer process, bauxite is digested in hot (175 °C) sodium hydroxide (NaOH) solution (Figure 11.14). This converts the alumina to aluminum hydroxide, $Al(OH)_3$, which dissolves in the hydroxide solution.

$$Al_2O_3 + 2\ OH^- + 3\ H_2O \rightarrow 2\ [Al(OH)_4]^-$$

The other components do not dissolve and are filtered off (Figure 11.14). The hydroxide solution is cooled, and the dissolved aluminum hydroxide precipitates, which when heated to 1050 °C is calcined into alumina.

$$2\ Al(OH)_3 \rightarrow Al_2O_3 + 3\ H_2O$$

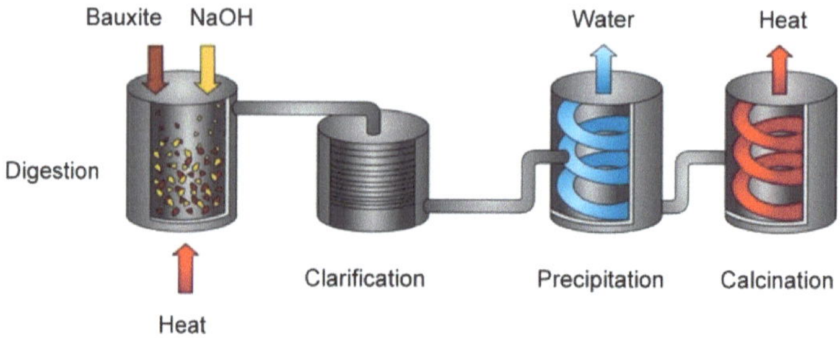


**Figure 11.14: Schematic representation of the Bayer process. Copyright: Andrew Perchard, Alcan Aluminium UK and courtesy of Glasgow University Archives (2007).**

Once a pure alumina is formed, it is dissolved in molten cryolite ($Na_3AlF_6$) and reduced to the pure metal at elevated temperatures (950 – 980 °C) using the Hall-Héroult process, developed independently Charles Hall (Figure 11.15) and Paul Héroult (Figure 11.16).

Both of the electrodes used in the electrolysis of aluminum oxide are carbon (Figure 11.17). The reaction at the cathode involves the reduction of the $Al^{3+}$,

$$Al^{3+} + 3\ e^- \rightarrow Al$$

**Figure 11.15: American inventor and engineer Charles Martin Hall (1863 - 1914).**

**Figure 11.16: French scientist Paul (Louis-Toussaint) Héroult (1863 - 1914).**

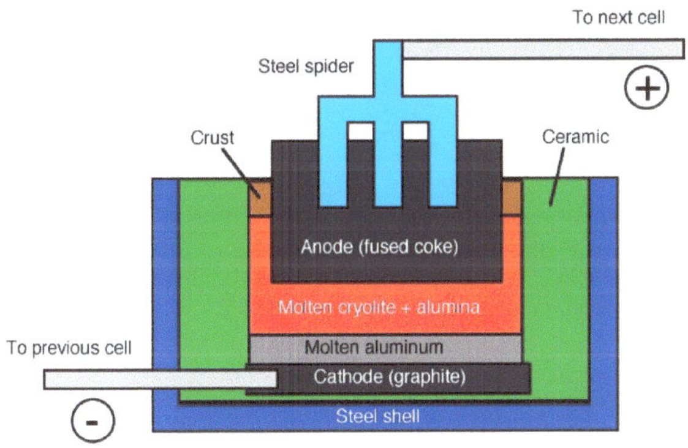


**Figure 11.17: Schematic of the Hall-Héroult cell.**

The aluminum metal then sinks to the bottom and is tapped off, where it is usually cast into large blocks called aluminum billets. At the anode oxygen is formed,

$$2\ O_2^- \rightarrow O_2 + 4\ e^-$$

where it reacts with the carbon anode is then oxidized to carbon dioxide,

$$C + O_2 \rightarrow CO_2$$

The anodes must be replaced regularly, since they are consumed. While the cathodes are not oxidized, they do erode due to electrochemical processes and metal movement. Although the Hall-Héroult process consumes a lot of energy, alternative processes have always found to be less economically and ecologically viable.

## Physical properties

Table 11.5 summarizes the physical properties of the Group 13 elements. While, aluminum, indium, and thallium have typical metal properties, gallium has the largest liquid range of any element. Boron exists as a molecular compound in the solid state, hence its high melting point.

| Element | Melting point (°C) | Boiling point (°C) | Density ($g/cm^3$) |
|---|---|---|---|
| B | 2300 | 3658 | 2.3 |
| Al | 661 | 2467 | 2.7 |
| Ga | 30 | 2403 | 5.9 (solid), 6.1 (liquid) |
| In | 156 | 2080 | 7.3 |
| Tl | 304 | 1457 | 11.9 |

**Table 11.5: Selected physical properties of the Group 13 elements.**

The element gallium was predicted, as eka-aluminum, by Mendeleev (Figure 11.7) in 1870, and subsequently discovered by Lecoq de Boisbaudran (Figure 11.8) in 1875; in fact, de Boisbaudran had been searching for the missing element for some years, based on his own independent theory. The first experimental indication of gallium came with the observation of two new violet lines in the spark spectrum of a sample deposited on zinc. Within a month of these initial results de Boisbaudran had isolated 1 g of the metal starting from several hundred kilograms of crude zinc blende ore. The new element was named in honor of France (Latin *Gallia*), and the striking similarity of its physical and chemical properties to those predicted by Mendeleev (Table 11.6) did much to establish the general acceptance of the periodic Law; indeed, when de Boisbaudran first stated that the density of Ga was 4.7 $g/cm^3$ rather than the predicted 5.9 $g/cm^3$, Mendeleev wrote to him suggesting that he redetermine the value (the correct value is 5.904 $g/cm^3$).

Gallium has a beautiful silvery blue appearance; it wets glass, porcelain, and most other surfaces (except quartz, graphite, and Teflon®) and forms a brilliant mirror when painted on to glass. The atomic radius and first ionization potential of gallium are almost identical with those of aluminum and the two elements frequently resemble each other in chemical properties. Both are amphoteric, but gallium is less electropositive as indicated by its lower electrode potential. Differences in the chemistry of the two elements can be related to the presence of a filled set of 3d orbitals in gallium.

| Property | Mendeleev's prediction (1871) for eka-aluminum, M | Observed properties of gallium (discovered 1875) |
|---|---|---|
| Atomic weight | ca. 68 | 69.72 |
| Density, g.cm$^{-3}$ | 5.9 | 5.904 |
| Melting point | Low | 29.78 °C |
| Vapor pressure | Non-volatile | $10^{-3}$ mmHg, 1000 °C |
| Valence | 3 | 3 |
| Oxide | $M_2O_3$ | $Ga_2O_3$ |
| Density of oxide (g/cm$^3$) | 5.5 | 5.88 |
| Properties of metal | M should dissolve slowly in acids and alkalis and be stable in air | Ga metal dissolves slowly in acids and alkalis and is stable in air |
| Properties of hydroxide | $M(OH)_3$ should dissolve in both acids and alkalis | $Ga(OH)_3$ dissolves in both acids and alkalis |
| Properties of salts | M salts will tend to form basic salts; the sulfate should form alums; $M_2S_3$ should be precipitated by $H_2S$ or $(NH_4)_2S$; an- hydrous $MCl_3$ should be more volatile than $ZnCl_2$ | Ga salts readily hydrolyze and form basic salts; alums are known; $Ga_2S_3$ can be precipitated under special conditions by $H_2S$ or $(NH_4)_2S$, anhydrous $GaCl_3$ is more volatile than $ZnCl_2$. |

**Table 11.6: Comparison of predicted and observed properties of gallium.**

## Trends for the Group 13 compounds

Boron is a non-metal with metalloidal tenancies. The higher ionization energies for boron than for its other Group homologs are far more than would be compensated by lattice energies, and thus, the $B^{3+}$ ion plays no part in the chemistry of boron, and its chemistry is dominated by the formation of covalent compounds. In contrast, the elements aluminum through thallium each has a low electronegativity and the chemistry of their compounds reflects this characteristic. Each of the Group 13 metals forms both covalent compounds and ionic coordination complexes.

All of the Group 13 (IIIA) elements have a valence shell electron configuration of $ns^2np^1$. As a consequence, all of the Group 13 elements for compounds in which they adopt a +3 oxidation state. While the lighter elements do form compounds with lower oxidation state, they are not the norm; however, the +1 oxidation state is more prevalent for the heavier elements in particular thallium. The rational for this is described as the inert pair effect. The inert pair effect is usually explained by the energy of the ns orbital is lower making it harder to ionize and stabilizing a $ns^2np^0$ valence shell. However, as may be seen from Table 11.7 the sum of the second and third ionization enthalpies is lower for indium (4501 kJ/mol), than for gallium (4916 kJ/mol), but with thallium intermediate (4820 kJ/mol). The true source of the inert pair effect is that the lower bond strengths observed for the heavier elements (due to more diffuse orbitals and therefore less efficient overlap) cannot compensate for the energy needed to promote the $ns^2$electrons. For example, the bond energies for gallium, indium, and thallium in $MCl_3$ are 242, 206, and 153 kJ/mol, respectively. It has also been suggested that relativistic effects make a contribution to the inert pair effect.

| Ionization enthalpy (kJ/mol) | Al | Ga | In | Tl |
|---|---|---|---|---|
| 1 | 576.4 | 578.3 | 558.1 | 589.0 |
| 2 | 1814.1 | 1969.3 | 1811.2 | 1958.7 |
| 3 | 2741.4 | 2950.0 | 2689.3 | 2868.8 |

**Table 11.7: Summary of first three ionization enthalpies for the Group 13 metals.**

In summary, it may be stated that while the chemistry of gallium, indium and thallium is very similar, that of aluminum is slightly different, while boron's chemistry is very different from the rest of the Group.

A second effect is noticed in the transition from aluminum, to gallium, to indium. Based upon their position in the Group it would be expected that the ionic radius and associated lattice parameters should follow the trend:

$$Al < Ga < In < Tl$$

However, as may be seen from Table 11.8 the values for gallium are either the same as, or smaller, than that of aluminum. In a similar manner, the covalent radius and covalent bond lengths as determined by X-ray crystallography for a range of compounds (Table 11.9).

| Element | Phosphide lattice parameter (Å) | Arsenide lattice parameter (Å) |
|---|---|---|
| Al | 5.4635 | 5.6600 |
| Ga | 5.4505 | 5.6532 |
| In | 5.8687 | 6.0583 |

**Table 11.8: Table. Lattice parameter (*a*) for zinc blende forms of the Group 13 phosphides and arsenides. Data from *Semiconductors: Group IV Elements and III-V Compounds*, Ed. O. Madelung, Springer-Verlag, Berlin (1991).**

| Element | M-C (Å) | M-N (Å) | M-O (Å) | M-Cl (Å) |
|---|---|---|---|---|
| Al | 1.96 - 2.02 | 2.03 - 2.19 | 1.74 - 1.93 | 2.09 - 2.11 |
| Ga | 1.97 - 2.01 | 1.95 - 2.12 | 1.89 - 1.94 | 2.12 - 2.23 |
| In | 2.14 - 2.17 | 2.23 - 2.31 | 2.19 - 2.20 | 2.39 - 2.47 |

**Table 11.9: Comparative crystallographically determined bond lengths.**

Gallium is significantly smaller than expected from its position within the Group 13 elements (Table 11.10). The rational for this may be attributed to an analogous effect as seen in the lanthanide contraction observed for the lanthanides and the 3$^{rd}$ row of transition elements. In multi-electron atoms, the decrease in radius brought about by an increase in nuclear charge is partially o set by increasing electrostatic repulsion among electrons. In particular, a shielding effect results when electrons are added in outer shells, electrons already present shield the outer electrons from nuclear charge, making them experience a lower effective charge on the nucleus. The shielding effect exerted by the inner electrons decreases in the order s > p > d > f. As a sub-shell is filled in a period the atomic radius decrease. This effect is particularly pronounced in the case of lanthanides, as the 4f sub-shell is not very effective at shielding the outer shell (n = 5 and n = 6) electrons. However, a similar, but smaller effect should be observed with the post-transition metal elements, i.e., gallium. This is indeed observed (Table 11.10).

| Element | Covalent radius (Å) | Ionic radius (Å) |
|---|---|---|
| Aluminum | 1.21 | 0.53 |
| Gallium | 1.22 | 0.62 |
| Indium | 1.42 | 0.80 |
| Iron (low spin) | 1.32 | 0.55 |
| Iron (high spin) | 1.52 | 0.64 |

**Table 11.10: Comparison of the covalent and ionic radii of Group 13 elements.**

The anomalous size of gallium has two positive effects:

- The similarity in size of aluminum and gallium means that their Group 15 derivatives have near identical lattice parameters (Table 11.7). This allows for both epitaxial growth of one material on the other, and also the formation of ternary mixtures (i.e., $Al_xGa_{1-x}As$) with matched lattice parameters. The ability to grow hetrojunction structures of Group 13-15 compounds (III-V) is the basis for the fabrication of a wide range of important optoelectronic devices, including: LEDs and laser diodes.
- The similarity in size of gallium(III) to iron(III) (Table 11.10) means that gallium can substitute iron in a range of coordination compounds without alteration of the structure. Because of a similar size and charge as $Fe^{3+}$, $Ga^{3+}$ is widely used as a non-redox-active $Fe^{3+}$ substitute for studying metal complexation in proteins and bacterial populations.

## Borides

The non-metallic nature of boron means that it makes a number of binary compounds with elements more electropositive than itself (i.e., metals). These compounds are called, borides, and some are also formed with metalloid elements as well (e.g., arsenic). In this regard, borides may be considered similar to carbides, silicides, and some phosphides.

Borides are prepared in a number of ways; however, direct combination of the elements is the simplest.

$$M + x B \rightarrow MB_x$$

Other routes include, electrolysis of the fused salts, and the reduction of the metal oxide with a mixture of carbon and boron carbide.

| Element | Melting point (°C) | Boride | Melting point (°C) |
|---|---|---|---|
| Ti | 1725 | $TiB_2$ | 3225 |
| Zr | 1855 | $ZrB_2$ | 2990 |
| Hf | 2233 | $HfB_2$ | 3100 |

**Table 11.11: The melting points of Group 4 metals and their borides.**

Metal borides are generally refractory in character and chemically inert, while they often have properties better that of the constituent elements. For example, the thermal conductivity of $TiB_2$ is about ten times greater than that of titanium, and the melting point is significantly higher (Table 11.11).

The structures of metal borides depend on the M:B ratio. Borides with an isolated boron atom have a low B:M ratio: $M_4B$, $M_3B$, $M_2B$, $M_5B_2$, and $M_7B_3$. In such compounds the boron atom is normally in a triangular-prismatic or square-antiprismatic hole in a metal lattice. Borides with equal or near equal metal and boron ratio have structures with either pairs of boron atoms (as in $V_3B_2$), single boron chains (seen in all MB compounds), or double boron chains (observed for many $M_3B_4$ compounds). Increasing the boron content results in two-dimensional structures. For example, $MB_2$ usually consists of alternate hexagonal layers of metal and boron (Figure 11.18). Finally, boron rich borides (e.g., $MB_4$, $MB_6$, and $MB_{12}$) all have three-dimensional structures.

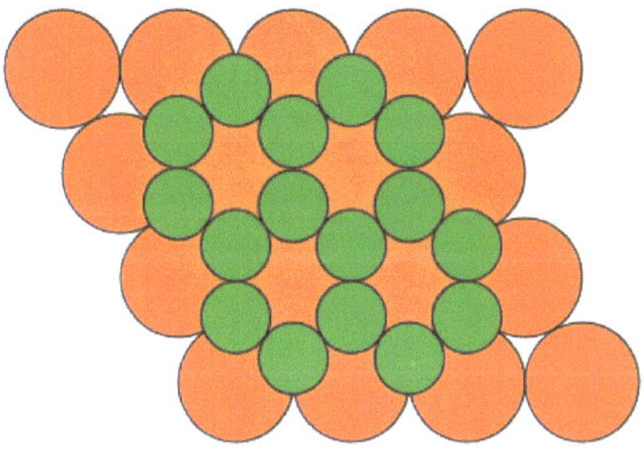

**Figure 11.18: Alternate layers of metal atoms (large circles) and boron (small circles) in $MB_2$.**

# Boron hydrides

## Borane and diborane

Borane ($BH_3$) formed in the gaseous state from decomposition of other compounds,

$$H_3B\text{-}PF_3 \xrightarrow{\Delta} BH_3 + PF_3$$

but cannot be isolated except as a Lewis acid-base complex,

$BH_3 + PPh_3 \rightarrow H_3B\text{-}PPh_3$

As such many borane adducts are known.

In the absence of a Lewis base the dimeric diborane ($B_2H_6$) is formed. Diborane is generally synthesized by the reaction if $BF_3$ with a hydride source, such as $NaBH_4$ or $LiAlH_4$.

$3\ NaBH_4 + 4\ BF_3 \rightarrow 2\ B_2H_6 \uparrow + 3\ NaBF_4$

$3\ LiAlH_4 + 4\ BF_3 \rightarrow 2\ B_2H_6 \uparrow + 3\ LiAlF_4$

The structure of diborane (Figure 11.19a) is considered to be electron deficient and has been confirmed by IR spectroscopy and electron diffraction. The four terminal B-H bonds are normal covalent bonds; however, the bridging B-H-B unit consists of two three-centered two-electron bonds, each ordinarily considered to be formed by the combination of two boron $sp^3$ orbitals and one hydrogen s orbital (Figure 11.19b). However, a consideration of the H-B-H bond angle associated with the terminal hydrides (120°) it is perhaps better to consider the $BH_2$ fragment to be $sp^2$ hybridized, and the B-H-B bridging unit to be a linear combination of one $sp^2$ orbital and one p orbital from each boron atom with the two hydrogen s orbitals. Diborane represents the archetypal electron deficient dimeric compound, of which $Al_2Me_3$ is also a member of this class of electron deficient molecules.

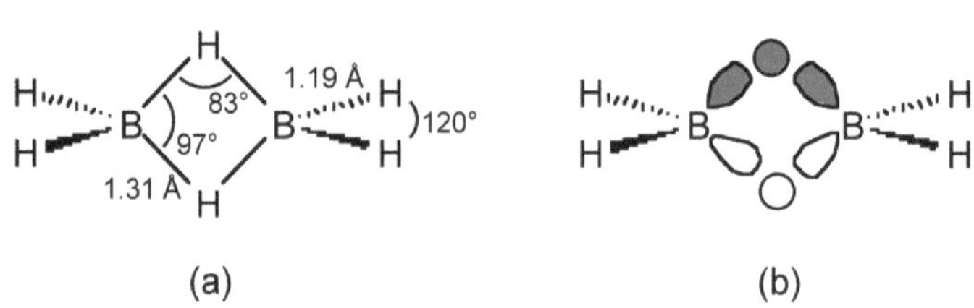


**Figure 11.19: The structure (a) and the typical view of the three-centered two-electron bonds (b) in diborane, $B_2H_6$.**

$B_2H_6$ is spontaneously and highly exothermically in flammable above 25 °C ($\Delta H$ = -2137.7 kJ/mol),

$B_2H_6 + 3\ O_2 \rightarrow B_2O_3 + 3\ H_2O$

It is often used as one of a range of solvate forms for safety for both its flammability and toxicity.

Most reactions of diborane involve the cleavage of the dimeric structure. Hydrolysis of diborane yields boric acid,

$$B_2H_6 + 6\,H_2O \rightarrow 2\,B(OH)_3 + 6\,H_2\uparrow$$

while alcoholysis yields the appropriate borate ester,

$$B_2H_6 + 6\,ROH \rightarrow 2\,B(OR)_3 + 6\,H_2\uparrow$$

Diborane reacts with Lewis bases to form the appropriate Lewis acid-base complex,

$$B_2H_6 + 6\,NR_3 \rightarrow 2\,H_3B\text{-}NR_3$$

*Borohydride*

The borohydride anion (or more properly the tetrahydridoborate anion), $BH_4^-$, can be considered as the Lewis acid-base complex between borane and $H^-$. A typical synthesis involves the reaction of a borate ester with a hydride source.

$$4\,NaH + B(OMe)_3 \rightarrow NaBH_4 + 3\,NaOMe$$

Sodium borohydride is a stable white crystalline solid that is stable in dry air and is non-volatile. The boron in borohydride ($BH_4^-$) is tetrahedral. Although it is insoluble in $Et_2O$, it is soluble in water (in which it reacts slowly), THF, ethylene glycol, and pyridine. Interestingly, $NaBH_4$ reacts rapidly with MeOH, but dissolves in EtOH. Sodium borohydride has extensive uses in organic chemistry as a useful reducing agent in which it donates a hydride ($H^-$).

## Higher boranes

Higher boron hydrides contain, in addition to the bridging B-H-B unit, one or more B-B bonds. The higher boranes are usually formed by the thermal decomposition of diborane,

$$2\ B_2H_6 \rightarrow B_4H_{10} + H_2$$

$$5\ B_2H_6 \rightarrow 2\ B_5H_9 + 6\ H_2$$

These higher boranes have 'open' cluster structures, e.g., Figure 11.20 - Figure 11.22. Tetraborane, or to be more precise tetraborane(10) or arachno-$B_4H_{10}$, is a foul-smelling toxic gas. Pentaborane(9) is a toxic liquid (with a distinctive garlic odor) that can detonate in air, and like decaborane(14) was at one time considered as a potential rocket fuel. Because simple boron compounds burn with a characteristic green flame, the nickname for these fuels in the US military was Green Dragon. Problems with using boranes as a fuel included their toxicity and the characteristic of bursting into flame on contact with the air; furthermore, the exhaust would also be toxic. The US program resulted in a stockpile of borane fuels, in particular pentaborane(9), which was not destroyed until 2000. The system for the destroying the boranes was appropriately known as Dragon Slayer.

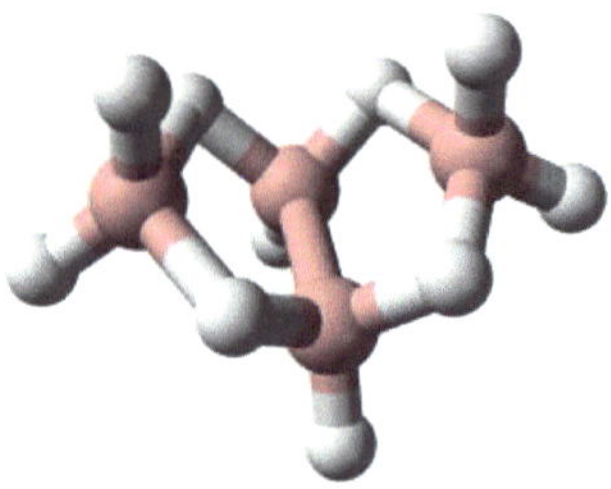

**Figure 11.20: The molecular structure of tetraborane(10), $B_4H_{10}$. Boron atoms are represented by pink spheres, and hydrogen by white spheres.**

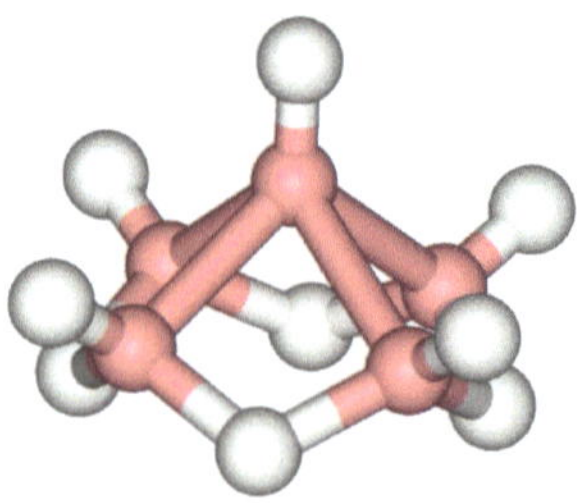

**Figure 11.21: The molecular structure of pentaborane(9), $B_5H_9$. Boron atoms are represented by pink spheres, and hydrogen by white spheres.**

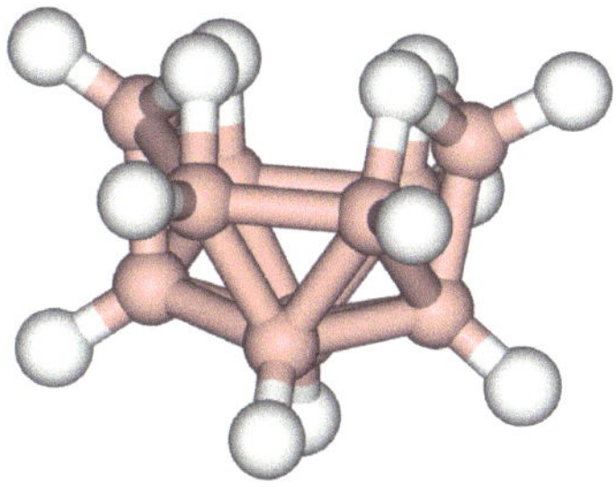

**Figure 11.22: The molecular structure of decaborane(14), $B_{10}H_{14}$. Boron atoms are represented by pink spheres, and hydrogen by white spheres.**

*Wade's Rules*

Ken Wade (Figure 11.23) developed a method for the prediction of shapes of borane clusters (Figure 11.24); however, it may be used for other classes of cluster compounds. Wade's rules are used to rationalize the shape of borane clusters by calculating the total number of skeletal electron pairs (SEP) available for cluster bonding.

**Figure 11.23: British chemist Ken Wade FRS (1932 - 2014).**

The general methodology to be followed when applying Wade's rules is as follows:

1. Determine the total number of valence electrons from the chemical formula, i.e., 3 electrons per B, and 1 electron per H.
2. Subtract 2 electrons for each B-H unit (or C-H in a carborane).

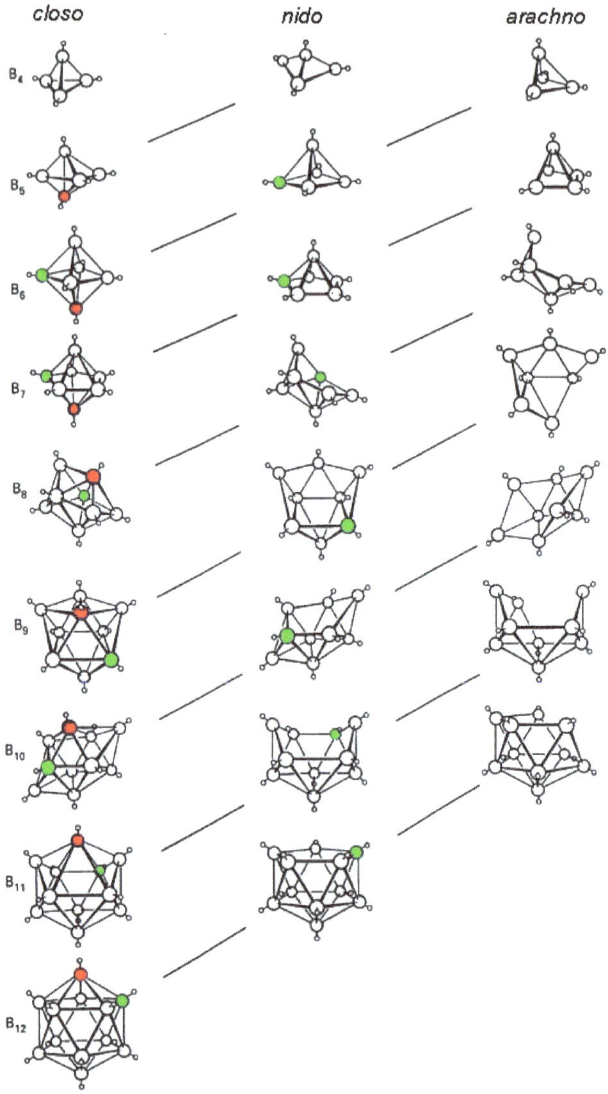


**Figure 11.24: Structural relationship between *closo*, *nido*, and *arachno* boranes (and hetero-substituted boranes). The diagonal lines connect species that have the same number of skeletal electron pairs (SEP). Hydrogen atoms except those of the B-H framework are omitted. The red atom is omitted first, the green atom removed second. Adapted from R. W. Rudolph, Boranes and heteroboranes: a paradigm for the electron requirements of clusters? *Acc. Chem. Res.*, 1976, 9, 446. Copyright: American Chemical Society (1976).**

3. Divide the number of remaining electrons by 2 to get the number of skeletal electron pairs (SEP).
4. A cluster with n vertices (i.e., n boron atoms) and n+1 SEP for bonding has a closo structure.
5. A cluster with n-1 vertices (i.e., n-1 boron atoms) and n+1 SEP for bonding has a nido structure.
6. A cluster with n-2 vertices (i.e., n-2 boron atoms) and n+1 SEP for bonding has an arachno structure.
7. A cluster with n-3 vertices (i.e., n-3 boron atoms) and n+1 SEP for bonding has an hypho structure.
8. If the number of boron atoms (i.e., n) is larger than n+1 SEP then the extra boron occupies a capping position on a triangular phase.

For example, the structure of $B_5H_{11}$ can be predicted as follows:

1. Total number of valence electrons = (5×B)+(11×H) = (5×3)+(11×1) = 26
2. Number of electrons for each B-H unit = (5×2) = 10
3. Number of skeletal electrons = 26-10 = 16
4. Number SEP = 16/2 = 8
5. If n+1 = 8 and n-2 = 5 boron atoms, then n=7

Structure of n = 7 is pentagonal bipyramid (Figure 11.24), therefore $B_5H_{11}$ is an *arachno* based upon a pentagonal bipyramid with two apexes missing (Figure 11.25).

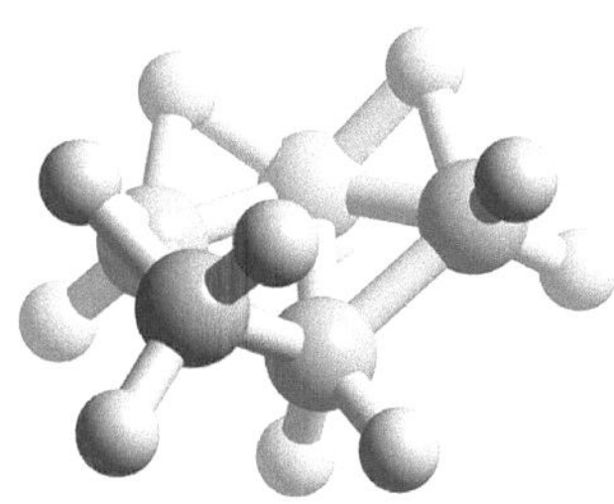

**Figure 11.25: Ball and stick representation of the structure of $B_5H_{11}$.**

Table 11.12 provides a summary of borane cluster with the general formula $B_nH_n^{x-}$ and their structures as defined by Wade's rules.

| Type | Basic formula | Example | # of verticies | # of vacancies | # of e- in B + charge | # of bonding MOs |
|---|---|---|---|---|---|---|
| *Closo* | $B_nH_n^{2-}$ | $B_6H_6^{2-}$ | n | 0 | 3n + 2 | n+1 |
| *Nido* | $B_nH_n^{4-}$ | $B_5H_9$ | n+1 | 1 | 3n + 4 | n+2 |
| *Arachno* | $B_nH_n^{6-}$ | $B_4H_{10}$ | n+2 | 2 | 3n + 6 | n+3 |
| *Hypho* | $B_nH_n^{8-}$ | $B_5H_{11}^{2-}$ | n+3 | 3 | 3n + 8 | n+4 |

**Table 11.12: Wade's rules for boranes.**

# Oxides, hydroxides, and oxygen donor ligands

## Trends for the oxides of the Group 13 elements

All of the Group 13 elements form a trivalent oxide ($M_2O_3$). The chemical properties of the oxides follow the trend acidic to basic going down the Group (Table 11.13). The physical properties are consistent with the electronegativities and covalent character in the M-O bonds. Thallium oxide is unique in that it decomposes above 100 °C to yield the thalium(I) oxide, $Tl_2O$. The other oxides are all stable to high temperatures.

| Oxide | Color | Chemical property | Melting point (°C) |
|---|---|---|---|
| $B_2O_3$ | White/colorless | Weak acid | 450 (trigonal), 510 (tetrahedral) |
| $Al_2O_3$ | White/colorless | Amphoteric | 2072 (α) |
| $Ga_2O_3$ | White/colorless | Amphoteric | 1900 (α), 1725 (β) |
| $In_2O_3$ | Yellow | Weakly basic | 1910 |
| $Tl_2O_3$ | Brown-black | Basic, oxidizing | 100 (decomposes) |

**Table 11.13: Properties of the Group 13 oxides.**

## Boron oxides, hydroxides, and oxyanions

*Oxides*

Boron oxide, $B_2O_3$, is made by the dehydration of boric acid,

$$2\ B(OH)_3 \overset{\Delta}{\rightleftharpoons} B_2O_3 + 3\ H_2O$$

It is a glassy solid with no regular structure but can be crystallized with extreme difficulty. The structure consists of infinite chains of triangular $BO_3$ unit (Figure 11.26). Boron oxide is acidic and reacts with water to reform boric acid.

The reaction of $B_2O_3$ with hydroxide yields the metaborate ion,

$$B_2O_3 + 6\ KOH \rightarrow 2\ K_3[B_3O_6] + 3\ H_2O$$

whose planar structure (Figure 11.27) is related to metaboric acid. Boron oxide fuses with a wide range of metal and non-metal oxides to give borate glasses.

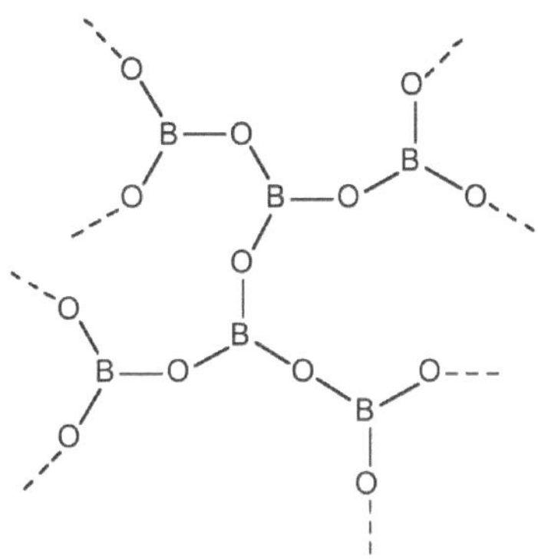

**Figure 11.26: Glassy structure of $B_2O_3$.**

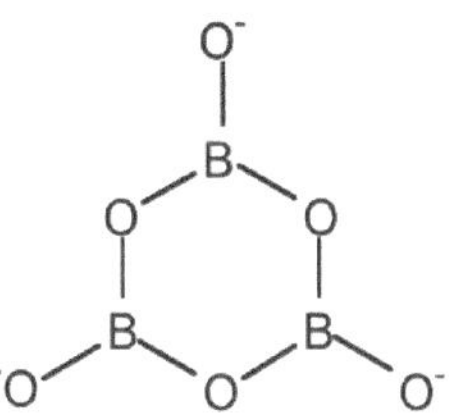

**Figure 11.27: Structure of the metaborate anion, $[B_3O_6]^{3-}$.**

*Boric acid*

Boric acid, $B(OH)_3$, usually obtained from the dissolution of borax, $Na_2[B_4O_5(OH)_4]$, is a planar solid with intermolecular hydrogen bonding forming a near hexagonal layered structure, broadly similar to graphite (Figure 11.28). The inter layer distance is 3.18 Å.

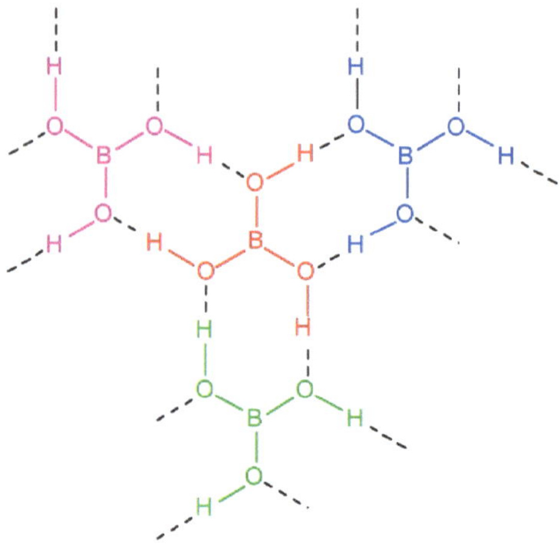

**Figure 11.28: The hydrogen bonded structure of boric acid.**

A summary of reactions of the reactivity of boric acid is shown in Figure 11.29.

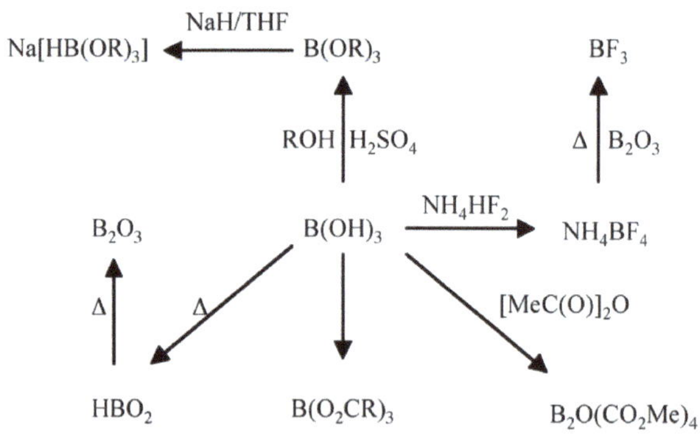


**Figure 11.29: Selected reactions of boric acid. Adapted from F. A. Cotton and G. Wilkinson, *Advanced Inorganic Chemistry*, 4 Ed. Wiley Interscience (1980).**

Upon dissolution of boric acid in water, boric acid does not act as a proton acid, but instead reacts as a Lewis base,

$$B(OH)_3 + 2\,H_2O \rightleftharpoons B(OH)_4^- + H_3O^+$$

The reaction may be followed by $^{11}B$ NMR spectroscopy from the change in the chemical shift (Figure 11.30).

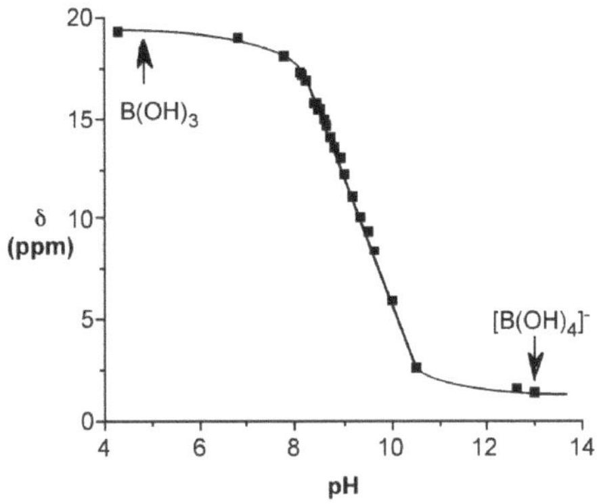


**Figure 11.30: The $^{11}B$ NMR chemical shift (ppm) of boric acid as a function of pH. Adapted from M. Bishop, N. Shahid, J. Yang, and A. R. Barron, Determination of the mode and efficacy of the cross-linking of guar by borate using MAS $^{11}B$ NMR of borate cross-linked guar in combination with solution $^{11}B$ NMR of model systems. *Dalton Trans.*, 2004, 2621. Copyright: Royal Society of Chemistry (2004).**

Since the $^{11}B$ NMR shift is directly proportional to the mole fraction of the total species present as the borate anion (e.g., $[B(OH)_4]^-$) the $^{11}B$ NMR chemical shift at a given temperature, $\delta_{(obs)}$, may be used to calculate both the mole fraction of boric acid and the borate anion, i.e.,

$$\chi_{(acid)} = \frac{\delta_{(obs)} - \delta_{(anion)}}{\delta_{(acid)} - \delta_{(anion)}}$$

$$\chi_{(anion)} = \frac{\delta_{(acid)} - \delta_{(obs)}}{\delta_{(acid)} - \delta_{(anion)}}$$

Using these equations, the relative speciation as a function of pH may be calculated for both boric acid (Figure 11.31). The pH at which a 50:50 mixture of acid and anion for boric acid is ca. 9.4.

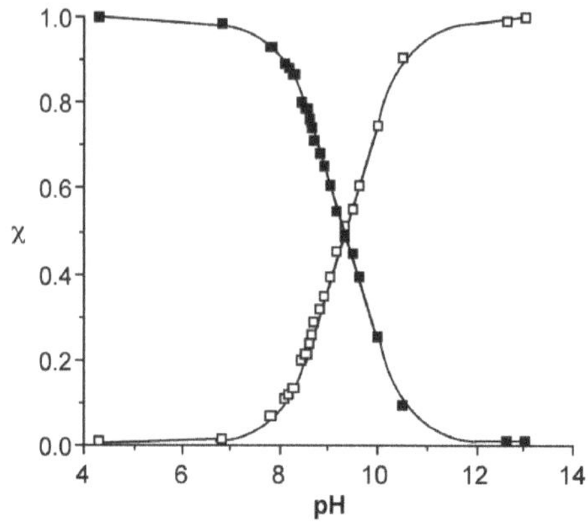


**Figure 11.31: The mole fraction of acid (black squares) and anion (white squares) forms of boric acid as a function of pH. Adapted from M. Bishop, N. Shahid, J. Yang, and A. R. Barron, Determination of the mode and efficacy of the cross-linking of guar by borate using MAS $^{11}$B NMR of borate cross-linked guar in combination with solution $^{11}$B NMR of model systems. *Dalton Trans.*, 2004, 2621. Copyright: Royal Society of Chemistry (2004).**

In concentrated solutions, the borate ion reacts further to form polyborate ions. The identity of the polyborate is dependent on the pH. With increasing pH, $B_5O_6(OH)_4^-$, $B_3O_3(OH)_4^-$, and $B_4O_5(OH)_4^{2-}$, are formed.

$$4\ B(OH)_3 + B(OH)_4^- \rightleftharpoons B_5O_6(OH)_4^- + 6\ H_2O$$

$$2\ B(OH)_3 + B(OH)_4^- \rightleftharpoons B_3O_3(OH)_4^- + 3\ H_2O$$

$$2\ B(OH)_3 + 2\ B(OH)_4^- \rightleftharpoons B_4O_5(OH)_4^{2-} + 5\ H_2O$$

The structure of each borate is shown in Figure 11.42. Once the ratio of $B(OH)_3$ to $B(OH)_4^-$ is greater than 50%, only the mono-borate is observed.

Borax, the usual mineral form of boric acid, is the sodium salt, $Na_2[B_4O_5(OH)_4]$, which upon dissolution in water re-equilibrates to $B(OH)_3$.

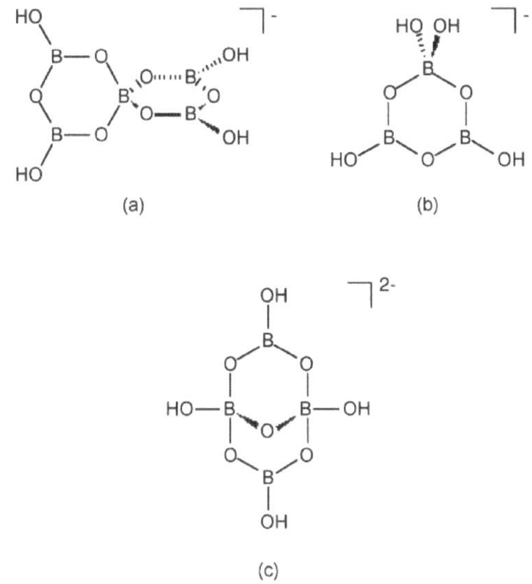


**Figure 11.32: The structures of the borate anions (a) $B_5O_6(OH)_4^-$, (b) $B_3O_3(OH)_4^-$, and (c) $B_4O_5(OH)_4^{2-}$.**

*Enough to make your hair curl*

In 1906, a German hairdresser, Charles Nessler who was living in London, decided to help his sister who was fed-up with having to put her straight hair in curlers. While looking for a solution, Nessler noticed that a clothesline contracted in a wavy shape when it was wet. Nessler wound his sister's hair on cardboard tubes; then he covered the hair with borax paste. After wrapping the tubes with paper (to exclude air) he heated the entire mass for several hours. Removing the paper and tubes resulted in curly hair. After much trial and error (presumably at his sister's discomfort) Nessler perfected the method by 1911 and called the process a permanent wave (or perm). The process involved the alkaline borax softening the hair sufficiently to be remodeled, while the heating stiffened the borax to hold the hair into shape. The low cost of borax meant that Nessler's methods was an immediate success.

*Metaboric acid*

Heating boric acid results in the partial dehydration to yield metaboric acid, $HBO_2$,

$$B(OH)_3 \underset{H_2O}{\overset{\Delta}{\rightleftharpoons}} HBO_2 \underset{H_2O}{\overset{\Delta}{\rightleftharpoons}} B_2O_3$$

Metaboric acid is also formed from the partial hydrolysis of $B_2O_3$. If the heating is carried out below 130 °C, $HBO_2$-III is formed in which $B_3O_3$ rings are joined by hydrogen bonding to the hydroxide on each boron atom (Figure 11.33). Continued heating to 150 °C results in $HBO_2$-II, whose structure consists of $BO_4$ tetrahedra and $B_2O_5$ groups chain linked by hydrogen bonding. Finally, heating above 150 °C yields cubic $HBO_2$-I with all the boron atoms tetrahedral.

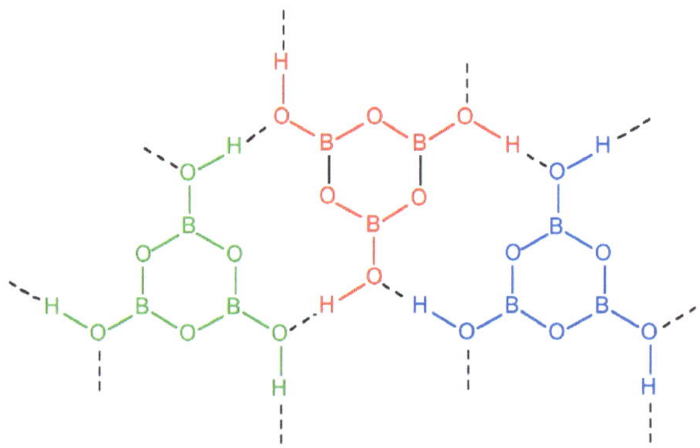

**Figure 11.33: Structure of the layered structure of $HBO_2$-III.**

*Borate esters*

Boric acid reacts with alcohols in the presence of sulfuric acid to form $B(OR)_3$ (Figure 11.29). This is the basis for a simple flame test for boron. Treatment of a compounds with methanol/sulfuric acid, followed by placing the reaction product in a flame results in a green flame due to $B(OMe)_3$. In the presence of diols, polyols, or polysaccharides, boric acid reacts to form alkoxide complexes. In the case of diols, the mono-diol $[B(OH)_2L]^-$ (Figure 11.34a) and the bis-diol $[BL_2]^-$ (Figure 11.34b) complexes.

Originally it was proposed that diols react with the anion rather than boric acid,

$$[B(OH)_4]^- \overset{+\,LH_2}{\rightleftharpoons} [B(OH)_2L]^- \overset{+\,LH_2}{\rightleftharpoons} [BL_2]^-$$

(a) (b)

**Figure 11.34: Structure of (a) the $[B(OH)_2L]^-$ and (b) the $[BL_2]^-$ anions formed by the reaction of boric acid (or borate) with a diol.**

In contrast, it was suggested that the optimum pH for the formation of carboxylic acid complexes is under the condition where $pK_a$ (carboxylic acid) < pH < $pK_a$ (boric acid). Under these conditions it is the boric acid, $B(OH)_3$, not the borate anion, $[B(OH)_4]^-$, that reacts to form the complex.

The conversion of boric acid to borate is related to the $pK_a$ of water. The formation of $[B(OH)_2L]^-$ from $B(OH)_3$ would be expected to occur via a similar initial reaction (attack by $RO^-$ or deprotonation of coordinated ROH) followed by a subsequent elimination of $H_2O$ and the formation of a chelate coordination, and would therefore be related to the $pK_a$ of the alcohol. Thus, the pH at which $[B(OH)_2L]^-$ is formed relative to $[B(OH)_4]^-$ will depend on the relative acidity of the alcohol. The $pK_a$ of a simple alcohol (e.g., MeOH = 15.5, EtOH = 15.9) are close to the value for water (15.7) and the lowest pH at which $[B(OH)_2L]^-$ is formed should be comparable to that at which $[B(OH)_4]^-$ forms. Thus, the formation of $[B(OH)_2L]^-$ as compared to $[B(OH)_4]^-$ is a competition between the reaction of $B(OH)_3$ with $RO^-$ and $OH^-$ (Figure 11.35), and as with carboxylic acids it is boric acid not borate that reacts with the alcohol,

$$B(OH)_3 \underset{}{\overset{+\,LH_2}{\rightleftharpoons}} [B(OH)_2L]^- \underset{}{\overset{+\,LH_2}{\rightleftharpoons}} [BL_2]^-$$

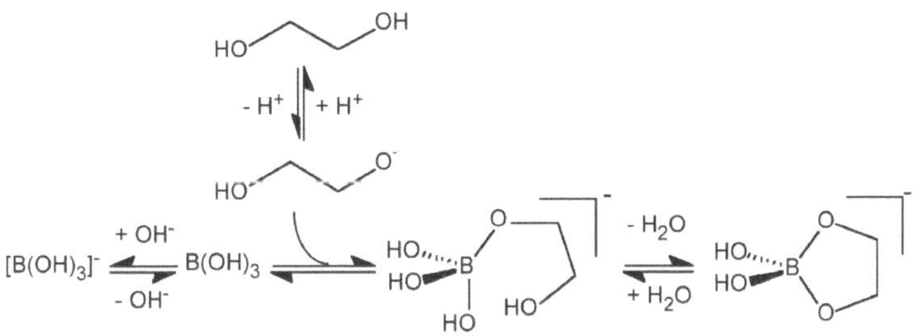


**Figure 11.35: Competition between hydroxide and 1,2-diol for complexation to boron in the aqueous boric acid/diol system.**

A key issue in the structural characterization and understanding of the diol/boric acid system is the assignment of the $^{11}B$ NMR spectral shifts associated with various complexes. A graphical representation of the observed shift ranges for boric acid chelate systems is shown in Figure 11.36.

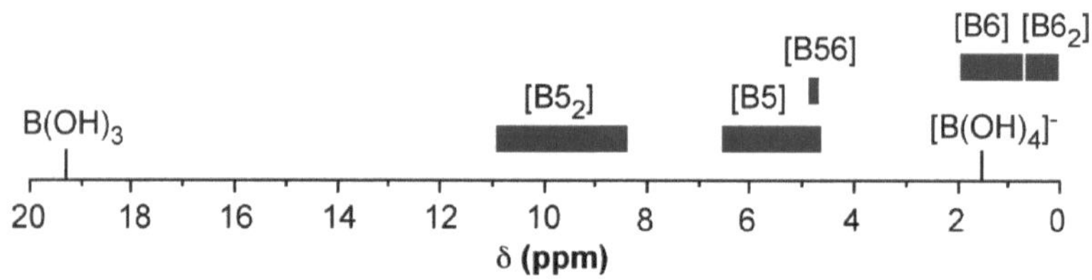


**Figure 11.36: $^{11}B$ NMR spectroscopic shifts of chelate alkoxide compounds in comparison with boric acid and borate anion. The acronyms [B5] and [B6] are used to denote a borate complex of a 1,2-diol and 1,3-diol forming a chelate 5-membered and 6-membered ring cycle, respectively. Similarly, [B5$_2$] and [B6$_2$] denotes a borate complex of two 1,2-diols or 1,3-diols both forming [BL$_2$]$^-$. Adapted from M. Bishop, N. Shahid, J. Yang, and A. R. Barron, Determination of the mode and efficacy of the cross-linking of guar by borate using MAS $^{11}B$ NMR of borate cross-linked guar in combination with solution $^{11}B$ NMR of model systems. *Dalton Trans.*, 2004, 2621. Copyright: Royal Society of Chemistry (2004).**

*Boric acid cross-linking of guar gum for hydraulic fracturing fluids*

Thick gels of guar gum cross-linked with borax or a transition metal complex are used in the oil well drilling industry as hydraulic fracturing fluids. The polysaccharide guaran (M ≈ $10^6$ Da) is the major (>85 wt%) component of guar gum and consists of a (1→4)-β-D-mannopyranosyl backbone with α-D-galactopyranosyl side chain units attached via (1→6) linkages. Although the exact ratio varies between different crops of guar gum the general structure is consistent with about one galactose to every other mannose (Figure 11.37).

The synthesis of a typical boric acid cross-linked guar gel, with viscosity properties suitable for use as a fracturing fluid, involves mixing guar gum, water and boric acid, $B(OH)_3$, in a 10:2000:1 wt/wt ratio. Adjusting the pH to between 8.5 and 9 results in a viscous gel. The boron:guaran ratio corresponds to almost 2 boron centers per 3 monosaccharide repeat units. Such a large excess of boric acid clearly indicates that the system is not optimized for cross-linking, i.e., a significant fraction of the boric acid is ineffective as a cross-linking agent.

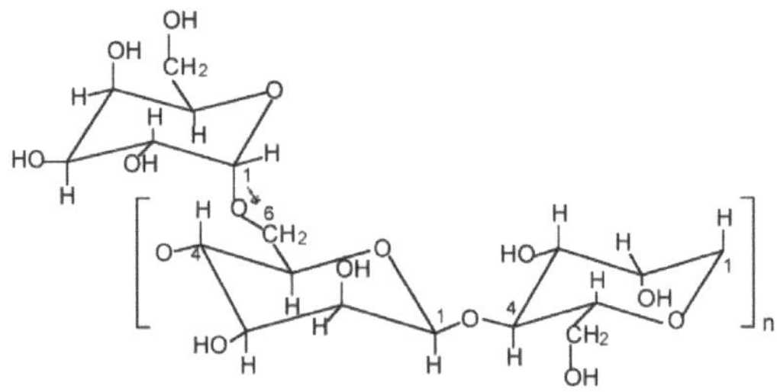

**Figure 11.37: The repeat unit of guaran, a naturally occurring polysaccharide, consisting of (1→4)-β- D-mannopyranosyl units with α-D-galactopyranosyl side chains attached to about every other mannose via (1→6) linkages.**

The reasons for the inefficiency of boric acid to cross-link guaran (almost 2 borate ions per 3 monosaccharide repeat unit are required for a viscous gel suitable as a fracturing fluid): the most reactive sites on the component saccharides (mannose and galactose) are precluded from reaction by the nature of the guar structure; the comparable acidity ($pK_a$) of the remaining guaran alcohol substituents and the water solvent, results in a competition between cross-linking and borate formation; a significant fraction of the boric acid is ineffective in cross-linking guar due to the modest equilibrium ($K_{eq}$).

## Aluminum oxides, hydroxides, and hydrated oxides

The many forms of aluminum oxides and hydroxides are linked by complex structural relationships. Bauxite has the formula $Al_x(OH)_{3-2x}$ ($0 < x < 1$) and is thus a mixture of $Al_2O_3$ (α-alumina), $Al(OH)_3$ (gibbsite), and AlO(OH) (boehmite). The latter is an industrially important compound that is used in the form of a gel as a pre-ceramic in the production of fibers and coatings, and as a re-retarding agent in plastics.

Knowledge of microstructural evolution in ceramic systems is important in determining their end-use application. In this regard alumina has been the subject of many studies in which the phase, morphology, porosity and crystallinity are controlled by physical and chemical processing. The transformation from boehmite [γ-Al(O)(OH)] to corundum (α-$Al_2O_3$) has been well characterized and is known to go through the following sequence:

$$\gamma\text{-Al(O)(OH)} \xrightarrow{\approx 500\ ^\circ\text{C}} \gamma\text{-Al}_2\text{O}_3 \xrightarrow{\approx 850\ ^\circ\text{C}} \delta\text{-Al}_2\text{O}_3 \xrightarrow{\approx 1000\ ^\circ\text{C}} \theta\text{-Al}_2\text{O}_3 \xrightarrow{>1100\ ^\circ\text{C}} \alpha\text{-Al}_2\text{O}_3$$

The phase changes from boehmite through θ-$Al_2O_3$ are known to be topotactic (i.e., changes in crystal structure are accomplished without changes in crystalline morphology); however, each phase change is ac- companied by a change in porosity. The θ- to α-$Al_2O_3$ phase transition occurs through nucleation and growth of the θ-$Al_2O_3$ crystallites. The α-$Al_2O_3$ phase transition temperature can be altered by the addition of certain additives. For example, because the α-$Al_2O_3$ phase occurs by nucleation, the addition of small seed crystals can lower the transition temperature between 100 and 200 °C. The addition of certain transition metals (chromium, manganese, iron, cobalt, nickel, and copper) has also been shown to decrease the transition temperature, while lanthanum or rare earth metals tend to increase the temperature. Finally, the addition of metal oxides has also shown to affect the growth rate in α-$Al_2O_3$. A third form of $Al_2O_3$ forms on the surface of the clean aluminum metal,

$$4\,Al + 3\,O_2 \rightarrow 2\,Al_2O_3$$

This oxide skin is rapidly self-repairing because its heat of formation is so large ($\Delta H = 3351$ kJ/mol). The thin, tough, transparent oxide layer is the reason for much of the usefulness of aluminum.

*Ceramic processing of alumina*

While aluminum is the most abundant metal in the earth's crust (ca. 8%) and aluminum compounds such as alum, $K[Al(SO_4)_2].12(H_2O)$, were known throughout the world in ancient times, it was not until the isolation of aluminum in the late eighteenth century by the Danish scientist H. C. Öersted that research into the chemistry of the Group 13 elements began in earnest. Initially, metallic aluminum was isolated by the reduction of aluminum trichloride with potassium or sodium; however, with the advent of inexpensive electric power in the late 1800's, it became economically feasible to extract the metal via the electrolysis of alumina ($Al_2O_3$) dissolved in cryolite, $Na_3AlF_6$, (the Hall-Heroult process). Today, alumina is prepared by the Bayer process, in which the mineral bauxite (named for Les Baux, France, where it was first discovered) is dissolved with aqueous hydroxides, and the solution is filtered and treated with $CO_2$ to precipitate alumina. With availability of both the mineral and cheap electric power being the major considerations in the economical production of aluminum, it is not surprising that the leading producers of aluminum are the United States, Japan, Australia, Canada, and the former Soviet Union.

*Aluminum oxides and hydroxides*

The many forms of aluminum oxides and hydroxides are linked by complex structural relationships. Bauxite has the formula $Al_x(OH)_{3-2x}$ ($0 < x < 1$) and is thus a mixture of $Al_2O_3$ (α-alumina), $Al(OH)_3$ (gibbsite), and AlO(OH) (boehmite). The latter is an industrially important compound which is used in the form of a gel as a pre-ceramic in the production of fibers and coatings, and as a fire retarding agent in plastics.

Heating boehmite and diaspore to 450 °C causes dehydration to yield forms of alumina which have structures related to their oxide-hydroxide precursors. Thus, boehmite produces the low-temperature form γ-alumina, while heating diaspore will give α-alumina (corundum). γ-alumina converts to the hcp structure at 1100 °C. A third form of $Al_2O_3$ forms on the surface of the clean aluminum metal. The thin, tough, transparent oxide layer is the reason for much of the usefulness of aluminum. This oxide skin is rapidly self-repairing because its heat of formation is so large (ΔH = -3351 kJ/mol).

$$4\,Al + 3\,O_2 \rightarrow 2\,Al_2O_3$$

*Ternary and mixed-metal oxides*

A further consequence of the stability of alumina is that most if not all of the naturally occurring aluminum compounds are oxides. Indeed, many precious gemstones are actually corundum doped with impurities. Replacement of aluminum ions with trace amounts of transition-metal ions transforms the formerly colorless mineral into ruby (red, $Cr^{3+}$), sapphire (blue, $Fe^{2+/3+}$, $Ti^{4+}$), or topaz (yellow, $Fe^{3+}$). The addition of stoichiometric amounts of metal ions causes a shift from the α-$Al_2O_3$ hcp structure to the other common oxide structures found in nature. Examples include the perovskite structure for $ABO_3$ type minerals (e.g., $CeTiO_7$ or $LaAlO_3$) and the spinel structure for $AB_2O_4$ minerals (e.g., beryl, $BeAl_2O_4$).

Aluminum oxide also forms ternary and mixed-metal oxide phases. Ternary systems such as mullite ($Al_6Si_2O_{13}$), yttrium aluminum garnet (YAG, $Y_3Al_5O_{12}$), the β-aluminas (e.g., $NaAl_{11}O_{17}$) and aluminates such as hibonite ($CaAl_{12}O_{19}$) possessing β-alumina or magnetoplumbite-type structures can offer advantages over those of the binary aluminum oxides.

Applications of these materials are found in areas such as engineering composite materials, coatings, technical and electronic ceramics, and catalysts.

For example, mullite has exceptional high temperature shock resistance and is widely used as an infrared-transparent window for high temperature applications, as a substrate in multilayer electronic device packaging, and in high temperature structural applications. Hibonite and other hexaluminates with similar structures are being evaluated as interfacial coatings for ceramic matrix composites due to their high thermal stability and unique crystallographic structures. Furthermore, aluminum oxides doped with an alkali, alkaline earth, rare earth, or transition metal are of interest for their enhanced chemical and physical properties in applications utilizing their unique optoelectronic properties.

*Synthesis of aluminum oxide ceramics*

In common with the majority of oxide ceramics, two primary synthetic processes are employed for the production of aluminum oxide and mixed metal oxide materials:

1. The traditional ceramic powder process.
2. The solution-gelation, or "sol-gel" process.

The environmental impact of alumina and alumina-based ceramics is in general negligible; however, the same cannot be said for these methods of preparation. As practiced commercially, both of the above processes can have a significant detrimental environmental impact.

*Traditional ceramic processing*

Traditional ceramic processing involves three basic steps generally referred to as powder-processing, shape- forming, and densification, often with a final mechanical finishing step. Although several steps may be energy intensive, the most direct environmental impact arises from the shape-forming process where various binders, solvents, and other potentially toxic agents are added to form and stabilize a solid ("green") body (Table 11.14).

| Function | Composition | Volume (%) |
|---|---|---|
| Powder | Alumina ($Al_2O_3$) | 27 |
| Solvent | 1,1,1-Trichloroethane/ethanol | 58 |
| Deflocculant | Menhaden oil | 1.8 |
| Binder | Poly(vinyl butyrol) | 4.4 |
| Plasticizer | Poly(ethylene glycol)/octyl phthalate | 8.8 |

**Table 11.14: Typical composition of alumina green body.**

The component chemicals are mixed to a slurry, cast, then dried and red. In addition to any innate health risk associated with the chemical processing these agents are subsequently removed in gaseous form by direct evaporation or pyrolysis. The replacement of chlorinated solvents such as 1,1,1-trichloroethylene (TCE) must be regarded as a high priority for limiting environmental pollution. The United States Environmental Protection Agency (EPA) included TCE on its 1991 list of 17 high-priority toxic chemicals targeted for source reduction. The plasticizers, binders, and alcohols used in the process present a number of potential environmental impacts associated with the release of combustion products during ring of the ceramics, and the need to recycle or discharge alcohols which, in the case of discharge to waterways, may exert high biological oxygen demands in the receiving communities. It would be desirable, therefore, to be able to use aqueous processing; however, this has previously been unsuccessful due to problems associated with batching, milling, and forming. Nevertheless, with a suitable choice of binders, etc., aqueous processing is possible. Unfortunately, in many cast-parts formed by green body processing the liquid solvent alone consists of over 50% of the initial volume, and while this is not directly of an environmental concern, the resultant shrinkage makes near net shape processing di cult.

### *Sol-gel*

Whereas the traditional sintering process is used primarily for the manufacture of dense parts, the solution- gelation (sol-gel) process has been applied industrially primarily for the production of porous materials and coatings.

Sol-gel involves a four-stage process: dispersion, gelation, drying, and ring. A stable liquid dispersion or sol of the colloidal ceramic precursor is initially formed in a solvent with appropriate additives. By changing the concentration (aging) or pH, the dispersion is "polymerized" to form a solid dispersion or gel. The excess liquid is removed from this gel by drying and the final ceramic is formed by ring the gel at higher temperatures.

The common sol-gel route to aluminum oxides employs aluminum hydroxide or hydroxide-based material as the solid colloid, the second phase being water and/or an organic solvent, however, the strong interactions of the freshly precipitated alumina gels with ions from the precursor solutions makes it di cult to prepare these gels in pure form. To avoid this complication, alumina gels are also prepared from the hydrolysis of aluminum alkoxides, $Al(OR)_3$.

$$Al(OR)_3 + H_2O/H^+ \rightarrow Al\text{-gel}$$

$$Al\text{-gel} \xrightarrow{\Delta} Al_2O_3$$

The exact composition of the gel in commercial systems is ordinarily proprietary, however, a typical composition will include an aluminum compound, a mineral acid, and a complexing agent to inhibit premature precipitation of the gel, e.g., Table 11.15.

| Function | Composition |
|---|---|
| Boehmite precursor | ASB [aluminum sec-butoxide, $Al(OC_4H_9)_3$] |
| Electrolyte | $HNO_3$ 0.07 mole/mole ASB |
| Complexing agent | Glycerol ca. 10 wt.% |

**Table 11.15: Typical composition of an alumina sol-gel for slip cast ceramics.**

The principal environmental consequences arising from the sol-gel process are those associated with the use of strong acids, plasticizers, binders, solvents, and *sec*-butanol formed during the reaction. Depending on the ring conditions, variable amounts of organic materials such as binders and plasticizers may be released as combustion products. $NO_x$'s may also be produced in the off gas from residual nitric acid or nitrate salts. Moreover, acids and solvents must be recycled or disposed of. Energy consumption in the process entails upstream environmental emissions associated with the production of that energy.

# Boron compounds with nitrogen donors

The B-N unit is isoelectronic (3 + 5 valence electrons) to the C-C unit (4 + 4 valence electrons). The two moieties are also isolobal, and as such there are many of the compound types formed by carbon have analogous derivatives in the chemistry of boron-nitrogen.

## Lewis acid-base addition compounds

Boron compounds, $BX_3$, are strong Lewis acids and as such form stable addition compounds with Lewis bases, in particular those with nitrogen donor ligands.

$BF_3 + NMe_3 \rightarrow F_3B\text{-}NMe_3$

In principle these Lewis acid-base complexes should be similar to their isolobal hydrocarbon analogs, however, whereas the dipole in ethane is zero (by symmetry) the dipole in $H_3NBH_3$ is 5.2 D as a consequence of the difference in the Pauling electronegativities (i.e., B = 2.04 and N = 3.04). It is this dipole that generally differentiates the B-N compounds from their C-C analogs.

Homolytic cleavage of the C-C bond in ethane will yield two neutral methyl radicals,

$H_3C\text{-}CH_3 \rightarrow \cdot CH_3 + \cdot CH_3$

In contrast, heterolytic cleavage will result in the formation of two charged species,

$H_3C\text{-}CH_3 \rightarrow CH_3^+ + \cdot CH_3^-$

Thus, the products either have a net spin or a net charge. By contrast, cleavage of the B-N bond in $H_3N\text{-}BH_3$ either yields products with both spin and charge or neither. Heterolytic cleavage of the B-N bond yields neutral compounds,

$H_3N\text{-}BH_3 \rightarrow NH_3 + BH_3$

while hemolytic cleavage results in the formation of radical ions,

$H_3N\text{-}BH_3 \rightarrow NH_3^{+\cdot} + BH_3^{-\cdot}$

The difference in bond strength between $H_3N\text{-}BH_3$ and ethane is reflected in the difference in bond lengths (Table 11.16).

| Compound | Bond length (Å) | Bond strength (kcal/mol) |
|---|---|---|
| $H_3C\text{-}CH_3$ | 1.533 | 89 |
| $H_3N\text{-}BH_3$ | 1.658 | 31 |

**Table 11.16: A comparison of bonding in $H_3E\text{-}E'H_3$.**

# Aminoboranes

The group $R'_2N\text{-}BR_2$ is isoelectronic and isolobal to the olefin sub-unit $R'_2C=CR_2$, and there is even appreciable π-bonding character (Figure 11.38). A measure of the multiple bond character can be seen from a comparison of the calculated B-N bond in $H_2NBN_2$ (1.391 Å) as compared to a typical olefin (1.33 Å). It is interesting that a consideration of the possible resonance form (Figure 11.38) suggest the dipole in the σ-bond is in the opposite direction of that of the π-bond.

**Figure 11.38: Resonance forms for $X_2N\text{-}BX'_2$.**

Unlike olefins, borazines oligomerize to form dimmers and trimers (Figure 11.39) in the absence of significant steric hindrance. Analogous structures are also observed for the other Group 13-15 homologs ($R_2AlNR'_2$, $R_2GaPR'_2$, etc.).

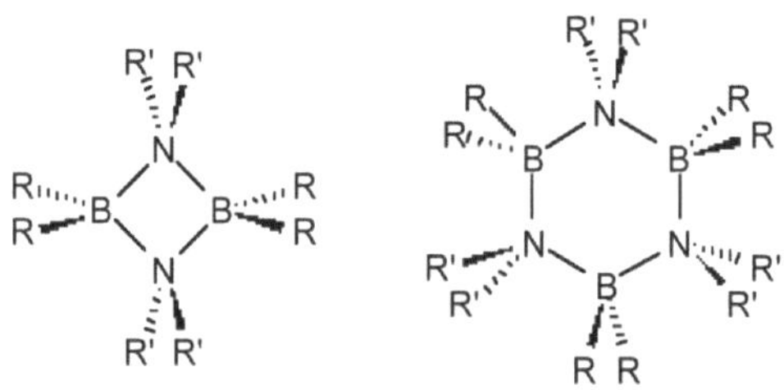

**Figure 11.39: Typical structures of aminoboranes.**

# Borazines

The condensation of boron hydride with ammonia results in the formation of a benzene analog: borazine,

$$BH_3 + NH_3 \rightleftharpoons H_3B\text{-}NH_3 \xrightarrow[-H_2]{\Delta} [H_2B\text{-}NH_2]_n \xrightarrow[-H_2]{\Delta} [HBNH]_6$$

Substituted derivatives are formed by the reaction with primary amines.

Despite the cyclic structure (Figure 11.40), borazine is not a true analog of benzene. Despite all the B-N bond distances being equal (1.44 Å) consistent with a delocalized structure, the difference in electronegativity of boron (2.04) and nitrogen (3.04) (Table 1.5) results in a polarization of the bonds (i.e., $B^{\delta+}$-$N^{\delta-}$) and hence a limit to the delocalization. The molecular orbitals of the π-system in borazine are lumpy in appearance (Figure 11.41a) compared to benzene (Figure 11.41b). This uneven distribution makes borazine prone to addition reactions, making it as a molecule less stable than benzene.

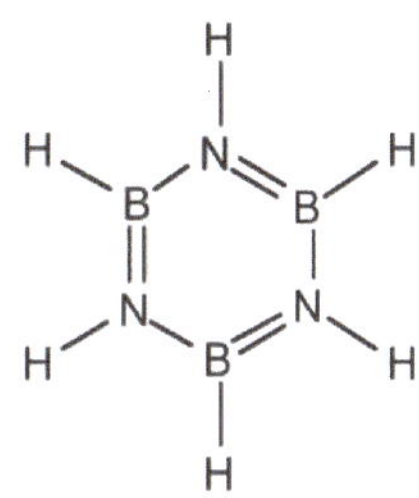

**Figure 11.40: Structure of borazine, the inorganic analog to benzene.**

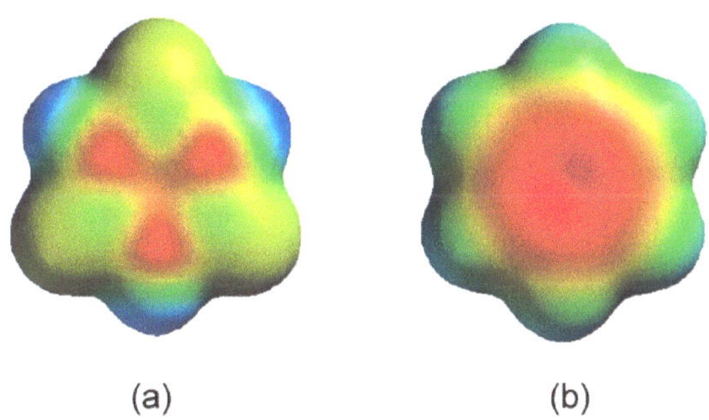


**Figure 11.41: Probability distributions for the π-bond orbitals in (a) borazine and (b) benzene.**

## Iminoboranes: analogs of acetylene

Iminoboranes, RB=NR', are analogs of alkynes, but like aminoboranes are only isolated as monomers with sterically hindered subsistent. In the absence of sufficient steric bulk oligomerization occurs, forming substituted benzene analogs.

## Boron nitrides: analogs of elemental carbon

The fusion of borax, $Na_2[B_4O_5(OH)_4]$ with ammonium chloride ($NH_4Cl$) results in the formation of hexagonal boron nitride (h-BN). Although h-BN has a planar, layered structure consisting of six-member rings similar to graphite (Figure 11.42), it is a white solid. The difference in color is symptomatic of the more localized bonding in BN than in graphite.

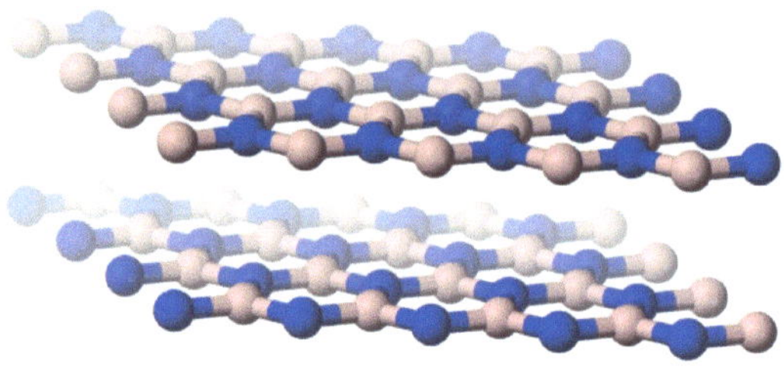

**Figure 11.42: Structure of h-BN.**

As is found for its carbon analog, hexagonal boron nitride (h-BN or α-BN) is converted at high temperatures (600 - 2000 °C) and pressures (50 - 200 kbar) to a cubic phase (c-BN or β-BN). In a similar manner to diamond, cubic-BN is very hard being actually able to cut diamond, and as a consequence its main use is as an industrial grinding agent. The cubic form has the sphalerite crystal structure (Figure 11.43). Finally, a wurtzite form of boron nitride (w-BN) is known that has similar structure as *lonsdaleite*, rare hexagonal polymorph of carbon. Table 11.17 shows a comparison of the properties of the hexagonal and cubic phases of BN with their carbon analogs.

| Phase | Carbon | Boron nitride |
|---|---|---|
| Cubic | Colorless, hard, melting point = 3550 °C. C-C = 1.514 Å | Colorless, hard, B-N = 1.56 Å |
| Hexagonal | Black solid, planar layers, conductor, melting point = 3652 - 3697 °C (sublimes), C-C = 1.415 Å | White solid, planar layers, semi- conductor ($E_g$ = 5.2 eV), melting point = 2973 °C (sublimes), B-N = 1.45 Å |

**Table 11.17: Comparison of structural and physical properties of carbon and boron nitride analogs.**

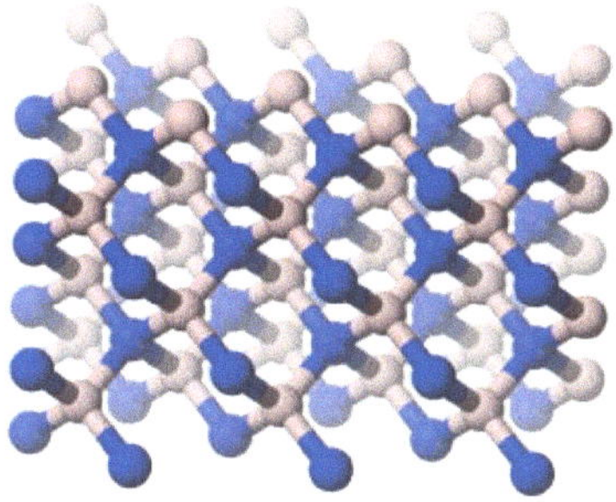

**Figure 11.43: Structure of c-BN.**

The partly ionic structure of BN layers in h-BN reduces covalency and electrical conductivity, whereas the interlayer interaction increases resulting in higher hardness of h-BN relative to graphite.

## Gallium arsenide

Gallium's main use is in semiconductor technology. For example, GaAs and related compounds can convert electricity directly into coherent light (laser diodes) and is employed in electroluminescent light-emitting diodes (LED's); it is also used for doping other semiconductors and in solid-state devices such as hetero-junction bipolar transistors (HBTs) and high power high speed metal semiconductor field effect transistors (MESFETs). The compound $MgGa_2O_4$ is used in ultraviolet-activated powders as a brilliant green phosphor used in Xerox copying machines.

Undoubtedly the binary compounds of gallium with the most industrial interest are those of the Group 15 (V) elements, GaE (E = N, P, As, Sb). The compounds which gallium forms with nitrogen, phosphorus, arsenic, and antimony are isoelectronic with the Group 14 elements. There has been considerable interest, particularly in the physical properties of these compounds, since 1952 when Welker first showed that they had semiconducting properties analogous to those of silicon and germanium.

Gallium phosphide, arsenide, and antimonide can all be prepared by direct reaction of the elements; this is normally done in sealed silica tubes or in a graphite crucible under hydrogen. Phase diagram data is hard to obtain in the gallium-phosphorus system because of loss of phosphorus from the bulk material at elevated temperatures. Thus, GaP has a vapor pressure of more than 13.5 atm. at its melting point; as compared to 0.89 atm. for GaAs. The physical properties of these three compounds are compared with those of the nitride in Table 11.18. All three adopt the zinc blende crystal structure and are more highly conducting than gallium nitride.

| Property | GaN | GaP | GaAs | GaSb |
|---|---|---|---|---|
| Melting point (°C) | > 1250 (dec) | 1350 | 1240 | 712 |
| Density (g/cm$^3$) | ca. 6.1 | 4.138 | 5.3176 | 5.6137 |
| Crystal structure | würtzite | zinc blende | zinc blende | zinc blende |
| Cell dimensions (Å)[a] | $a$ = 3.187, $c$ = 5.186 | $a$ = 5.4505 | $a$ = 5.6532 | $a$ = 6.0959 |
| Refractive index[b] | 2.35 | 3.178 | 3.666 | 4.388 |
| k (ohm$^{-1}$cm$^{-1}$) | $10^{-9}$ - $10^{-7}$ | $10^{-2}$ - $10^{2}$ | $10^{-6}$ - $10^{3}$ | 6 - 13 |
| Band gap (eV)[c] | 3.44 | 2.24 | 1.424 | 0.71 |

**Table 11.18: Physical properties of 13-15 compound semiconductors. [a]Values given for 300 K. [b]Dependent on photon energy; values given for 1.5 eV incident photons. [c]Dependent on temperature; values given for 300 K.**

## Gallium arsenide versus silicon

Gallium arsenide is a compound semiconductor with a combination of physical properties that has made it an attractive candidate for many electronic applications. From a comparison of various physical and electronic properties of GaAs with those of Si (Table 11.19) the advantages of GaAs over Si can be readily ascertained.

| Properties | GaAs | Si |
|---|---|---|
| Formula weight | 144.63 | 28.09 |
| Crystal structure | zinc blende | diamond |
| Lattice constant | 5.6532 | 5.43095 |
| Melting point (°C) | 1238 | 1415 |
| Density (g/cm$^3$) | 5.32 | 2.328 |
| Thermal conductivity (W/cm.K) | 0.46 | 1.5 |
| Band gap (eV) at 300 K | 1.424 | 1.12 |
| Intrinsic carrier conc. (cm$^{-3}$) | $1.79\times10^{6}$ | $1.45\times10^{10}$ |
| Intrinsic resistivity (ohm.cm) | $10^{8}$ | $2.3\times10^{5}$ |
| Breakdown field (V/cm) | $4\times10^{5}$ | $3\times10^{5}$ |
| Minority carrier lifetime (s) | $10^{-8}$ | $2.5\times10^{-3}$ |
| Mobility (cm$^2$/V.s) | 8500 | 1500 |

**Table 11.19: Comparison of properties of GaAs and Si.**

*Band gap*

The band gap of GaAs is 1.42 eV; resulting in photon emission in the infra-red range. Alloying GaAs with Al to give $Al_xGa_{1-x}As$ can extend the band gap into the visible red range. Unlike Si, the band gap of GaAs is direct, i.e., the transition between the valence band maximum and conduction band minimum involves no momentum change and hence does not require a collaborative particle interaction to occur. Photon generation by inter-band radiative recombination is therefore possible in GaAs. Whereas in Si, with an indirect band gap, this process is too inefficient to be of use. The ability to convert electrical energy into light forms the basis of the use of GaAs, and its alloys, in optoelectronics; for example, in light emitting diodes (LEDs), solid state lasers (light amplification by the stimulated emission of radiation).

A significant drawback of small band gap semiconductors, such as Si, is that electrons may be thermally promoted from the valence band to the conduction band. Thus, with increasing temperature the thermal generation of carriers eventually becomes dominant over the intentionally doped level of carriers. The wider band gap of GaAs gives it the ability to remain 'intentionally' semiconducting at higher temperatures; GaAs devices are generally more stable to high temperatures than similar Si devices.

*Carrier density*

The low intrinsic carrier density of GaAs in a pure (undoped) form indicates that GaAs is intrinsically a very poor conductor and is commonly referred to as being semi-insulating. This property is usually altered by adding dopants of either the p- (positive) or n- (negative) type. This semi-insulating property allows many active devices to be grown on a single substrate, where the semi-insulating GaAs provides the electrical isolation of each device; an important feature in the miniaturization of electronic circuitry, i.e., VLSI (very-large-scale-integration) involving over 100,000 components per chip (one chip is typically between 1 and 10 mm square).

*Electron mobility*

The higher electron mobility in GaAs than in Si potentially means that in devices where electron transit time is the critical performance parameter, GaAs devices will operate with higher response times than equivalent Si devices. However, the fact that hole mobility is similar for both GaAs and Si means that devices relying on cooperative electron and hole movement, or hole movement alone, show no improvement in response time when GaAs based.

*Crystal growth*

The bulk crystal growth of GaAs presents a problem of stoichiometric control due the loss, by evaporation, of arsenic both in the melt and the growing crystal (>600 °C). Melt growth techniques are, therefore, designed to enable an overpressure of arsenic above the melt to be maintained, thus preventing evaporative losses. The loss of arsenic also negates diffusion techniques commonly used for wafer doping in Si technology; since the diffusion temperatures required exceed that of arsenic loss.

The thermal gradient and, hence, stress generated in melt grown crystals have limited the maximum diameter of GaAs wafers because with increased wafer diameters the thermal stress generated dislocation (crystal imperfections) densities eventually becomes unacceptable for device applications. Gallium arsenide single crystals are very brittle, requiring that considerably thicker substrates than those employed for Si devices.

## Electronic grade gallium arsenide

The synthesis and purification of bulk polycrystalline semiconductor material represents the first step towards the commercial fabrication of an electronic device. This polycrystalline material is then used as the raw material for the formation of single crystal material that is processed to semiconductor wafers.

In contrast to electronic grade silicon (EGS), whose use is a minor fraction of the global production of elemental silicon, gallium arsenide (GaAs) is produced exclusively for use in the semiconductor industry. However, arsenic and its compounds have significant commercial applications. The main use of elemental arsenic is in alloys of Pb, and to a lesser extent Cu, while arsenic compounds are widely used in pesticides and wood preservatives and the production of bottle glass. Thus, the electronics industry represents a minor user of arsenic. In contrast, although gallium has minor uses as a high-temperature liquid seal, manometric fluids and heat transfer media, and for low temperature solders, its main use is in semiconductor technology.

*Synthesis and purification of gallium arsenide.*

Gallium arsenide can be prepared by the direct reaction of the elements,

$$\text{Ga(liquid)} + \text{As(vapor)} \xrightarrow{>1240\ ^\circ\text{C}} \text{GaAs(solid)}$$

However, while conceptually simple the synthesis of GaAs is complicated by the different vapor pressures of the reagents and the highly exothermic nature of the reaction. Furthermore, since the synthesis of GaAs at atmospheric pressure is accompanied by its simultaneous decomposes due to the loss by sublimation, of arsenic, the synthesis must be carried out under an overpressure of arsenic in order to maintain a stoichiometric composition of the synthesized GaAs.

In order to overcome the problems associated with arsenic loss, the reaction is usually carried out in a sealed reaction tube. However, if a stoichiometric quantity of arsenic is used in the reaction a constant temperature of 1238 °C must be employed in order to maintain the desired arsenic overpressure of 1 atm. Practically, it is easier to use a large excess of arsenic heated to a lower temperature. In this situation the pressure in the tube is approximately equal to the equilibrium vapor pressure of the volatile component (arsenic) at the lower temperature. Thus, an over pressure of 1 atm arsenic may be maintained if within a sealed tube elemental arsenic is heated to 600-620 °C while the GaAs is maintained at 1240-1250 °C.

Figure 11.41 shows the sealed tube configuration that is typically used for the synthesis of GaAs. The tube is heated within a two-zone furnace. The boats holding the reactants are usually made of quartz, however, graphite is also used since the latter has a closer thermal expansion match to the GaAs product. If higher purity is required, then pyrolytic boron nitride (PBN) is used. One of the boats is loaded with pure gallium the other with arsenic. A plug of quartz wool may be placed between the boats to act as a diffuser. The tube is then evacuated and sealed. Once brought to the correct reaction temperatures (Figure 11.41), the arsenic vapor is transported to the gallium, and they react to form GaAs in a controlled manner. Table 11.20 gives the typical impurity concentrations found in polycrystalline GaAs.

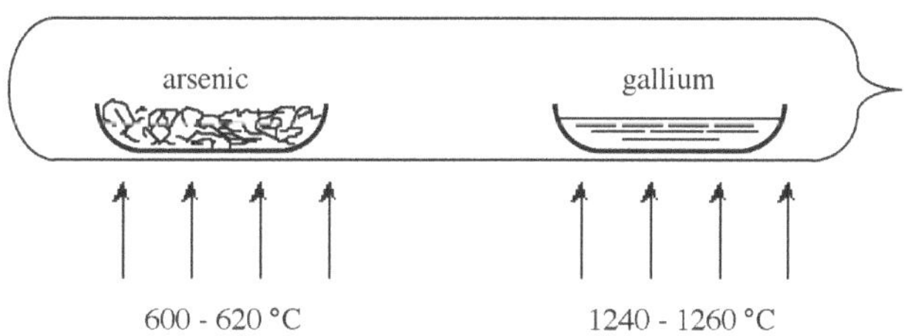


**Figure 11.41: Schematic representation of a sealed tube synthesis of GaAs.**

| Element | Concentration (ppm) | Element | Concentration (ppm) |
|---|---|---|---|
| boron | 0.1 | silicon | 0.02 |
| carbon | 0.7 | phosphorus | 0.1 |
| nitrogen | 0.1 | sulfur | 0.01 |
| oxygen | 0.5 | chlorine | 0.08 |
| fluorine | 0.2 | nickel | 0.04 |
| magnesium | 0.02 | copper | 0.01 |
| aluminum | 0.02 | zinc | 0.05 |

**Table 11.20: Impurity concentrations found in polycrystalline GaAs.**

Polycrystalline GaAs formed in from the direct reaction of the elements is often used as the starting material for single crystal growth via Bridgeman or Czochralski crystal growth. It is also possible to prepare single crystals of GaAs directly from the elements using in-situ, or direct, compounding within a high-pressure liquid encapsulated Czochralski (HPLEC) technique.

*Growth of gallium arsenide crystals*

When considering the synthesis of Group 13-15 compounds for electronic applications, the very nature of semiconductor behavior demands the use of high purity single crystal materials. The polycrystalline materials synthesized above are, therefore, of little use for 13-15 semiconductors but may, however, serve as the starting material for melt grown single crystals. For GaAs, undoubtedly the most important 13-15 (III - V) semiconductor, melt grown single crystals are achieved by one of two techniques: the Bridgman technique, and the Czochralski technique.

The Bridgman technique requires a two-zone furnace, of the type shown in Figure 11.42. The left-hand zone is maintained at a temperature of ca. 610 °C, allowing sufficient overpressure of arsenic within the sealed system to prevent arsenic loss from the gallium arsenide. The right-hand side of the furnace contains the polycrystalline GaAs raw material held at a temperature just above its melting point (ca. 1240 °C). As the furnace moves from left to right, the melt cools and solidifies.

If a seed crystal is placed at the left-hand side of the melt (at a point where the temperature gradient is such that only the end melts), a specific orientation of single crystal may be propagated at the liquid-solid interface eventually to produce a single crystal.

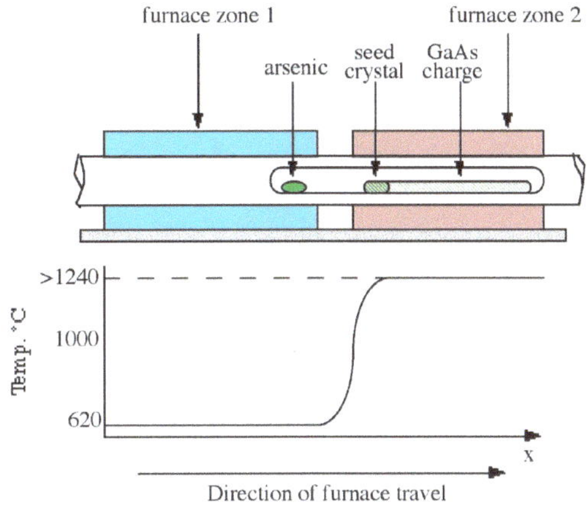


**Figure 11.42: A schematic diagram of a Bridgman two-zone furnace used for melt growths of single crystal GaAs.**

The Czochralski technique, which is the most commonly used technique in industry, is shown in Figure 11.43. The process relies on the controlled withdrawal of a seed crystal from a liquid melt. As the seed is lowered into the melt, partial melting of the tip occurs creating the liquid solid interface required for crystal growth. As the seed is withdrawn, solidi cation occurs, and the seed orientation is propagated into the grown material. The variable parameters of rate of withdrawal and rotation rate can control crystal diameter and purity. As shown in Figure 11.43 the GaAs melt is capped by boron trioxide ($B_2O_3$). The capping layer, which is inert to GaAs, prevents arsenic loss when the pressure on the surface is above atmospheric pressure. The growth of GaAs by this technique is thus termed liquid encapsulated Czochralski (LEC) growth.

While the Bridgman technique is largely favored for GaAs growth, larger diameter wafers can be obtained by the Czochralski method. Both of these melt techniques produce materials heavily contaminated by the crucible, making them suitable almost exclusively as substrate material. Another disadvantage of these techniques is the production of defects in the material caused by the melt process.

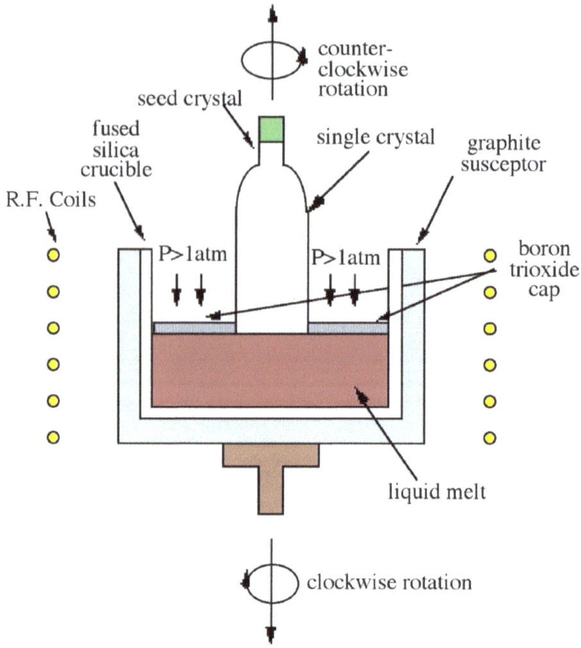


**Figure 11.43: A schematic diagram of the Czochralski technique as used for growth of GaAs single crystal bond.**

## Chalcogenides of aluminum, gallium, and indium

The only stable chalcogenides of aluminum are $Al_2S_3$ (white), $Al_2Se_3$ (grey), and $Al_2Te_3$ (dark grey). They are each prepared by the direct reaction of the elements (100 °C) and hydrolyze rapidly in aqueous solution,

$$Al_2E_3 + 6\,H_2O \rightarrow 2\,Al(OH)_3 + 3\,H_2E$$

All the chalcogenides have a hexagonal ZnS structure in which $^2/_3$ of the metal sites are occupied. The chalcogenides of gallium and indium are more numerous than those of aluminum and are listed in Table 11.21 and Table 11.22 along with selected physical properties.

| Com-pound | Structural type | Crystallographic system | Cell parameters (Å, °) | Band Gap (eV)[a] |
|---|---|---|---|---|
| GaS | | Hexagonal | $a = 3.587$, $c = 15.492$ | 3.05 (dir.), 2.593 (ind.) |
| GaS | ZnS or NaCl | Cubic | $a = 5.5$ | 4.0 (opt.) |
| β-GaSe | GaS | Hexagonal | $a = 3.742$, $c = 15.919$ | 2.103 (dir.), 2.127 (ind.) |
| γ-GaSe | GaS | Rhombohedral | $a = 3.755$, $c = 23.92$ | |
| δ-GaSe | GaS | Hexagonal | $a = 3.755$, $c = 31.99$ | |
| β-GaTe | GaS | Hexagonal | $a = 4.06$, $c = 16.96$ | |
| GaTe | GaS | Monoclinic | $a = 17.44$, $b = 4.077$, $c = 10.456$, $\beta = 104.4$ | 1.799 (dir.) |
| α-$Ga_2S_3$ | Wurtzite | Cubic | $a = 5.181$ | |
| α-$Ga_2S_3$ | Wurtzite | Monoclinic | $a = 12.637$, $b = 6.41$, $c = 7.03$, $\beta = 131.08$ | 3.438 (opt.) |
| β-$Ga_2S_3$ | Defect wurtz-ite | Hexagonal | $a = 3.685$, $c = 6.028$ | 2.5 - 2.7 (opt.) |
| α-$Ga_2Se_3$ | Sphalerite | Cubic | $a = 5.429$ | 2.1 (dir.), 2.04 (ind.) |
| α-$Ga_2Te_3$ | Sphalerite | Cubic | $a = 5.886$ | 1.22 (opt.) |

**Table 11.21: Stoichiometries, structures and selected physical properties of the crystalline chalcogenides of gallium. [a]dir = direct, ind = indirect, opt = optical.**

The hexagonal β-form of $Ga_2S_3$ is isostructural with the aluminum analogue; however, while the α-phase was proposed to be hexagonal it was later shown to be monoclinic. A cubic α-phase has been reported. Cubic sphalerite structures are found for $Ga_2Se_3$, $Ga_2Te_3$, and $In_2Te_3$, in which the structure is based on a cubic close packing of the chalcogenides and the metal atoms occupying 1/3 of the tetrahedral sites. These structures are all formed with rapid crystallization; slow crystallization and/or thermal annealing leads to ordering and the formation of more complex structures. The indium sulfides, and selenides derivatives are spinel (γ-$Al_2O_3$), and defect würtzite, respectively.

Unlike the chalcogenides of aluminum, those of gallium and indium also form subvalent compounds, i.e., those in which the metal is formally of an oxidation state less than +3. Of these subvalent chalcogenides the (formally) divalent materials are of the most interest. The thermodynamically stable phase of GaS has a hexagonal layer structure (Figure 11.44) with Ga-Ga bonds (2.48 Å). The compound can, therefore, be considered as an example of Ga(II). Each Ga is coordinated by three sulfur atoms and one gallium, and the sequence of layers along the *z*-axis is ···S-Ga-Ga-S···S-Ga-Ga-S···.

| Compound | Structural type | Crystallographic system | Cell parameter (Å, °) | Band gap (eV)[a] |
|---|---|---|---|---|
| β-InS | GaS | Orthorhombic | $a = 3.944$, $b = 4.447$, $c = 10.648$ | 2.58 (dir.), 2.067 (ind.) |
| InS[b] | $Hg_2Cl_2$ | Tetragonal | $a = 3.823$, $c = 10.868$ | |
| InSe | GaS | Rhombohedral | $a = 4.00$, $c = 25.32$ | 1.3525 (dir.), 1.32 (ind.) |
| β-InSe | GaS | Hexagonal | $a = 4.05$, $c = 16.93$ | |
| InTe | TlSe | Tetragonal | $a = 8.437$, $c = 7.139$ | Metallic |
| InTe[b] | NaCl | Cubic | $a = 6.18$ | |
| α-$In_2S_3$ | γ-$Al_2O_3$ | Cubic | $a = 5.36$ | |
| β-$In_2S_3$ | Spinel | Tetragonal | $a = 7.618$, $c = 32.33$ | 2.03 (dir.), 1.1 (ind.) |
| α-$In_2Se_3$ | Defect wurtzite | Hexagonal | $a = 16.00$, $c = 19.24$ | |
| β-$In_2Se_3$ | Defect wurtzite | Rhombohedral | $a = 4.025$, $c = 19.222$ | 1.2 - 1.5 (ind.) |
| α-$In_2Te_3$ | Sphalerite | Cubic | $a = 6.158$ | 0.92 - 1.15 (opt.) |
| $In_6Te_7$ | | Monoclinic | $a = 9.090$, $b = 3.887$, $c = 17.705$, $\beta = 108.20$ | 0.89 (dir.), 0.7 (ind.) |
| $In_6Se_7$ | $In_6Te_7$ | Monoclinic | $a = 9.430$, $b = 4.063$, $c = 18.378$, $\beta = 109.34$ | 0.86 (dir.), 0.34 (ind.) |
| $In_4Se_3$ | | Orthorhombic | $a = 15.297$, $b = 12.308$, $c = 4.081$ | 0.64 (dir.) |
| $In_4Te_3$ | $In_4Se_3$ | Orthorhombic | $a = 15.630$, $b = 12.756$, $c = 4.441$ | 0.48 (dir.) |

**Table 11.22: Stoichiometries, structures and selected physical properties of the crystalline chalcogenides of indium. [a]dir = direct, ind = indirect, opt = optical. [b]High pressure phase.**

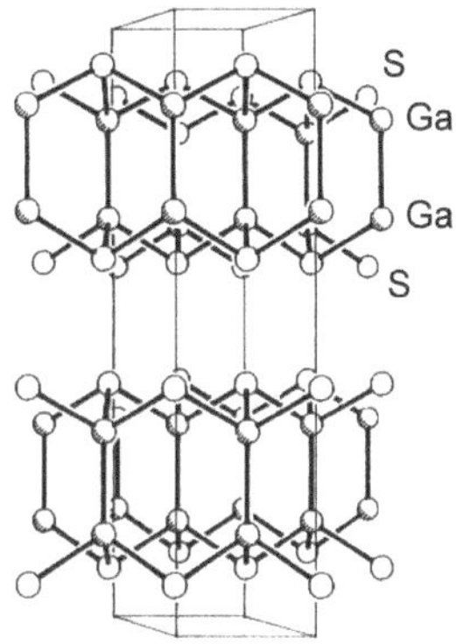


**Figure 11.44: The S-Ga-Ga-S···S-Ga-Ga-S structure of hexagonal GaS. Gallium atoms are shown shaded.**

The structures of β-GaSe, and β-InSe are similar to hexagonal GaS. The layered structure of GaTe is similar in that it consists of ···TeGaGaTe··· layers, but is monoclinic, while InS is found in both a (high pressure) tetragonal phase (Figure 11.45a) as well as an orthorhombic phase (Figure 11.45b). By contrast to these M-M bonded layered compounds InTe (Figure 11.46) has a structure formalized as In(I)[In(III)$Te_2$]; each In(III) is tetrahedrally coordinated to four Te and these tetrahedra are linked via shared edges; the In(I) centers lying between these chains.

(a)

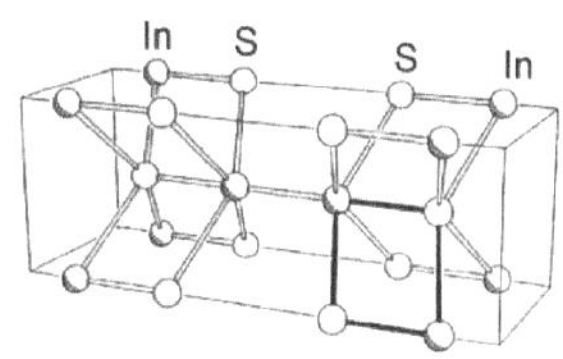


(b)

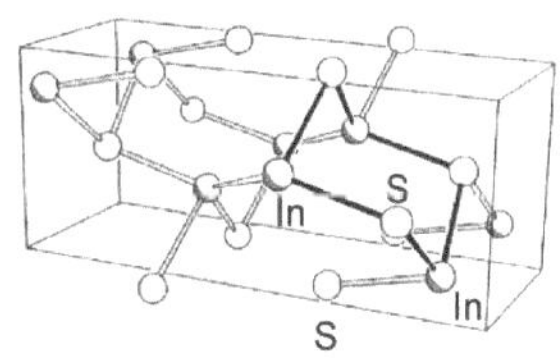


**Figure 11.53: Unit cell of (a) tetragonal InS and (b) orthorhombic InS. Indium atoms are shown shaded, and the solid bonds represent the smallest cyclic structural fragment.**

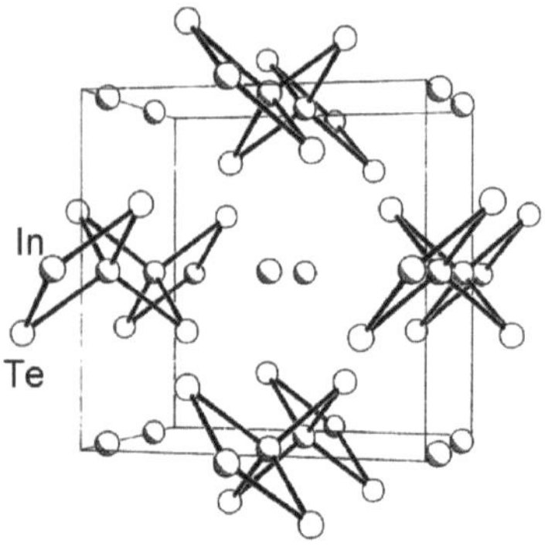


**Figure 11.46: Unit cell of tetragonal InTe. Indium atoms are shown shaded, and the solid bonds represent the $[InTe_2]_\infty$ chains.**

Further sub-chalcogenides are known for indium, e.g.; $In_4Se_3$, which contains $[In(III)_3Se_2]^{5+}$ groups (Figure 11.47). While the formally In(I) molecule $In_2S$ has been detected in the gas phase, it is actually a mixture of In and InS in the solid state.

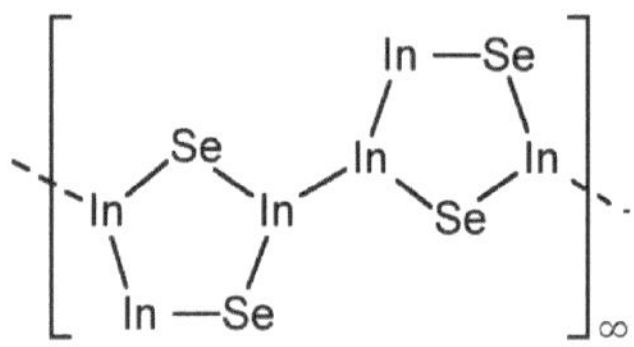


**Figure 11.47: Structure of $[In(III)_3Se_2]^{5+}$ groups in $In_4Se_3$.**

# Group 13 halides

## Trihalides

As shown in Table 11.23 all the combinations of Group 13 element (M) and halogen (X) exist for the trihalides ($MX_3$), except thallium(III) iodide. It should be noted that while there is a compound with the general formula TlI, it is actually a thallium(I) compound of $I_3^-$.

The trihalides of boron are all monomers with a coordination number of 3 (Table 11.24), as evidence from their low melting points (Table 11.23). In contrast, the fluorides and chlorides of the heavier Group 13 elements (except $GaCl_3$) are generally ionic or have a high ionic character, with a coordination number of 6 (Table 11.24, Figures 11.48 and 11.49). The bromides and iodides (except $InBr_3$) are generally dimeric with a coordination number of 4 (Table 11.24) and have molecular structures involving halide bridging ligands (Figure 11.50 and Table 11.25). $AlCl_3$ is unusual in that in the solid state it has an ionic structure, but it is readily sublimed, and in the vapor phase (and liquid phase) it has a dimeric structure (Figure 11.50).

| Element | Melting point (°C) | Boiling point (°C) |
|---|---|---|
| $BF_3$ | -126.8 | -100.3 |
| $BCl_3$ | -107.3 | 12.6 |
| $BBr_3$ | -46.3 | 91.3 |
| $BI_3$ | 49.9 | 210 |
| $AlF_3$ | 1291 | - |
| $AlCl_3$ | 192.4 (anhydrous), 0.0 (hexahydrate) | 120 (hexahydrate) |
| $AlBr_3$ | 97.8 | 265 |
| $AlI_3$ | 189.4 (anhydrous), 185 dec. (hexahydrate) | 300 subl. |
| $GaF_3$ | 800 | 1000 |
| $GaCl_3$ | 77.9 | 201 |
| $GaBr_3$ | 121.5 | 278.8 |
| $GaI_3$ | 212 | 345 |
| $InF_3$ | 1172 | - |
| $InCl_3$ | 586 | 800 |
| $InBr_3$ | 220 | - |
| $InI_3$ | 210 subl. | - |
| $TlF_3$ | 300 dec. | - |
| $TlCl_3$ | 40 dec. | - |
| $TlBr_3$ | 40 dec. | - |

**Table 11.23: Selected physical properties of the Group 13 trihalides, $MX_3$.**

| Element | Fluoride | Chloride | Bromide | Iodide |
|---|---|---|---|---|
| B | 3 | 3 | 3 | 3 |
| Al | 6 | 6 (4) | 4 | 4 |
| Ga | 6 | 4 | 4 | 4 |
| In | 6 | 6 | 6 | 4 |
| Tl | 6 | 6 | 4 | - |

**Table 11.24: The Group 13 element coordination numbers for the Group 13 trihalides, $MX_3$.**

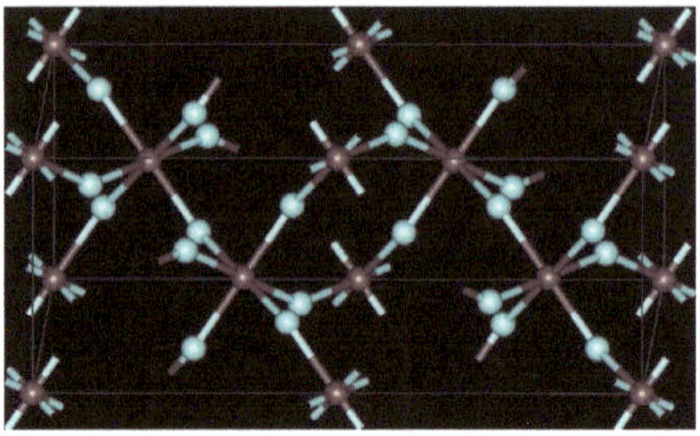

**Figure 11.48: Solid state structure of $MF_3$ (M = Al, Ga, In). Grey spheres represent the metal, while pale blue spheres represent fluorine.**

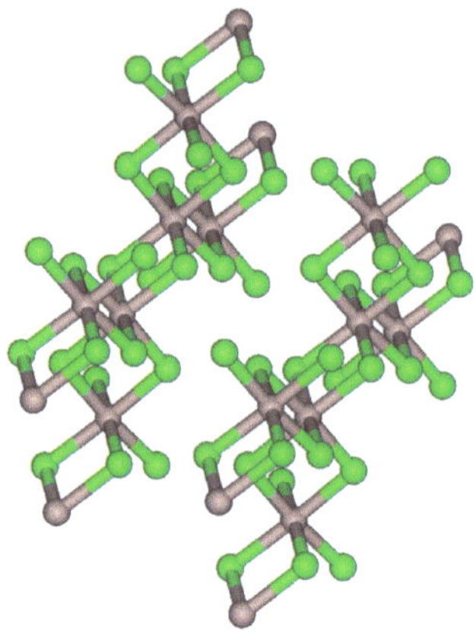

**Figure 11.49: Solid state structure of $MCl_3$ (M = Al, In, Tl) and $MB_3$ (M = In). Grey spheres represent the metal, while green spheres represent the halogen.**

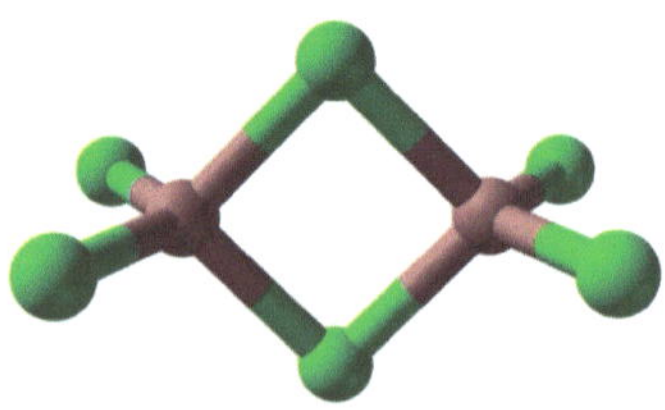

**Figure 11.50: Structure of $GaCl_3$, $MBr_3$ (M = Al, Ga, Tl), and $MI_3$ (M = Al, Ga, In). Grey spheres represent the metal, while green spheres represent the halogen.**

| Compound | M-$X_t$ (Å)[a] | M-$X_b$ (Å)[a] | $X_t$-M-$X_t$ (°)[a] | $X_b$-M-$X_b$ (°)[a] | M-X-M (°) |
|---|---|---|---|---|---|
| $Al_2Br_6$ | 2.21 | 2.33 | 115 | 93 | 87 |
| $In_2I_6$ | 2.64 | 2.84 | 125.1 | 93.7 | 86.3 |

**Table 11.25: Selected bond lengths and angles for dimeric $M_2X_6$ compounds. [a]$X_t$ and $X_b$ indicate terminal and bridging halides, respectfully.**

*Synthesis*

Boron trifluoride ($BF_3$) is manufactured commercially by the reaction of boron oxides with hydrogen fluoride,

$$B_2O_3 + 6\ HF \rightarrow 2\ BF_3 + 3\ H_2O$$

The HF is produced in-situ from sulfuric acid and fluorite ($CaF_2$). On smaller scales, $BF_3$ is prepared by the thermal decomposition of diazonium salts,

$$PhN_2[BF_4] \rightarrow PhF + BF_3 + N_2$$

Boron trichloride is also made from boron oxide, but in the presence of carbon,

$$B_2O_3 + 3\ C + 3\ Cl_2 \rightarrow 2\ BCl_3 + 3\ CO$$

Many of the trihalides are readily prepared by either the direct reaction of the metal with the appropriate halogen,

$$2\ Al + 3\ Cl_2 \rightarrow 2\ AlCl_3$$

$$2\ Al + 3\ Br_2 \rightarrow 2\ AlBr_3$$

$$2\ Al + 3\ I_2 \rightarrow 2\ AlI_3$$

or the acid,

$$2\ Al + 6\ HCl \rightarrow 2\ AlCl_3 + 3\ H_2$$

$$2\ Al + 6\ HBr \rightarrow 2\ AlBr_3 + 3\ H_2$$

Thallium tribromide can be prepared in $CH_3CN$ by treating a solution of the monobromide with bromine gas,

$$TlBr + Br_2 \rightarrow TlBr_2$$

*Reactivity*

The reaction chemistry of the Group 13 trihalides tends to fall into two categories:

- Lewis acid-base complex formation.
- Hydrolysis.

There are, however, a number of reactions involving halide exchange reactions. Aluminum tribromide reacts with carbon tetrachloride at 100 °C to form carbon tetrabromide,

$$4\ AlBr_3 + 3\ CCl_4 \rightarrow 4\ AlCl_3 + 3\ CBr_4$$

and with phosgene yields carbonyl bromide and aluminum chlorobromide,

$$AlBr_3 + COCl_2 \rightarrow COBr_2 + AlCl_2Br$$

Group 13 halides are used as synthons for their organometallic derivatives,

$$MX_3 + 3\ RMgX \rightarrow MR_3 + MgX_2$$

$$2\ MR_3 + MX_3 \rightarrow 3\ MXR_2$$

*Lewis acid-base complexes*

All of the trihalides are strong Lewis acids, and as such react with Lewis base compounds to form Lewis acid-base complexes,

$$MX_3 + L \rightleftharpoons X_3M\text{-}L$$

The extent of the equilibrium is dependent on the Lewis acidity of the trihalide and the basicity of the Lewis base. For example, with $BCl_3$, oxygen donor ligands result in approximately 50:50 ratio of $BCl_3$ and $BCl_3L$, while for nitrogen donor ligands the equilibrium is shifted to the formation of the complex.

The general structure of the Lewis acid-base complexes is such that the Group 13 element is close to tetrahedral (Figure 11.51). However, for aluminum and the heavier Group 13 elements, more than one ligand can coordinate (Figure 11.52) up to a maximum of six.

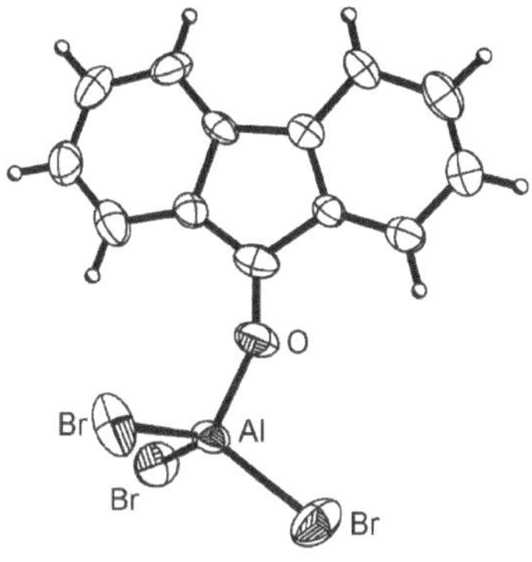


**Figure 11.59: Molecular structure of a typical Group 13 metal trihalide Lewis acid-base complex: $AlBr_3$(9-fluorenone). Adapted from C. S. Branch, S. G. Bott, and A. R. Barron, Group 13 trihalide complexes of 9-fluorenone: a comparison of methods for assigning relative Lewis acidity. *J. Organomet. Chem.*, 2003, 666, 23. Copyright: Elsevier (2003).**

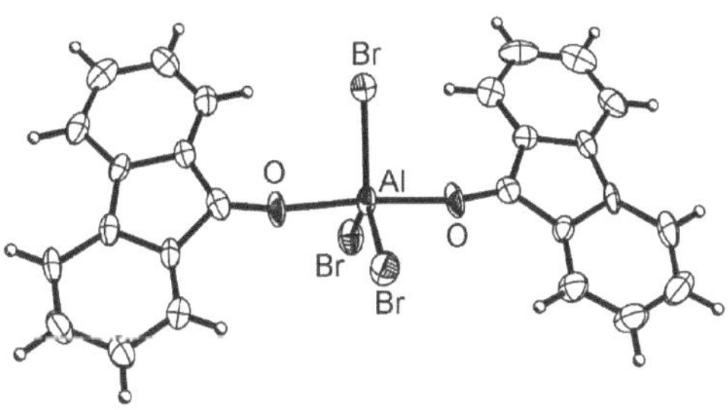


**Figure 11.59: Molecular structure of $AlBr_3$(9-fluorenone)$_2$. Adapted from C. S. Branch, S. G. Bott, and A. R. Barron, Group 13 trihalide complexes of 9-fluorenone: a comparison of methods for assigning relative Lewis acidity. *J. Organomet. Chem.*, 2003, 666, 23. Copyright: Elsevier (2003).**

It should be noted that the dimeric form of $MX_3$ (Figure 11.50) can be thought of as mutual Lewis acid-base complexes, in which a Lewis basic lone pair of a halide on one $MX_3$ unit donates to the Lewis acidic metal on another $MX_3$ unit.

*Hydrolysis*

Generally, the fluorides are insoluble in water while the heavier halides are more soluble. However, $BF_3$, $BCl_3$, and $BBr_3$ all decompose in the presence of water,

$$BX_3 + H_2O \rightarrow B(OH)_3 + 3\ HX$$

In the case of the fluoride, the HF formed reacts with $BF_3$ to form fluoroboric acid,

$$HF + BF_3 \rightarrow HBF_4$$

However, there is also a minor equilibrium (2-3%) resulting in the formation of the $BF_3$ complex of $OH^-$ and $H_3O^+$,

$$BF_3 + H_2O \rightleftharpoons F_3B\text{-}OH_2 \rightleftharpoons [BF_3OH][BF_3(OH_3)]$$

While the boron compounds (and $AlBr_3$) decompose even in moist air, $AlCl_3$ reacts more slowly to make aluminum chlorohydrate (ACH) which has the general formula $Al_nCl_{3n-m}(OH)_m$. While ACH has been proposed to exist as a number of cluster species, it is actually a range of nanoparticles. ACH is also known as polyaluminum chloride (PAC). The latter name is often used in water purification, where ACH is preferred over alum derivatives ($Al_2(SO_4)_3$). The combination of ACH and a high molecular weight quaternized ammonium polymer (e.g., dially dimethyl ammonium chloride (DADMAC)), has been known as an effective combination as a flocculant in water treatment process to remove dissolved organic matter and colloidal particles present in suspension.

Aluminum chlorohydrate (ACH) and aluminum-zirconium compounds, are frequently used as the active ingredient in antiperspirants. The mode of action of most aluminum-based compounds involves the dramatic change in the particle size from nano to micro as a function of pH and electrolyte changes on the skin (as compared to the antiperspirant stick or suspension) and hence forming a gel plug in the duct of the sweat gland. The plugs prevent the gland from excreting liquid and are removed over time by the natural sloughing of the skin. A further mechanism of action involves the absorption of 3-methyl-2-hexenoic acid (Figure 11.53). Human perspiration is odorless until bacteria ferment it. Bacteria thrive in hot, humid environments such as the underarm. When adult armpits are washed with alkaline pH soaps, the skin loses its acid mantel (pH = 4.5 - 6), raising the skin pH and disrupting the skin barrier. The bacteria thrive in the basic environment, and feed on the sweat from the apocrine glands and on dead skin and hair cells, releasing 3-methyl-2-hexenoic acid, which is the primary cause of body odor. As with all carboxylic acids, 3-methyl-2-hexenoic acid, reacts in a facile manner with the surface of the alumina nanoparticles. Aluminum chloride salts also have a slight astringent effect on the pores; causing them to contract, further preventing sweat from reaching the surface of the skin.

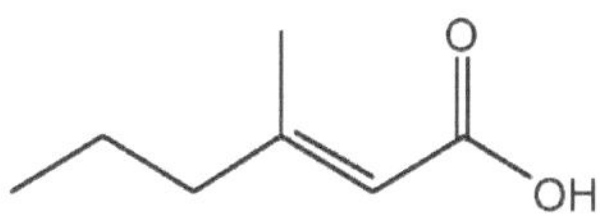


**Figure 11.53: The structure of 3-methyl-2-hexenoic acid.**

*Boron trihalides: a special case*

The lighter boron trihalides, $BX_3$ (X = F, Cl, Br) form stable adducts with common Lewis bases. Their relative Lewis acidities can be evaluated in terms of the relative exothermicity of the adduct-forming reaction:

$BF_3 < BCl_3 < BBr_3$ (strongest Lewis acid)

This trend is opposite to that expected based upon the electronegativity of the halogens. The best explanation of this trend takes into account the extent of π-donation that occurs between the filled lone pair orbital on the halogens and the empty p-orbital on the planar boron (Figure 11.54). As such, the greater the π-bonding the more stable the planar $BX_3$ configuration as opposed to the pyramidalization of the $BX_3$ moiety upon formation of a Lewis acid-base complex.

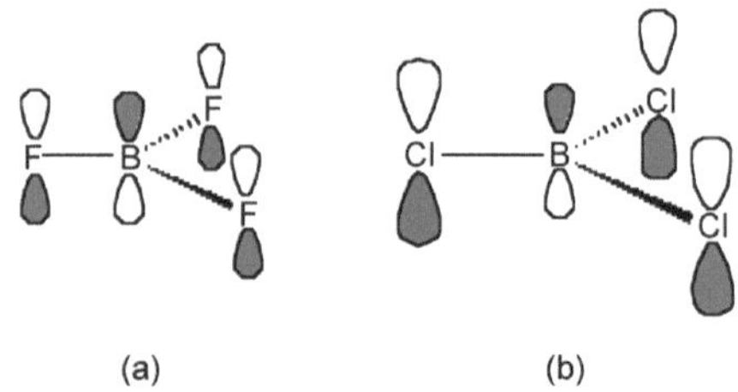


**Figure 11.54: Schematic representation of π-donation from filled halogen p-orbitals into empty p-orbital in the halides $BX_3$.**

The criteria for evaluating the relative strength of π-bonding are not clear, however, one suggestion is that the F atom is small compared to the Cl atom, and the lone pair electron in $p_z$ of F is readily and easily donated and overlapped to empty $p_z$ orbital of boron (Figure 11.54a). In contrast, the overlap for the large (diffuse) p-orbitals on the chlorine is poorer (Figure 11.54b). As a result, the π-donation of F is greater than that of Cl. Interestingly, as may be seen from Table 11.26, any difference in B-X bond length does not follow the expected trend associated with shortening of the B-X bond with stronger π-bonding. In fact, the B-Br distance is the most shortened as compared to that expected from the covalent radii (Table 11.26).

| Compound | B-X (Å) | X covalent radius (Å) | Sum of covalent radii (Å)[a] | Δ (Å) |
|---|---|---|---|---|
| $BF_3$ | 1.313 | 0.57(3) | 1.41 | 0.097 |
| $BCl_3$ | 1.75 | 1.02(4) | 1.86 | 0.11 |
| $BBr_3$ | 1.898 | 1.20(3) | 2.04 | 0.142 |
| $BI_3$ | 2.125 | 1.39(3) | 2.23 | 0.105 |

**Table 11.26: The B-X bond distances in the boron trihalides, $BX_3$, as compared to the sum of the covalent radii. [a]Covalent radius of B = 0.84(3) Å.**

At the simplest level the requirements for bonding to occur based upon the molecular or atomic orbital are:

- Directional relationship of the orbitals.
- Relative symmetry of the orbitals.
- Relative energy of the orbitals.
- Extent of orbital overlap

In the case of the boron trihalides, the direction (parallel) and symmetry (p-orbitals) are the same, and the only significant difference will be the relative energy of the donor orbitals (i.e., the lone pair on the halogen) and the extent of the overlap. The latter will be dependent on the B-X bond length (the shorter the bond the greater the potential overlap) and the diffusion of the orbitals (the less diffuse the orbitals the better the overlap). Both of these factors will benefit B-F over B-Cl and B-Br. Thus, the extent of potential overlap would follow the order. Despite these considerations, it is still unclear of the exact details of the rationalization of the low Lewis basicity of $BF_3$ as compared to $BCl_3$ and $BBr_3$.

## Anionic halides

The trihalides all form Lewis acid-base complexes with halide anions,

$$MX_3 + X^- \rightarrow MX_4^-$$

and as such salts of $BF_4^-$, $AlCl_4^-$, $GaCl_4^-$, and $InCl_4^-$ are common. In the case of gallium, the $Ga_2Cl_7^-$ anion (Figure 11.55) is formed from the equilibrium:

$$2\ GaCl_4^- \rightleftharpoons Ga_2Cl_7^- + Cl^-$$

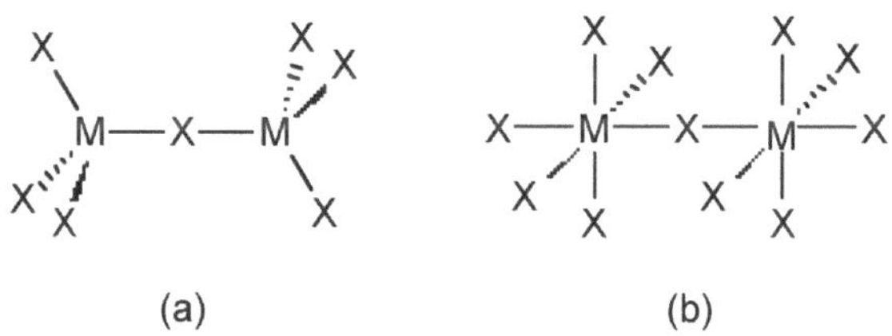


**Figure 11.55: Structures of the (a) $M_2Cl_7$ and (b) $M_2Cl_9$ anions.**

As a consequence of its larger size indium forms a wide range of anionic halides addition compounds with trigonal bipyramidal, square pyramidal, and octahedral coordination geometries. For example, salts of $InCl_5^{2-}$, $InBr_5^{2-}$, $InF_6^{3-}$, $InCl_6^{3-}$ and $InBr_6^{3-}$ have all been made. The $InCl_5^{2-}$ ion has been found to be square pyramidal in the salt $[NEt_4]_2InCl_5$ but is trigonal bipyramidal in the acetonitrile solvate of $[Ph_4P]_2InCl_5$. The oligomeric anionic halides $In_2X_7^-$ and $In_2X_9^{3-}$ (X = Cl and Br) contain binuclear anions with tetrahedral and octahedrally coordinated indium atoms, respectively (Figure 11.55).

# Low valent halides

*Oxidation state +2*

Boron forms a series of low oxidation halides containing B-B bonds a formal oxidation state of +2. Passing an electric discharge through $BCl_3$ using mercury electrodes results in the synthesis of $B_2Cl_4$,

$$2\ BCl_3 + 2\ Hg \rightarrow B_2Cl_4 + Hg_2Cl_2$$

An alternative route is by the co-condensation of copper as a reducing agent with $BCl_3$,

$$2\ BCl_3 + 2\ Cu \rightarrow B_2Cl_4 + 2\ CuCl$$

$B_2F_4$ has a planar structure (Figure 11.56 with $D_{2h}$ symmetry, while $B_2Cl_4$ has the same basic structure it has a staggered geometry (Figure 6.57). The energy for bond rotation about the B-B bond is very low (5 kJ/mol) that can be compared to ethane (12.5 kJ/mol). The bromide, $B_2Br_4$, is also observed to be staggered in the solid state. The staggered conformation is favorable on steric grounds, however, for $B_2F_4$ the planar geometry is stabilized by the smaller size of the halide, and more importantly the presence of strong delocalized π-bonding.

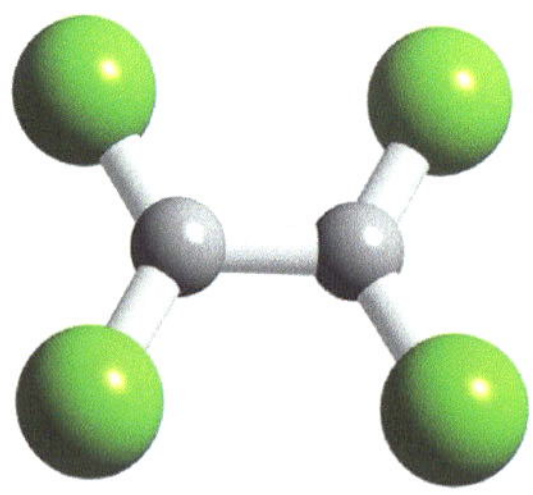

**Figure 11.56: The planar structure of $B_2F_4$.**

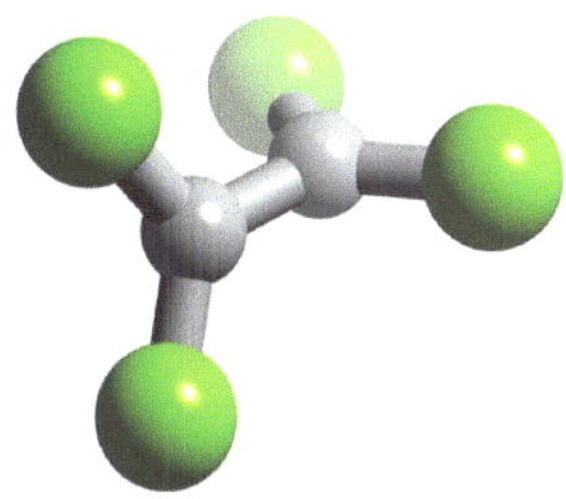

**Figure 11.57: The staggered structure of $B_2Cl_4$.**

*Oxidation state +1*

Boron forms a number of halides with cluster structures, $B_nCl_n$ where n = 4 (Figure 11.58), 8, 9, 10, 11, and 12. Each compound is made by the decomposition of $B_2Cl_4$. For gallium, none of the monohalides are stable at room temperature, but GaCl and GaBr have been produced in the gas form from the reaction of HX and molten gallium. The stability of thallium(I) as compared to thallium(III), results in the monohalides, TlCl, TlBr, and TlI being stable. Each compound is insoluble in water, and photosensitive.

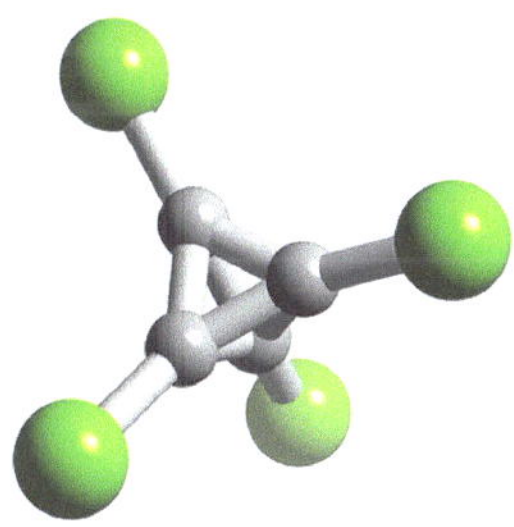

**Figure 11.58: Structure of $B_4Cl_4$.**

*Intermediate halides*

The dihalides ($MX_2$) of gallium, indium, and thallium do not actually contain the metal in the +2 oxidation state. Instead they are actually mixed valence compound, i.e., $M^+[MX_4]^-$. The dihalides of gallium are unstable in the presence of water disproportionating to gallium metal and gallium(III) entities. They are soluble in aromatic solvents, where arene complexes have been isolated and the arene is $\eta^6$-coordinated to the $Ga^+$ ion. $InBr_2$ and $InI_2$ are greenish and yellow crystalline solids, respectively, which are formulated $In(I)[In(III)X_4]$. $TlCl_2$ and $TlBr_2$ both are of similar formulations.

$Ga_2X_3$ (X = Br, I) and $In_2Br_3$ are formulated $M(I)_2[M(II)_2X_6]$. Both anions contain a M-M bond where the metal has a formal oxidation state of +2. The $Ga_2Br_6^{2-}$ anion is eclipsed like the $In_2Br_6^{2-}$ anion, whereas the $Ga_2I_6^{2-}$ anion is isostructural with $Si_2Cl_6$ with a staggered conformation. In Cl is colorless and is formulated $In(I)_3[In(III)Cl_6]$.

$Ga_3Cl_7$ contains the $Ga_2Cl_7^-$ ion, which has a structure similar to the dichromate, $Cr_2O_7^{2-}$, ion with two tetrahedrally coordinated gallium atoms sharing a corner (Figure 11.59). The compound can be formulated as gallium(I) heptachlorodigallate(III), $Ga(I)[Ga(III)_2Cl_7]$.

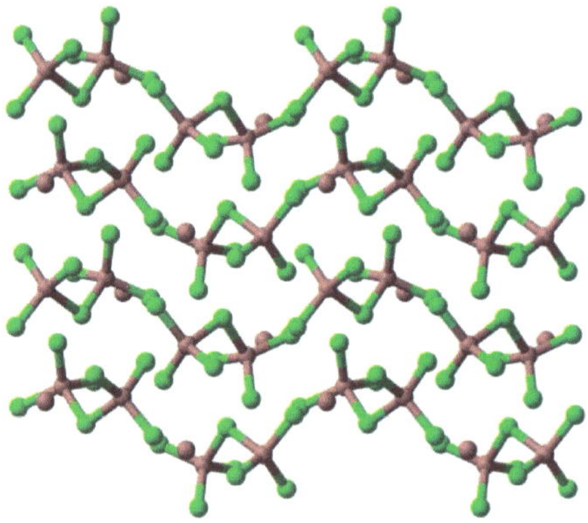

**Figure 11.59: The crystal structure of $Ga_3Cl_7$, better described as gallium(I) heptachlorodigallate(III), $Ga(I)[Ga(III)_2Cl_7]$. Grey spheres represent the gallium, while green spheres represent the chlorine.**

$In_4Br_7$ is light sensitive (like TlCl and TlBr) decaying to $InBr_2$ and In metal. It is a mixed salt containing the $InBr_4^-$ and $InBr_6^{3-}$ anions balanced by $In^+$ cations. It is formulated In(I) $[In(III)Br_4]_2[In(III)Br_6]$. $In_5Br_7$ is a pale yellow solid formulated as $In(I)_3[In(II)_2Br_6]Br$. The $In(II)_2Br_6^{2-}$ anion has an eclipsed ethane like structure with an In-In bond length of 2.70 Å. $In_5Cl_9$ is formulated $In(I)_3[In(III)_2Cl_9]$, with the $In_2Cl_9^{2-}$ anion having two 6 coordinate indium atoms with 3 bridging chlorine atoms, face sharing bioctahedra. Finally, $In_7Cl_9$ and $In_7Br_9$ have a structure formulated as $InX_6[In(III)X_6]X_3$.

# Chapter 12: Group 14

## The elements

The group was once also known as the *tetrels* (from Greek *tetra* meaning four), stemming from the earlier naming convention of this group as Group IVA. Table 12.1 lists the derivation of the names of the Group 14 elements.

| Element | Sym-bol | Name |
|---|---|---|
| Carbon | C | From the Latin *carbo* meaning *coal* |
| Silicon | Si | From the Latin *silicis* meaning *flints* |
| Germa-nium | Ge | From the Latin *Germania* for Germany |
| Tin | Sn | From the Anglo-Saxon and from the Latin *stannum* meaning *melts easily* |
| Lead | Pb | From the Anglo-Saxon, and from the Latin *plumbum* meaning *soft metal* |

**Table 12.1: Derivation of the names of each of the Group 14 elements.**

## Discovery

### *Carbon*

Carbon was known in prehistory in the form of soot; while charcoal was made in Roman times (by heating wood while exclude air) and diamonds were known as early as 2500 BC in China. In 1772, Antoine Lavoisier (Figure 7.4) showed that diamonds were a form of carbon, when he burned samples of carbon and diamond and showed that both formed the same amount of carbon dioxide per gram of material. Carl Scheele (Figure 9.5) showed that graphite was a form of carbon rather a form of lead.

A new allotrope of carbon, fullerene, was discovered in 1985 by Robert Curl, Harry Kroto, and Richard Smalley (Figure 12.1) who subsequently shared the Nobel Prize in Chemistry in 1996. Fullerenes have been revealed to include nanostructured forms such as Buckyballs and nanotubes. The renewed interest in new forms lead to the discovery of further exotic allotropes, including glassy carbon, and the realization that amorphous carbon is not amorphous.

**Figure 12.1: Rice University chemists Richard E. Smalley (1943 - 2005) and Robert F. Curl (1933 - ).**

*Silicon*

Silicon was first identified by Antoine Lavoisier (Figure 7.4) in 1787 as a component of flints and was later mistaken by Humphry Davy (Figure 8.5) for a compound rather than an element. In 1824, Berzelius (Figure 8.3) prepared amorphous silicon by the reaction of potassium with silicon tetra fluoride,

$$SiCl_4 + 4\,K \rightarrow Si + 4\,KCl$$

*Germanium*

In 1869 Dmitri Mendeleev (Figure 11.7) predicted the existence of several unknown elements, including ekasilicon (Es) between silicon and tin.

In 1885 a new mineral (named argyrodite because of its high silver content) was found in a mine near Freiberg, Saxony. Clemens Winkler (Figure 12.2) isolated Mendeleev's missing element. He originally was going to name neptunium because like this element, because like ekasilicon, the planet Neptune had been preceded by mathematical prediction of its existence. However, the name neptunium had already been given to an element and so Winkler named the new metal germanium in honor of his fatherland.

**Figure 12.2: German chemist Clemens Alexander Winkler (1838 - 1904).**

Winkler was able to isolate sufficient germanium from 500 kg of ore to determine a number of properties, including an atomic weight of 72.32 g/mol by analyzing pure germanium tetrachloride ($GeCl_4$). Winkler prepared several new compounds of germanium, including the fluorides, chlorides, sulfides, germanium dioxide, and tetraethylgermane ($Ge(C_2H_5)_4$). The physical data from these compounds, corresponded with Mendeleev's predictions (Table 12.2).

| Property | Ekasilicon | Germanium |
|---|---|---|
| Atomic mass | 72 | 72.59 |
| Density (g/cm$^3$) | 5.5 | 5.35 |
| Melting point (°C) | High | 947 |
| Color | Gray | Gray |
| Oxide type | Refractory dioxide | Refractory dioxide |
| Oxide density (g/cm$^3$) | 4.7 | 4.7 |
| Oxide activity | Feebly basic | Feebly basic |
| Chloride boiling point (°C) | Under 100 | 86 ($GeCl_4$) |
| Chloride density (g/cm$^3$) | 1.9 | 1.9 |

**Table 12.2: Properties predicted for ekasilicon compared those determined for germanium.**

*Tin*

Tin is one of the earliest metals known. When the addition of about 5% tin to molten copper produced an alloy (bronze) that was easier to work and much harder than copper, as such it revolutionized civilization. The widespread use of bronze to make tools and weapons became part of what archaeologists call the Bronze Age. The Bronze Age arrived in Egypt, Mesopotamia and the Indus Valley culture by around 3000 BC.

*Lead*

Lead has been commonly used for thousands of years because of its ease of extraction, and its ease of smelting. Lead beads dating back to 6400 BC have been found in Çatalhöyük in modern-day Turkey, while lead was used during the Bronze Age.

## Abundance

Carbon and silicon are amongst the most abundant elements (Table 12.3). Silicon is the second most abundant element (after oxygen) in the Earth's crust, making up 28% of the crust. Carbon is the fourth most abundant chemical element in the universe after hydrogen, helium, and oxygen. In combination with oxygen in carbon dioxide, carbon is found in the Earth's atmosphere (in quantities of approximately 810 gigatons) and dissolved in all water bodies (approximately 36,000 gigatons). Around 1,900 gigatons are present in the biosphere. Hydrocarbons (such as coal, petroleum, and natural gas) contain carbon amounts to around 900 gigatons. Natural diamonds occur in the rock kimberlite, found in ancient volcanic "necks," or "pipes". Most diamond deposits are in Africa but there are also deposits in Canada, the Russian Arctic, Brazil, and Australia.

| Element | Terrestrial abundance (ppm) |
|---|---|
| C | 480 (Earth's crust), 28 (sea water), 350 (atmosphere $CO_2$), 1.6 (atmosphere, $CH_4$), 0.25 (atmosphere, CO) |
| Si | 28,000 (Earth's crust), 2 (sea water) |
| Ge | 2 (Earth's crust), 1 (soil), $5\times10^{-7}$ (sea water) |
| Sn | 2 (Earth's crust), 1 (soil), $4\times10^{-6}$ (sea water) |
| Pb | 14 (Earth's crust), 23 (soil), $2\times10^{-6}$ (sea water) |

**Table 12.3: Abundance of Group 14 elements.**

## Isotopes

Table 12.4 summarizes the naturally occurring isotopes of the Group 14 elements.

| Isotope | Natural abundance (%) |
|---|---|
| Carbon-12 | 98.9 |
| Carbon-13 | 1.1 |
| Carbon-14 | trace |
| Silicon-28 | 92.23 |
| Silicon-29 | 4.67 |
| Silicon -30 | 3.1 |
| Germanium-70 | 21.23 |
| Germanium-72 | 27.66 |
| Germanium-73 | 7.73 |
| Germanium-74 | 35.94 |
| Germanium-76 | 7.44 |
| Tin-112 | 0.97 |
| Tin-114 | 0.66 |
| Tin-115 | 0.34 |
| Tin-116 | 14.54 |
| Tin-117 | 7.68 |
| Tin-118 | 24.22 |
| Tin-119 | 8.59 |
| Tin-120 | 32.58 |
| Tin-122 | 4.63 |
| Tin-124 | 5.79 |
| Lead-204 | 1.4 |
| Lead-24.1 | 24.1 |
| Lead-207 | 22.1 |
| Lead-208 | 52.4 |

**Table 12.4: Abundance of the major isotopes of the Group 14 elements.**

Although radioactive, carbon-14 is formed in upper layers of the troposphere and the stratosphere, at altitudes of 9 - 15 km. Thermal neutrons produced by cosmic rays collide with the nuclei of nitrogen-14, forming carbon-14 and a proton. Because of its relatively short half-life of 5730 years, carbon-14 is absent in ancient rocks, but is incorporated in living organisms.

*Carbon dating*

Carbon dating is a process whereby the age of a material that contains carbon can be determined by comparing the decay rate of that material with that of living material.

Carbon-14 has a half-life ($t_{1/2}$) of $5.73 \times 10^3$ years for its decay to nitrogen-14 by the loss of a β particle,

$$^{14}_{6}C \rightarrow\ ^{14}_{7}N +\ ^{0}_{-1}e$$

The rate of radioactive decay can be expressed as a rate constant (k):

$$k = \frac{\ln[2]}{t_{1/2}} = \frac{0.693}{t_{1/2}}$$

For carbon-14, using

$$k = \frac{0.693}{5.73 \times 10^3} = 1.21 \times 10^{-4}\ \text{year}^{-1}$$

In 1947 samples of the Dead Sea Scrolls were analyzed by carbon dating. It was found that the carbon-14 present had an activity of d/min.g (where d = disintegration); by contrast in living material the activity is 14 d/min.g. Thus,

$$\ln \frac{14}{11} = (1.21 \times 10^{-4})\, t$$

$$t = \frac{\ln 1.272}{1.21 \times 10^{-4}} = 2.0 \times 10^3 \text{ years}$$

From the measurement performed in 1947 the Dead Sea Scrolls were determined to be 2000 years old giving them a date of 53 BC and confirming their authenticity. This discovery is in contrast to the carbon dating results for the Turin Shroud that was supposed to have wrapped Jesus' body. Carbon dating has shown that the cloth was made between 1260 and 1390 AD. Thus, the Turin Shroud is clearly a fake having been made over a thousand years after its supposed manufacture.

# Industrial production

Due to the industrial importance of carbon and silicon, as well as the wide range of fullerene materials, the production of these elements is discussed elsewhere. However, the diamond supply chain is controlled by a limited number of commercial concerns, the largest of which is DeBeers in London (Figure 12.3). Diamonds make up only a very small fraction of ore bearing rock. The ore is crushed and subsequently the particles are sorted by density. Diamonds are located in the diamond-rich fraction by X-ray fluorescence, after which the final sorting steps are done by hand.

**Figure 12.3: The Charterhouse Street entrance of the offices of De Beers in London, are the venue for the 10 sights that are held during the year. Sightholders, both manufacturers (of polished diamonds) and dealers (in rough diamonds), meet in London, once every 5 weeks, 10 times a year, to inspect their diamond allocations.**

Following oxygen (46%), silicon is the most abundant element in the earth's crust (28%). However, silicon does not occur in its elemental form, but as its oxide ($SiO_2$) or as silicates. Sand, quartz, amethyst, agate, flint, and opal are some of the forms in which the oxide appears. Granite, hornblende, asbestos, feldspar, clay and mica, etc. are a few of the numerous silicate minerals. With such boundless supplies of the raw material, the costs associated with the production of bulk silicon is not one of abstraction and conversion of the oxide(s), but of purification of the crude elemental silicon. While 98% elemental silicon, known as metallurgical-grade silicon (MGS), is readily produced on a large scale, the requirements of extreme purity for electronic device fabrication require additional purification steps in order to produce electronic-grade silicon (EGS).

Germanium ore concentrates are mostly sulfidic, e.g., as an impurity in zinc blende. They are converted to the oxides by heating under air (roasting),

$$GeS_2 + 3\,O_2 \rightarrow GeO_2 + 2\,SO_2$$

Part of the germanium ends up in the dust produced during this process, while the rest is converted to germanates which are leached together with the zinc by sulfuric acid. After neutralization the germanium and other metals are precipitated (leaving the $Zn^{2+}$ in solution). Germanium dioxide is obtained as a precipitate and converted with chlorine gas or hydrochloric acid to germanium tetrachloride, which has a low boiling point and can be purified by distillation.

$$GeO_2 + 4\,HCl \rightarrow GeCl_4 + 2\,H_2O$$

$$GeO_2 + 2\,Cl_2 \rightarrow GeCl_4 + O_2$$

The germanium tetrachloride is hydrolyzed to the give pure oxide ($GeO_2$), which is then converted to germanium glass for the semiconductor industry,

$$GeO_2 + 4\,H_2 \rightarrow Ge + 2\,H_2O$$

Germanium used in steel production and other applications that do not require the high purity is produced by reduction with carbon,

$$GeO_2 + C \rightarrow Ge + CO_2$$

Tin is mined and subsequently smelted, and its production has changed little. In contrast, lead-rich ores contain less than 10% lead, but ores containing as little as 3% lead can be economically exploited. Ores are crushed and concentrated to 70%. Sulfide ores are roasted, producing lead oxide and a mixture of sulfates and silicates of lead. Lead oxide from the roasting process is reduced in a coke- red blast furnace. This converts most of the lead to its metallic form. Metallic lead that results from the roasting and blast furnace processes still contains significant contaminants of arsenic, antimony, bismuth, zinc, copper, silver, and gold. The melt is treated in a reverberatory furnace (Figure 12.4) with air, steam, and sulfur, which oxidizes the contaminants except silver, gold, and bismuth.

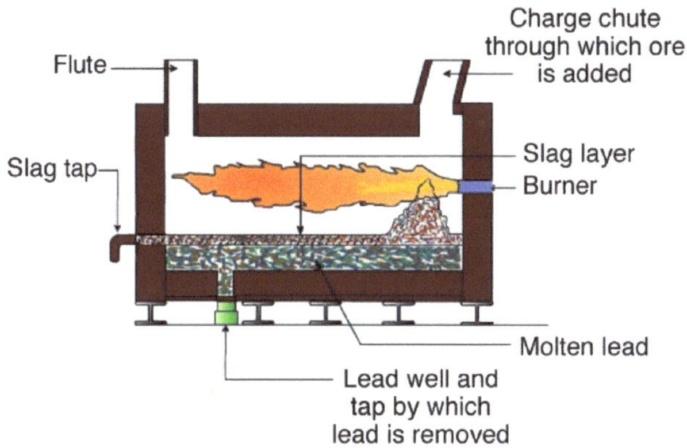


**Figure 12.4: Schematic of a reverberatory furnace.**

## Physical properties

Table 12.5 provides a summary of the physical properties of the Group 14 elements.

| Element | Melting point (°C) | Boiling point (°C) | Density (g/cm$^3$) |
|---|---|---|---|
| C | 642 (sublimes) | | 2.267 (graphite), 3.515 (diamond), 1.8 - 2.1 (amorphous) |
| Si | 1414 | 3265 | 2.3290 |
| Ge | 938 | 2833 | 5.323 |
| Sn | 232 | 2602 | 7.365 (white), 5.769 (gray) |
| Pb | 327 | 1749 | 11.34 |

**Table 12.5: Selected physical properties of the Group 14 elements.**

*Cubic structure*

The elements carbon through tin (in its α form) all exist in diamond cubic structure (Figure 12.5a), while lead crystallizes in a cubic close packed structure (Figure 12.5b). As expected, the lattice parameter ($a$) increases with increased atomic radius (Table 12.6). The switch from diamond cubic to close packed cubic may be rationalized by the relative atomic sizes. The diamond cubic structure comprises of two interpenetrating cubic close packed lattices. As the atomic size increases large interstitial vacancies would result, resulting in an unfavorable low-density structure.

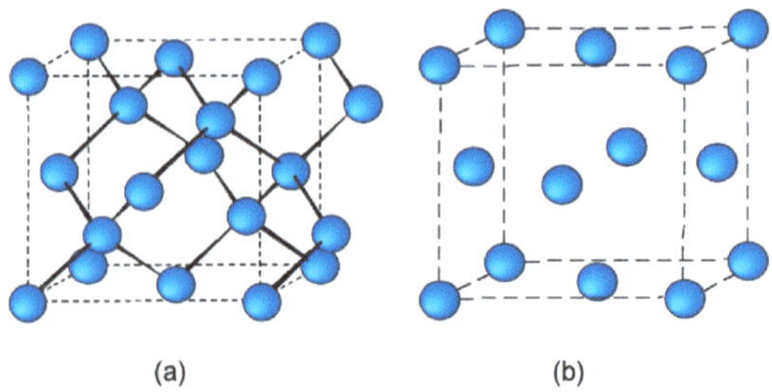

Figure 12.5: Unit cell structure of (a) the diamond cubic lattice showing the two-interpenetrating face-centered cubic lattices, and (b) the cell of cubic close packed structure for comparison.

| Element | Structure | *a* (Å) | Atomic radius (Å) |
|---|---|---|---|
| C | diamond cubic | 3.566 | 0.70 |
| Si | diamond cubic | 5.431 | 1.10 |
| Ge | diamond cubic | 5.657 | 1.25 |
| α-Sn (gray) | diamond cubic | 6.489 | 1.45 |
| Pb | cubic close packed | 4.951 | 1.80 |

**Table 12.6: Lattice parameter and crystal density for Group 14 elements.**

# Carbon

## Carbon black: from copying to communication

It is a commonly held fallacy that James Watt (Figure 12.6) was the inventor of the steam engine. This actually honor belongs to Thomas Newcomen (Figure 12.7). Watt's actual achievement was to improve Newcomen's design of a steam pump. While Watt was working in the repair shop at the University of Glasgow, he was fixing a Newcomen pump when he developed several key improvements on the original design.

The reason for Watt's success was that Britain was going through a major industrial boom and was in need of significant quantities of raw material including coal. Many of the coalmines, especially those in Devon were prone to flooding. Unfortunately, Newcomen's engine (which was actually a steam pump) could not pump the water out fast enough, whereas Watt's engine was powerful enough to drain the mines. As a result, Watt's career as a manufacturer was assured. However, Watt did not see the full potential of his invention, this was left to an employee of his, William Murdock (Figure 9.15), whose invention of the gearing to allow the steam engine to be used for powering machinery.

**Figure 12.6: Scottish inventor and mechanical engineer James Watt FRS (1736 - 1819).**

**Figure 12.7: English ironmonger and inventor of the steam engine Thomas Newcomen (1664 - 1729).**

One outcome that Watt found from his increased business was an increase in paperwork! While he was living in Redruth, Cornwall (Figure 9.13), close to where many of the mines were situated, he told a friend that he was having excessive difficulty in finding intelligent managing clerks. In 1780 Watt solved his paperwork problems by inventing the first method of making copies. This was the subject of a Patent entitled 'A new method of copying letters and writing expeditiously'. Watt's invention involved making ink out of gum Arabic and carbon black.

*Gum arabic*, also known as gum acacia, is a natural gum made of the hardened sap taken from two species of the acacia tree; *acacia senegal* and *acacia seyal*. Carbon black is a form of amorphous carbon that has a high surface area to volume ratio it is commonly produced by the incomplete combustion of heavy petroleum products such as coal tar.

Watt's ink would stay wet for 24 hours. Writing with the ink and then pressing the result against another piece of paper created a copy. Initially there was great resistance to the copy paper. In particular banks believed that this form of copying could result in forgeries. However, by the end of its first year on sale, Watt sold over 200 examples. The real commercial push came after Watt demonstrated his process to the House of Parliament. The resultant consternation amongst the Members of Parliament resulted in their having to be reminded that Parliament was still in session! By 1785 copies were in general use, however, Cyrus P. Dalkin made the biggest advance in 1823. By using a mixture of carbon black and hot parafin wax the back of a piece of paper was coated. The carbon black was transferred to another piece of paper underneath by the pressure of the pen. Dalkin had created carbon copies. Both Ralph Wedgwood in England and Pellegrino Turri in Italy had developed forms of carbon paper between 1806 and 1808, but it was Dalkin's version of carbon copy paper that found usage (Figure 12.8).

**Figure 12.8: Modern carbon paper.**

Initially there was almost no market for this new carbon copy paper, until in 1868 when American Lebbeus Rogers was talking part in a balloon ascent. Rogers was an owner of biscuit and greengrocer companies and was intrigued when a reporter for the Associated Press who was interviewing him used Dalkin's carbon paper. Rogers gave up his biscuit business and founded a firm to sell carbon paper. In 1873 he talked with E. Remington and Sons the typewriter manufacturer, and it was this application that created a success out of the carbon paper business.

The term 'CC' commonly used in e-mail programs today grew from the use of carbon paper and means *carbon copy*.

Interestingly, carbon black is still used today in modern photocopiers and laser printers. One of the attributes that allow its use is the ability for carbon black particles to become charged. This is related to another application of carbon black in early communications: its ability to conduct electric, and the changes that occur as a function of external pressure.

While Alexander Graham Bell was the first to be awarded a US patent for the electric telephone in March 1876, it was an invention by Thomas Edison (Figure 12.9) that provided significant improvements. Early telephones relied on a metal diaphragm that was attached to an electromagnet. Speaking into the metal diaphragm caused its vibration, which in turn vibrated the electromagnetic and hence created a current. Unfortunately, the current was very low and subsequent clarity was poor. Between 1877 and 1878 Edison investigated methods to improve the clarity of the signal. The key development was the carbon microphone (Figure 12.10).

**Figure 12.9: American inventor, scientist and businessman Thomas Alva Edison (1847 - 1931).**

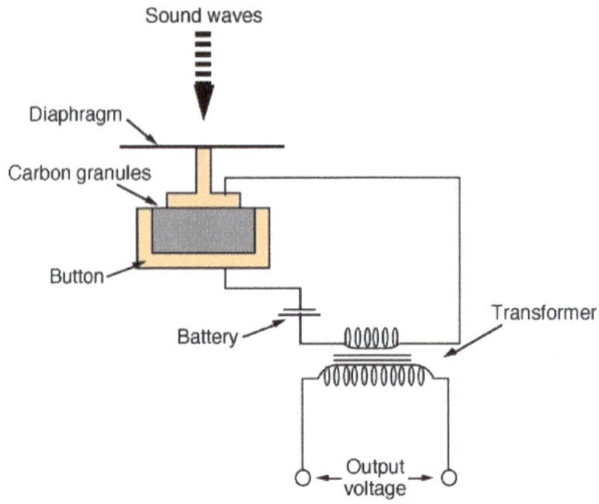


**Figure 12.10: A schematic of a simple carbon button microphone.**

The carbon microphone, also known as a carbon button microphone consisting of two metal plates separated by granules of carbon. One plate faces outward and acts as a diaphragm. When sound waves strike this plate, the pressure on the granules, changes, which in turn changes the electrical resistance between the plates. A direct current is passed from one plate to the other, and the changing resistance results in a changing current, which can be passed through a telephone system. The carbon microphone was used in all telephones until the 1980s.

**Figure 12.11: American chemist Edward Goodrich Acheson (1856 - 1931).**

It was one of Edison's researchers, Edward Acheson (Figure 12.11) whose discovery allowed for the extension of communication on a global scale. While trying to make artificial diamonds, Acheson began mixing clay and coke (carbon) at very high temperatures. In an electric furnace at high temperatures he found hexagonal crystals of silicon carbide (SiC) attached to the carbon electrode. He called this material carborundum. When, by mistake, he overheated the mixture to 4150 °C he found that the silicon evaporated (boiling point 3265 °C) leaving pure and highly crystalline carbon: graphite.

Graphite is a natural mineral normally associated with other minerals, although an enormous deposit of graphite was discovered in Cumbria, England, which the locals found very useful for marking sheep! Graphite was named by Abraham Werner (Figure 12.12) in 1789 from the Greek *graphein* meaning to draw/write due to its use in pencils. However, its use was limited due to the cost. The ability to manufacture high purity graphite resulted in its use in electrodes, dynamo brushes, and batteries. The most terrifying application of graphite was as a result of a material problem associated with the V-2 rocket built by Germany during the Second World War.

**Figure 12.12: German geologist Abraham Gottlob Werner (1749 - 1817).**

The V-2 rockets or Vergeltungswaffen-2 (vengeance weapon-2) was 47 feet long and reached 3,600 mph with an altitude of 300,000 feet (Figure 12.13). In order to control the direction of flight the V-2 was guided by four external rudders on the tail fins, and four internal vanes at the exit of the jet (Figure 12.40). These vanes were made of graphite, it being the only material that would survive the extreme temperatures.

**Figure 12.13: A V-2 rocket at the Peenemünde Museum.**

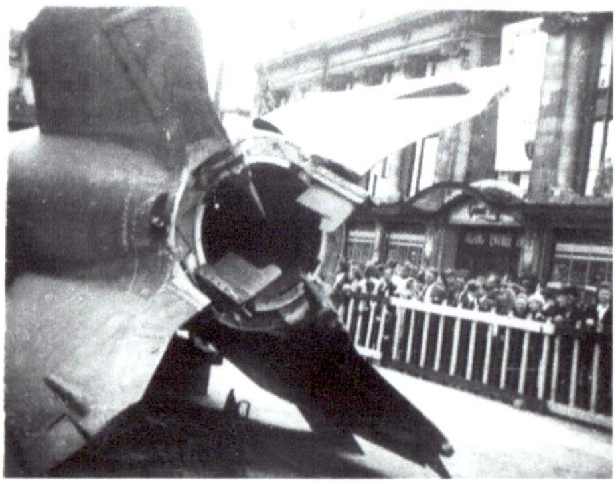

**Figure 12.14: View of the V-2 rocket showing the graphite vanes for directional control.**

Of course it was as a result of the same German scientists who worked on the V-2, working for NASA, that allowed rockets of sufficient power to escape the Earth's gravitational pull and send man to the moon, and position many of the satellites that are now vital for global communication.

# Carbon nanomaterials

*Fullerenes and nanotubes*

Although nanomaterials had been known for many years prior to the report of $C_{60}$ the field of nanoscale science was undoubtedly founded upon this seminal discovery. Part of the reason for this explosion in nanochemistry is that while carbon materials range from well-defined nano-sized molecules (i.e., $C_{60}$) to tubes with lengths of hundreds of microns, they do not exhibit the instabilities of other nanomaterials as a result of the very high activation barriers to their structural rearrangement. Despite this range of carbon nanomaterials possible they exhibit common reaction chemistry: that of organic chemistry.

Carbon-60 ($C_{60}$), commonly known as the Buckyball or Buckminsterfullerene, has a spherical shape comprising of highly pyramidalized $sp^2$ carbon atoms. The $C_{60}$ variant is often compared to the typical soccer football, hence *buckyball*; however, confusingly, this term is commonly used for higher derivatives. Fullerenes are similar in sheet structure to graphite, but they contain pentagonal (or sometimes heptagonal) rings that prevent the sheet from being planar. The unusual structure of $C_{60}$ led to the introduction of a new class of molecules known as fullerenes, which now constitute the third allotrope of carbon. Fullerenes are commonly defined as any of a class of closed hollow aromatic carbon compounds that are made up of twelve pentagonal and differing numbers of hexagonal faces.

Higher order fullerenes include carbon nanotubes that can be described as fullerenes that have been stretched along a rotational axis to form a tube. In addition, there have also been reports of nanohorns and nano fibers, however, these may be considered as variations on the general theme. It should be noted that fullerenes and nanotubes have been shown to be in flames produced by hydrocarbon combustion. Unfortunately, these naturally occurring varieties can be highly irregular in size and quality, as well as being formed in mixtures, making them unsuitable for both research and industrial applications.

Carbon-60 is probably the most studied individual type of nanomaterial. The spherical shape of $C_{60}$ is constructed from twelve pentagons and twenty hexagons and resembles a soccer ball (Figure 12.15a). The next stable higher fullerene is $C_{70}$ (Figure 12.15b) that is shaped like a rugby or American football. The progression of higher fullerenes continues, where the structural relationship between each involves the addition of six membered rings. Mathematically (and chemically) two principles define the existence of a stable fullerene, i.e., Euler's theorem and isolated pentagon rule (IPR). Euler's theorem states that for the closure of each spherical network, n (n ≥ 2) hexagons and 12 pentagons are required, while the IPR says no two pentagons may be connected directly with each other as destabilization is caused by two adjacent pentagons.

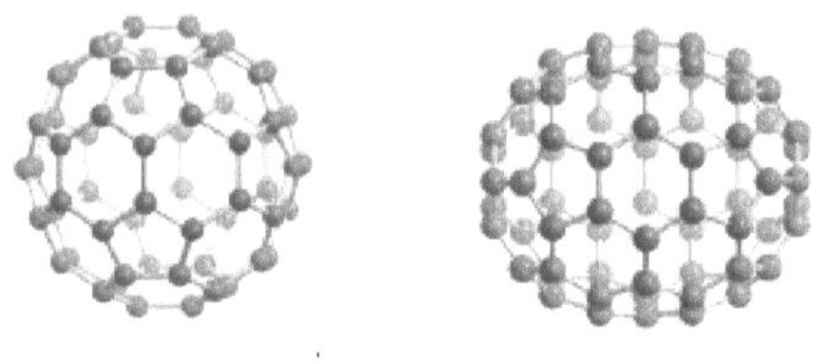


**Figure 12.15: Molecular structures of (a) $C_{60}$ and (b) $C_{70}$.**

Fullerenes are soluble in various common organic solvents. Due to their hydrophobic nature, fullerenes are most soluble in $CS_2$ ($C_{60}$ = 7.9 mg/mL) and toluene ($C_{60}$ = 2.8 mg/mL). Although fullerenes have a conjugated system, their aromaticity is distinctive from benzene that has all C-C bonds of equal lengths, in fullerenes two distinct classes of bonds exist. The shorter bonds are at the junctions of two hexagons ([6,6] bonds) and the longer bonds at the junctions of a hexagon and a pentagon ([5,6] bonds). This difference in bonding is responsible for some of the observed reactivity of fullerenes.

The first observation of fullerenes was in molecular beam experiments at Rice University. Subsequent studies demonstrated that it was relatively easy to produce grams of fullerenes. Although, purification remains a challenge and determines fullerene's commercial price. The first method of production of measurable quantities of fullerenes used laser vaporization of carbon in an inert atmosphere, but this produced microscopic amounts of fullerenes. Laboratory scales of fullerene are prepared by the vaporization of carbon rods in a helium atmosphere. Commercial production ordinarily employs a simple ac or dc arc. The fullerenes in the black soot collected are extracted in toluene and purified by liquid chromatography. The magenta $C_{60}$ comes off the column first, followed by the red $C_{70}$, and other higher fullerenes.

Endohedral fullerenes are fullerenes that have incorporated in their inner sphere atoms, ions or clusters. Endohedral fullerenes are generally divided into two groups: endohedral metallofullerenes and non-metal doped fullerenes. The first endohedral metallofullerenes was called La@$C_{60}$. The @ sign in the name reflects the notion of a small molecule trapped inside a shell. Doping fullerenes with metals takes place in-situ during the fullerene synthesis in an arc reactor or via laser evaporation. A wide range of metals have been encased inside a fullerene, i.e., Sc, Y, La, Ce, Ba, Sr, K, U, Zr, and Hf. Unfortunately, the synthesis of endohedral metallofullerenes is unspecific because in addition a high yield of unfilled fullerenes, compounds with different cage sizes are prepared (e.g., La@$C_{60}$ or La@$C_{82}$). A characteristic of endohedral metallofullerenes is that electrons will transfer from the metal atom to the fullerene cage and that the metal atom takes a position of-center in the cage. The size of the charge transfer is not always simple to determine, but it is usually between 2 and 3 units (e.g., $La_2$@$C_{80}$) but can be as high as 6 electrons (e.g., $La_3N$@$C_{80}$). These anionic fullerene cages are very stable molecules and do not have the reactivity associated with ordinary empty fullerenes. This lack of reactivity is utilized in a method to purify endohedral metallofullerenes from empty fullerenes.

Although fullerenes have a conjugated aromatic system all the carbons are quaternary (i.e., containing no hydrogen), which results in making many of the characteristic substitution reactions of planar aromatics impossible. Thus, two types of chemical transformations exist:

1. redox reactions
2. addition reactions.

Of these, addition reactions have the largest synthetic value. Another remarkable feature of fullerene addition chemistry is the thermodymics of the process. Since the $sp^2$ carbon atoms in a fullerene are paramidalized there is significant strain energy. For example, the strain energy in $C_{60}$ is ca 8 kcal/mol, which is 80% of its heat of formation. So, the relief of this strain energy leading to $sp^3$ hybridized C atoms is the major driving force for addition reactions (Figure 12.16). As a consequence, most additions to fullerenes are exothermic reactions.

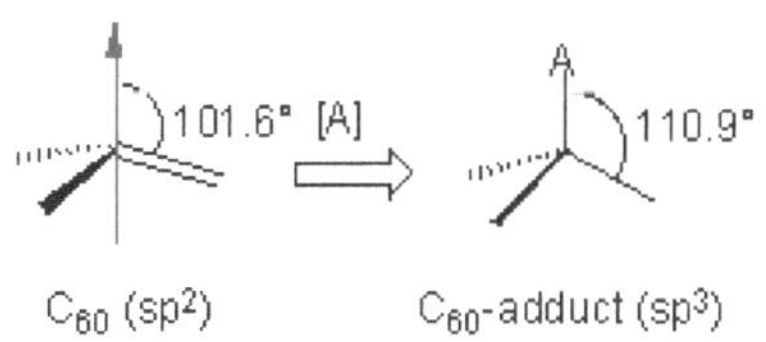


**Figure 12.16: Strain release after addition of reagent A to a pyramidalize carbon of $C_{60}$.**

Cyclic voltammetry (CV) studies show that $C_{60}$ can be reduced and oxidized reversibly up to 6 electrons with one-electron transfer processes. Fulleride anions can be generated by electrochemical method and then be used to synthesize covalent organofullerene derivatives. Alkali metals can chemically reduce fullerene in solution and solid state to form $M_xC_{60}$ (x = 3 - 6). $C_{60}$ can also be reduced by less electropositive metals like mercury to form $C_{60}^-$ and $C_{60}^{2-}$. In addition, salts can also be synthesized with organic molecules, for example $[TDAE^+][C_{60}^-]$ possesses interesting electronic and magnetic behavior.

Geometric and electronic analysis predicted that fullerene behaves live an electro-poor conjugated polyolefin. Indeed, $C_{60}$ and $C_{70}$ undergo a range of nucleophilic reactions with carbon, nitrogen, phosphorous and oxygen nucleophiles. $C_{60}$ reacts readily with organolithium and Grignard compounds to form alkyl, phenyl or alkanyl fullerenes. Possibly the most widely used additions to fullerene is the Bingel reaction (Figure 12.17), where a carbon nucleophile, generated by deprotonation of α-halo malonate esters or ketones, is added to form a cyclopropanation product.

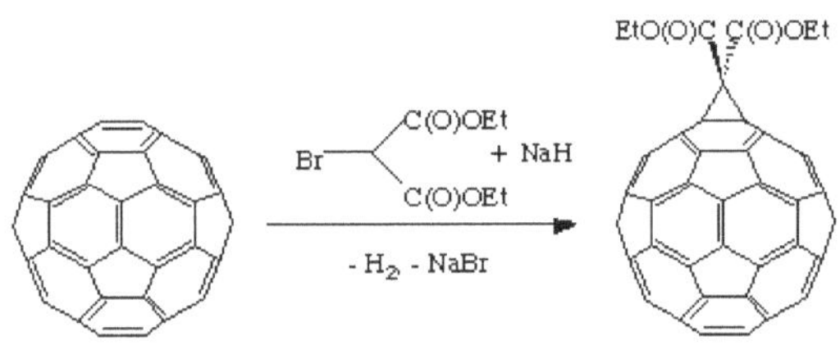


**Figure 12.17: Bingel reaction of $C_{60}$ with 2-bromoethylmalonate.**

Cycloaddition is another method to functionalize fullerenes, in particular because of its selectivity with the 6,6 bonds, limiting the possible isomers (Figure 12.18). The dienophilic feature of the [6,6] double bonds of $C_{60}$ enables the molecule to undergo various cycloaddition reactions in which the monoadducts can be generated in high yields.

The oxidation of fullerenes, such as $C_{60}$, has been of increasing interest with regard to applications in photoelectric devices, biological systems, and possible remediation of fullerenes. The oxidation of $C_{60}$ to $C_{60}O_n$ (n = 1, 2) may be accomplished by photooxidation, ozonolysis, and epoxidation. With each of these methods, there is a limit to the isolable oxygenated product, $C_{60}O_n$ with

$n < 3$. Highly oxygenated fullerenes, $C_{60}O_n$ with $3 \leq n \leq 9$, have been prepared by the catalytic oxidation of $C_{60}$ with $ReMeO_3/H_2O_2$.

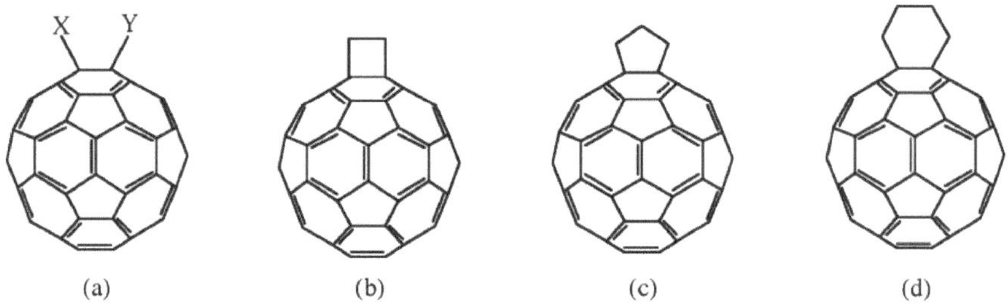

**Figure 12.18: Geometrical shapes built onto a [6,6] ring junction: a) open, b) four-membered ring, c) five-membered ring, and d) six-membered ring.**

*Carbon nanotubes*

A key breakthrough in carbon nanochemistry came in 1993 with the report of needle-like tubes made exclusively of carbon. This material became known as carbon nanotubes (CNTs). The first type discovered were multi walled tubes (MWCNTs) resembling many pipes nested within each other. Shortly afterwards single walled nanotubes (SWCNTs) were discovered. Single walled tubes resemble a single pipe that is potentially capped at each end. The properties of single walled and multi walled tubes are generally the same, although single walled tubes are believed to have superior mechanical strength and thermal and electrical conductivity; it is also more difficult to manufacture them.

Single walled carbon nanotubes (SWCNTs) consists of a graphene sheet rolled into a tube and capped by half a fullerene (Figure 12.19). The carbon atoms in a SWCNT are $sp^2$ hybridized. The structure of a nanotube is analogous to taking this graphene sheet and rolling it into a seamless cylinder. The different types of SWNTs are defined by their diameter and chirality. Most of the presently used single-wall carbon nanotubes have been synthesized by the pulsed laser vaporization method, however, increasingly SWCNTs are prepared by vapor liquid solid catalyzed growth.

SWCNTs have been shown to be stronger than steel as estimates for the Young's modulus approaches 1 Tpa. Their electrical conductance is comparable to copper with anticipate current densities of up to $10^{13}$ A/cm$^2$ and a resistivity as low as $0.34 \times 10^{-4}$ Ω.cm at room temperatures. Finally, they have a high thermal conductivity (3000 - 6000 W.m/K).

(a)

(b)

(c)

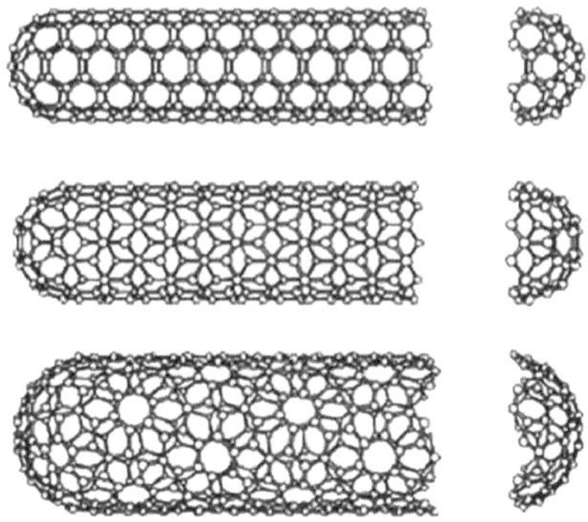

**Figure 12.25: Structure of single walled carbon nanotubes (SWCNTs) with (a) armchair, (b) zig-zag, and (c) chiral chirality.**

The electronic properties of a particular SWCNT structure are based on its chirality or twist in the structure of the tube which is defined by its *n,m* value. The values of *n* and *m* determine the chirality, or "twist" of the nanotube. The chirality in turn affects the conductance of the nanotube, its density, its lattice structure, and other properties. A SWCNT is considered metallic if the value *n-m* is divisible by three. Otherwise, the nanotube is semi-conducting.

Multi walled carbon nanotubes (MWCNTs) range from double walled NTs, through many-walled CNTs (Figure 12.20) to carbon nano fibers. Carbon nano fibers are the extreme of multi walled tubes (Figure 12.21) and they are thicker and longer than either SWCNTs or MWCNTs, having a cross-sectional of ca. 500 $Å^2$ and are between 10 to 100 μm in length.

A range of methodologies have been developed to produce nanotubes in sizeable quantities, including arc discharge, laser ablation, high pressure carbon monoxide (HiPco), and vapor liquid solid (VLS) growth.

The first observation of nanotubes was in the carbon soot formed during the arc discharge production of fullerenes. The high temperatures caused by the discharge caused the carbon contained in the negative electrode to sublime and the CNTs are deposited on the opposing electrode. Tubes produced by this method were initially multi walled tubes (MWCNTs). However, with the addition of cobalt to the vaporized carbon, it is possible to grow SWCNTs.

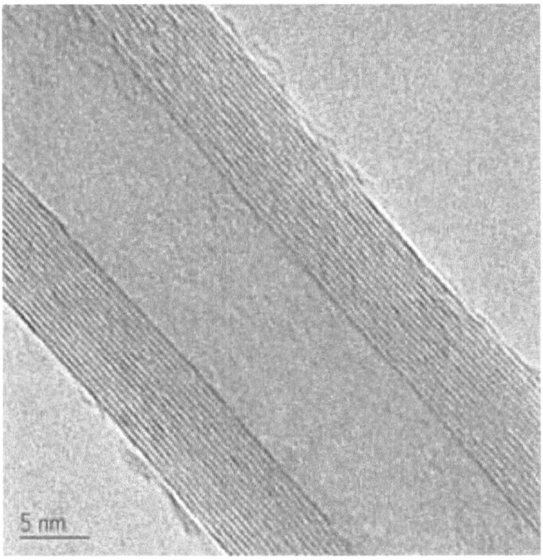


**Figure 12.20: TEM image of an individual multi walled carbon nanotube (MWNTs). Copyright of Nanotech Innovations.**

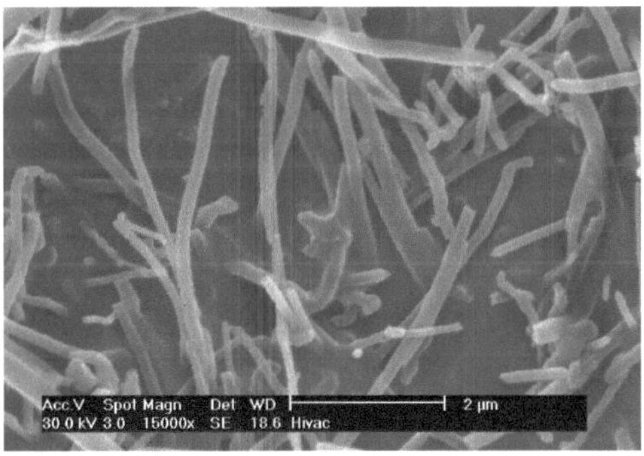


**Figure 12.21: SEM image of vapor grown carbon nano fibers.**

In order to overcome some of the difficulties of these high-energy processes, the chemical catalysis method was developed in which a hydrocarbon feedstock is used in combination with a metal catalyst. The catalyst is typically, but not limited to iron, colbalt, or iron/molybdenum, it is heated under reducing conditions in the presence of a suitable carbon feedstock, e.g., ethylene. This method can be used for both SWCNTs and MWCNTs; the formation of each is controlled by the identity of the catalyst and the reaction conditions. The use of CO as a feedstock, in place of a hydrocarbon, led to the development of the high-pressure carbon monoxide (HiPco) procedure for SWNT synthesis.

Initially developed for small-scale investigations of catalyst activity, vapor liquid solid (VLS) growth of nanotubes has been highly studied, and now shows promise for large-scale production of nanotubes. Recent approaches have involved the use of well-defined nanoparticle or molecular precursors and many different transition metals have been employed, but iron, nickel, and cobalt remain to be the focus of most research. The nanotubes grow at the sites of the metal catalyst; the carbon-containing gas is broken apart at the surface of the catalyst particle, and the carbon is transported to the edges of the particle, where it forms the nanotube. The length of the tube grown in surface supported catalyst VLS systems appears to be dependent on the orientation of the growing tube with the surface.

The limitation of using carbon nanotubes in any practical applications has been its solubility; for example, SWCNTs have little to no solubility in most solvent due to the aggregation of the tubes. Aggregation/roping of nanotubes occurs as a result of the high van der Waals binding energy of ca. 500 eV per mm of tube contact. The functionalization of nanotubes, i.e., the attachment of chemical functional groups provides the path to overcome these barriers. Functionalization can improve solubility as well as processability. It is important when discussing functionalization that a distinction is made between covalent and non-covalent functionalization.

Current methods for solubilizing nanotubes without covalent functionalization include highly aromatic solvents, super acids, polymers, or surfactants. Non-covalent functionalization is generally on the concept of supramolecular interactions between the SWCNT and some macromolecule as a result of various adsorption forces, such as van der Waals' and $\pi$-stacking interactions. The chemical speciation of the nanotube itself is not altered as a result of the interaction. In contrast, covalent functionalization relies on the chemical reaction at either the sidewall or end of the SWCNT.

Direct covalent sidewall functionalization is associated with a change of hybridization from $sp^2$ to $sp^3$ and a simultaneous loss of conjugation. An alternative approach to covalent functionalization involves the reaction of defects present (or generated) in the structure of the nanotube. Defect sites can be the open ends and holes in the sidewalls, and pentagon and heptagon irregularities in the hexagon graphene framework (often associated with bends in the tubes). All these functionalizations are exohedral derivatizations. Taking the hollow structure of nanotubes into consideration, endohedral functionalization of SWNTs is possible, i.e., the filling of the tubes with atoms or small molecules. It is important to note that covalent functionalization methods have one problem in common: extensive covalent functionalization modifies SWCNT properties by disrupting the continuous π system of SWCNTs. A variety of methods have been used to achieve the functionalization of nanotubes (Figure 12.22).

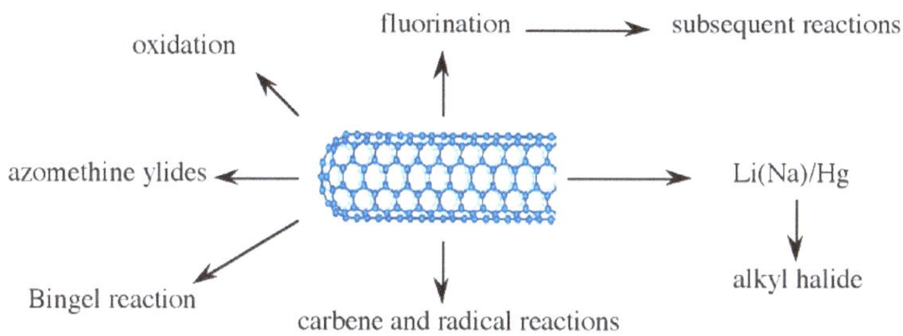


**Figure 12.22: Schematic description of various covalent functionalization strategies for SWCNTs.**

SWCNTs may be functionalized using 1,3 dipolar addition of azomethine ylides. The functionalized SWNTs are soluble in most common organic solvents. Organic functionalization through the use of alkyl halides, a radical pathway, on tubes treated with lithium in liquid ammonia offers a simple and flexible route to a range of functional groups. In this reaction, functionalization occurs on every 17 carbons. Most success has been found when the tubes are dodecylated. These tubes are soluble in chloroform, DMF, and THF. The addition of oxygen moieties to SWNT sidewalls can be achieved by treatment with acid or wet air oxidation, and ozonolysis. The direct epoxidation of SWNTs may be accomplished by the direct reaction with a peroxide reagent, or catalytically. Catalytic de-epoxidation (Figure 12.23) allows for the quantitative analysis of sidewall epoxide and led to the surprising result that previously assumed pure SWNTs actually contain ca. 1 oxygen per 250 carbon atoms.

One of the easiest functionalization routes, and a useful synthon for subsequent conversions, is the fluorination of SWCNTs, using elemental fluorine. Importantly, a C:F ratios of up to 2:1 can be achieved without disruption of the tubular structure. The fluorinated SWCNTs (F-SWCNTs) proved to be much more soluble than pristine SWCNTs in alcohols (1 mg/mL in iso-propanol), DMF and other selected organic solvents. Scanning tunneling microscopy (STM) revealed that the fluorine formed bands of approximately 20 nm, while calculations using DFT revealed 1,2 addition is more energetically preferable than 1,4 addition, which has been confirmed by solid state $^{13}C$ NMR. F-SWNTs make highly flexible synthons and subsequent elaboration has been performed with organolithium, Grignard reagents, and amines.

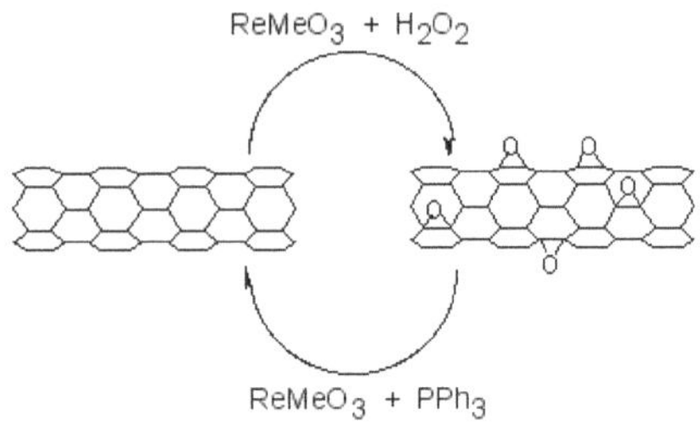


**Figure 12.23: Catalytic oxidation and de-epoxidation of SWCNTs.**

Functionalized nanotubes can be characterized by a variety of techniques. Changes in the Raman spectrum of a nanotube sample can indicate if functionalization has occurred. Pristine tubes exhibit two distinct bands. They are the radial breathing mode (230 $cm^{-1}$) and the tangential (G) mode (1590 $cm^{-1}$). When functionalized, a new band, called the disorder (D) band, appears at ca.1350 $cm^{-1}$. This band is attributed to $sp^3$-hybridized carbons in the tube. Unfortunately, while the presence of a significant D mode is consistent with sidewall functionalization and the relative intensity of D (disorder) mode versus the tangential G mode (1550 - 1600 $cm^{-1}$) is often used as a measure of the level of substitution. However, it has been shown that Raman is an unreliable method for determination of the extent of functionalization since the relative intensity of the D band is also a function of the substituents, distribution as well as concentration.

Recent studies suggest that solid state $^{13}C$ NMR are possibly the only definitive method of demonstrating covalent attachment of particular functional groups.

*Graphene*

Graphene is a one-atom-thick planar sheet of $sp^2$-bonded carbon atoms that are densely packed in a honeycomb crystal lattice (Figure 12.24). The name comes from graphite and alkene; graphite itself consists of many graphene sheets stacked together.

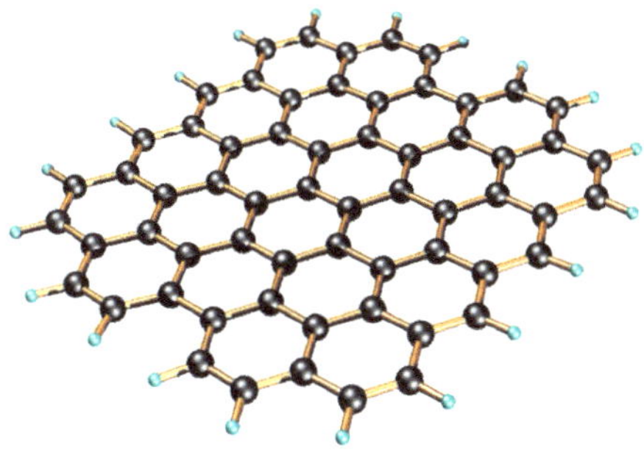

**Figure 12.24: Idealized structure of a single graphene sheet.**

Single-layer graphene nanosheets were first characterized in 2004, prepared by mechanical exfoliation (the scotch-tape method) of bulk graphite. Later graphene was produced by epitaxial chemical vapor deposition on silicon carbide and nickel substrates. Most recently, graphene nanoribbons (GNRs) have been prepared by the oxidative treatment of carbon nanotubes and by plasma etching of nanotubes embedded in polymer films.

Graphene has been reported to have a Young's modulus of 1 TPa and intrinsic strength of 130 GP; similar to single walled carbon nanotubes (SWNTs). The electronic properties of graphene also have some similarity with carbon nanotubes. Graphene is a zero-bandgap semiconductor. Electron mobility in graphene is extraordinarily high (15,000 $cm^2/V.s$ at room temperature) and ballistic electron transport is reported to be on length scales comparable to that of SWNTs.

The aforementioned methods of graphene preparation are effective; however, they are impractical for large-scale manufacturing. The most plentiful and inexpensive source of graphene is bulk graphite. Chemical methods for exfoliation of graphene from graphite provide the most realistic and scalable approach to graphene materials. Graphene layers are held together in graphite by enormous van der Waals forces. Overcoming these forces is the major obstacle to graphite exfoliation. To date, chemical efforts at graphite exfoliation have been focused primarily on intercalation, chemical derivatization, thermal expansion, oxidation-reduction, the use of surfactants, or some combination of these.

Probably the most common route to graphene involves the production of graphite oxide (GO) by extremely harsh oxidation chemistry. The structure of GO has been the subject of numerous studies; it is known to contain epoxide functional groups along the basal plane of sheets as well as hydroxyl and carboxyl moieties along the edges (Figure 12.25). In contrast to other methods for the synthesis of GO, the *m*-peroxybenzoic acid (m-CPBA) oxidation of microcrystalline synthetic graphite at room temperature yields graphite epoxide in high yield, without significant additional defects.

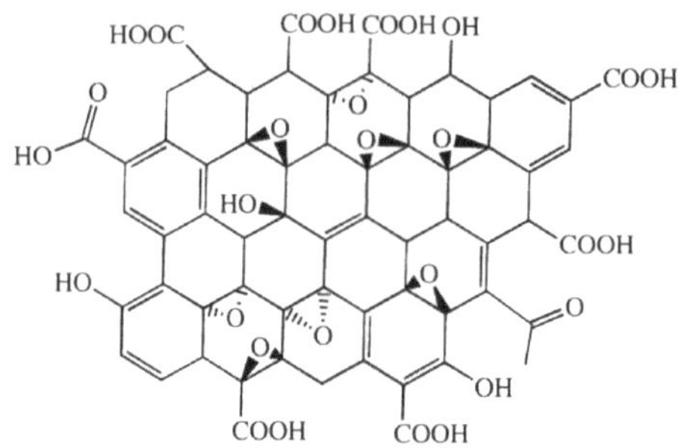

**Figure 12.25: Idealized structure proposed for graphene oxide (GO). Adapted from C. E. Hamilton, PhD Thesis, Rice University (2009).**

As graphite oxide is electrically insulating, it must be converted by chemical reduction to restore the electronic properties of graphene. Chemically converted graphene (CCG) is typically reduced by hydrazine or borohydride. The properties of CCG can never fully match those of graphene for two reasons:

1. Oxidation to GO introduces defects.
2. Chemical reduction does not fully restore the graphitic structure.

As would be expected, CCG is prone to aggregation unless stabilized. Graphene materials produced from pristine graphite avoid harsh oxidation to GO and subsequent (incomplete) reduction; thus, materials produced are potentially much better suited to electronics applications.

A catalytic approach to the removal of epoxides from fullerenes and SWNTs has been applied to graphene epoxide and GO. Treatment of oxidized graphenes with methyltrioxorhenium ($MeReO_3$, MTO) in the presence of $PPh_3$ results in the oxygen transfer, to form $O{=}PPh_3$ and allow for quantification of the C:O ratio.

An alternate approach to producing graphene materials involves the use of pristine graphite as starting material. The fundamental value of such an approach lies in its avoidance of oxidation to GO and subsequent (incomplete) reduction, thereby preserving the desirable electronic properties of graphene. It has been reported that N,N-dimethylformamide (DMF) dispersions of graphene are possible, but no detailed characterization of the dispersions were reported. Recently, dispersions of graphene have been reported in *ortho*-dichlorobenzene (ODCB) using a wide range of graphite sources. The choice of ODCB for graphite exfoliation was based on several criteria:

- ODCB is a common reaction solvent for fullerenes and is known to form stable SWNT dispersions.
- ODCB is a convenient high-boiling aromatic and is compatible with a variety of reaction chemistries.
- ODCB, being aromatic, is able to interact with graphene via π-π stacking. It has been suggested that good solvents for graphite exfoliation should have surface tension values of 40 50 mJ/m$^2$. ODCB has a surface tension of 36.6 mJ/m$^2$, close to the proposed range.

Graphite is readily exfoliated in ODCB with homogenization and sonication. Three starting materials were successfully dispersed: microcrystalline synthetic, thermally expanded, and highly ordered pyrolytic graphite (HOPG). Dispersions of microcrystalline synthetic graphite have a concentration of 0.03 mg/mL, determined gravimetrically. Dispersions from expanded graphite and HOPG are less concentrated (0.02 mg/mL).

High resolution transmission electron microscopy (HRTEM) shows mostly few-layer graphene ($n < 5$) with single layers and small flakes stacked on top (Figure 12.26). Large graphitic domains are visible; this is further supported by selected area electron diffraction (SAED) and fast Fourier transform (FFT) in selected areas. Atomic force microscope (AFM) images of dispersions sprayed onto silicon substrates shows extremely thin flakes with nearly all below 10 nm. Average height is 7 - 10 nm. The thinnest are less than 1 nm, graphene monolayers. Lateral dimensions of nanosheets range from 100 500 nm.

As-deposited films cast from ODCB graphene show poor electrical conductivity, however, after vacuum annealing at 400 °C for 12 hours the films improve vastly, having sheet resistances on the order of 60 Ω/sq. By comparison, graphene epitaxially grown on Ni has a reported sheet resistance of 280 Ω/sq.

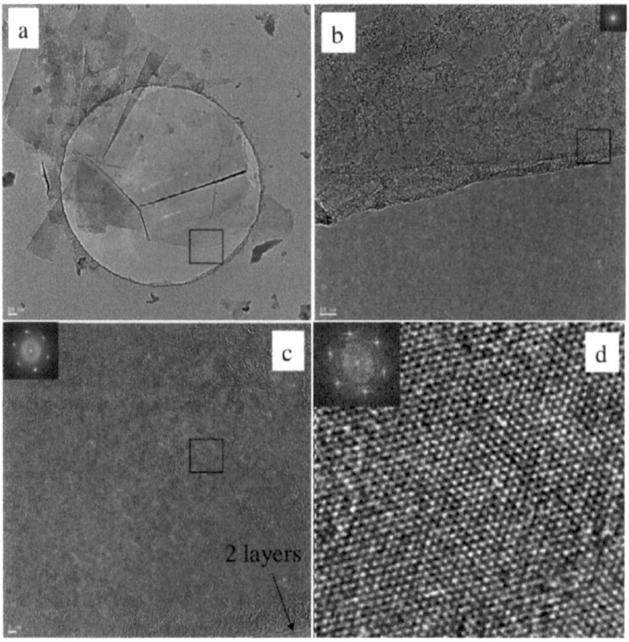


**Figure 12.26: TEM images of single layer graphene from HOPG dispersion. (a) monolayer and few layers of graphene stacked with smaller flakes; (b) selected edge region from (a), (c) selected area from (b) with FFT inset, (d) HRTEM of boxed region in (c) showing lattice fringes with FFT inset. Adapted from C. E. Hamilton, PhD Thesis, Rice University (2009).**

The chemical derivatization of graphene and GO is still relatively unexplored compared to fullerenes and CNTs. Some methods previously demonstrated for SWNTs have been adapted to GO or graphene. GO carboxylic acid groups have been converted into acyl chlorides followed by amidation with long-chain amines.

Additionally, the coupling of primary amines and amino acids via nucleophilic attack of GO epoxide groups has been reported. Yet another route coupled isocyanates to carboxylic acid groups of GO. Functionalization of partially reduced GO by aryldiazonium salts has also been demonstrated. The Billups reaction has been performed on the intercalation compound potassium graphite ($C_8K$), as well as graphite fluoride, and most recently GO. Graphene alkylation has been accomplished by treating graphite fluoride with alkyllithium reagents.

ODCB dispersions of graphene may be readily converted to covalently functionalize graphene. Thermal decomposition of benzoyl peroxide is used to initiate radical addition of alkyl iodides to graphene in ODCB dispersions.

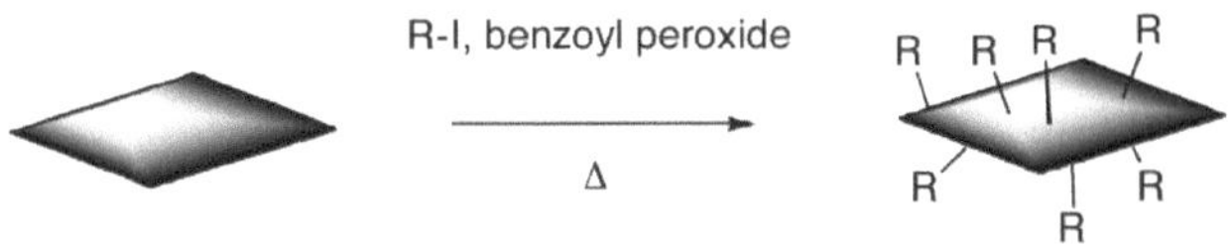


Additionally, functionalized graphene with nitrenes generated by thermal decomposition of aryl azides

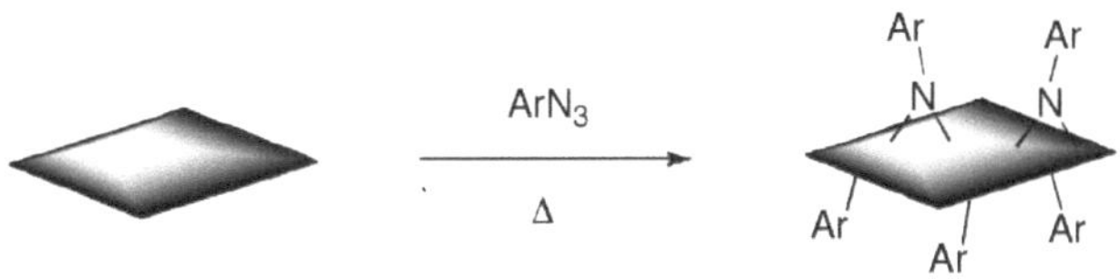


## Nitrogen compounds of carbon

There are a myriad of organic compounds containing carbon-nitrogen bonds, including: amines, imines, and nitriles. However, here we are concerned with the simplest carbon-nitrogen compounds.

### Cyanogen

Cyanogen, $(CN)_2$, may be considered the smallest molecular fragment containing carbon and nitrogen (Figure 12.27a). The reaction chemistry of cyanogen is related to that of the halogens, i.e., $F_2$, $Cl_2$, etc. Consequently, cyanogen is called a pseudo halogen.

$$N{\equiv}C{-}C{\equiv}N \qquad N{\equiv}C{-}C{\equiv}C{-}C{\equiv}N$$

(a) (b)

**Figure 12.27: The structures of (a) cyanogen, $(CN)_2$ and (b) dicyanoacetylene (carbon subnitride, $C_4N_2$).**

As shown in Table 12.7 the bonding in cyanogen is consistent with localization of the π-bonding between carbon and nitrogen given the similarity of the C-N bond distance in cyanogens and acetonitrile. However, there is clearly some π-delocalization associated with the C-C distance given it shortening as compared to ethane.

| Compound | Formula | C-C bond distance (Å) | C-N bond distance (Å) |
|---|---|---|---|
| Cyanogen | $(CN)_2$ | 1.393 | 1.163 |
| Hydrogen cyanide | HCN | - | 1.154 |
| Acetonitrile | $CH_3CN$ | 1.46 | 1.16 |
| Ethane | $C_2H6$ | 1.535 | - |
| Ethylene | $C_2H4$ | 1.339 | - |

**Table 12.7: A comparison of the bond distances in selected carbon nitrogen compounds.**

Cyanogen is produced by the reaction of a mixture of the cyanide and chloride of mercury,

$$Hg(CN)_2 + HgCl_2 \rightarrow (CN)_2 + 2\,Hg + Cl_2$$

Alternatively, the decomposition of unstable copper(II) cyanide, formed from a copper(II) salts with a Group 1 cyanide, yields cyanogens,

$$CuSO_4 + 2\,KCN \rightarrow Cu(CN)_2 + K_2SO_4$$

$$2\,Cu(CN)_2 \rightarrow (CN)_2 + 2\,CuCN$$

Cyanogen is a flammable gas (melting point = -28 °C and boiling point = -21 °C) that produces the second hottest natural flame (after carbon subnitride, $C_4N_2$) with a temperature of over 4525 °C when burnt in oxygen. Heating cyanogen in the absence of oxygen results self-polymerizes,

$$(CN)_2 \xrightarrow{\Delta}$$

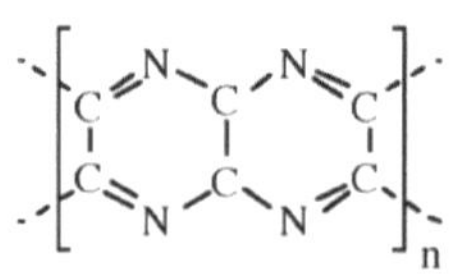

Hydrolysis of cyanogen results in addition across the carbon-nitrogen triple bonds and the formation of oxamide,

$(CN)_2 + 2\,H_2O \rightarrow$ 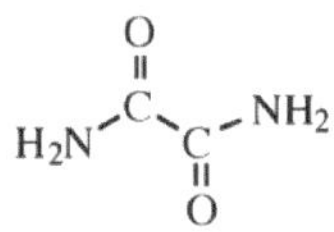

Cleavage of the C-C bond does occur in the presence of base (e.g., KOH), with the formation of cyanide ($CN^-$) and cyanate ($CNO^-$) salts,

$$(CN)_2 + 2\,OH^- \rightarrow CN^- + CNO^- + H_2O$$

## Dicyanoacetylene

Dicyanoacetylene (also known as carbon subnitride or by its IUPAC name but-2-ynedinitrile) has the structure shown in Figure 12.27b and may be thought of as a dicyanaide substituted acetylene. At room temperature, dicyanoacetylene is a clear liquid, however, solid dicyanoacetylene has been detected in the atmosphere of Titan (the largest moon of the planet Saturn) by infrared spectroscopy. Di-cyanoacetylene is an entropic explosive giving carbon powder and nitrogen gas. In the presence of oxygen, it burns with a bright blue-white flame at a temperature of 4990 °C.

## Hydrogen cyanide

Hydrogen cyanide (HCN) is a colorless, highly poisonous, gas (melting point = -13.5 °C and boiling point = 25.6 °C). Due to its original isolation from Prussian blue (hydrated ferric ferrocyanide), hydrogen cyanide is also known by the name of prussic acid.

The synthesis of hydrogen cyanide is accomplished commercially by the partial oxidation of methane in the presence of ammonia, using a platinum catalyst.

$$2\,CH_4 + 3\,O_2 + 2\,NH_3 \rightarrow 2\,HCN + 6\,H_2O$$

The heat to activate the reaction is derived from the partial combustion of the methane and ammonia. The resulting aqueous solution is dried by distilled from phosphorus pentoxide ($P_2O_5$) to yield anhydrous hydrogen cyanide.

Hydrogen cyanide may also be formed in the absence of oxygen; however, in this case the reaction must be heated externally.

$$CH_4 + NH_3 \rightarrow HCN + 3\ H_2$$

Small quantities of hydrogen cyanide for laboratory use may be prepared by the reaction of an acid with a cyanide salt (either potassium or sodium),

$$KCN + H^+ \rightarrow HCN + K^+$$

The structure of hydrogen cyanide is shown in Figure 12.28 along with its isomeric form, hydrogen isocyanide (HNC). While hydrogen cyanide is present in the pits of many fruits, and is generated by burnet moths and some millipedes, hydrogen isocyanide is only found in interstellar space. It is postulated, however, that along with HCN, HNC is an important building block for amino acids and hence life.

$$H{-}C{\equiv}N \qquad\qquad H{-}N^{+}{\equiv}C^{-}$$

(a) (b)

**Figure 12.28: The structures of (a) hydrogen cyanaide (HCN) and (b) its isomer hydrogen isocyanide (HNC).**

In the liquid state hydrogen cyanide forms strong hydrogen bonds (Figure 12.29). Hydrogen cyanide is a good solvent for polar compounds due to its high permittivity ($\epsilon_r$) and high dipole moment (2.98 D).

$$H^{\delta+}{-}C{\equiv}N^{\delta-} {-}{-}{-}{-}\ H^{\delta+}{-}C{\equiv}N^{\delta-}$$

**Figure 12.29: The hydrogen bonding in liquid hydrogen cyanide.**

In aqueous solution hydrogen cyanide is a weak acid and several salts are known,

$$HCN + H_2O \rightleftharpoons H_3O^+ + CN^-$$

However, HCN also reacts with water to give ammonium formate via formamide,

$$HCN + H_2O \rightarrow HC(O)NH_2 \xrightarrow{H_2O} HCO_2^- + NH_4^+$$

In a similar manner to cyanogen's relationship to the halogens, the cyanide anion ($CN^-$) is considered a pseudo halide (i.e., $F^-$, $Cl^-$, etc), and as such forms many coordination compounds, e.g., $[Fe(CN)_6]^{3-}$ and $[Ag(CN)_2]^-$.

*Assassination, execution, and the Holocaust*

Hydrogen cyanide is fatal to humans due to its inhibition of the enzyme cytochrome C oxidase by the cyanide ion ($CN^-$), which results in the halting of cellular respiration. A concentration of 300 mg/m$^3$ will kill within 10 minutes, while 3200 mg/m$^3$ (ca. 3500 ppm) will be fatal in about 1 minute.

The symptoms of cyanide poisoning appear similar to a heart attack and this has led to it being the poison of choice for both fictional murder mystery writers as well as the former KGB (Konitet gosudarstvennoy bezopasnosti or Committee for State Security) and its predecessor SMERSH (from the contraction *smert shpionam* meaning *death to spies*) in real life. Possibly the most famous use of hydrogen cyanide for assassination was the use of an atomizer mist gun by KGB agent Bohdan Stashynsky for the killing of the Ukranian political writer and anti-communist Lev Rebet in 1957, and later in 1959 that of fellow Ukranian, Stepan Bandera. In both cases the intention was to induce cardiac arrest and make it look like the victim had died of a heart attack.

Without doubt the most notorious use of hydrogen cyanide is in the form of the product Zyklon B, which was originally developed as an insecticide. Dr. Walter Heerdt found that hydrogen cyanide could be absorbed onto substrates such as absorbent pellets (e.g., silica), fibers, or diatomaceous earth (Figure 12.30). While stable in an airtight container, once opened the hydrogen cyanide is released. The B in the trade name comes from the German name for prussic acid (the common name for hydrogen cyanide), i.e., Blausäure meaning *blue acid*.

The first wholesale use of Zyklon B was actually in the US where it was used as early as 1929 to disinfect the freight trains and cloths of Mexican immigrants entering the US. The first use of Zyklon B in the concentration camps during World War II was for a similar purpose (in particular for delousing to control typhus); however, for its use as an insecticide, Zyklon B contained a warning odorant.

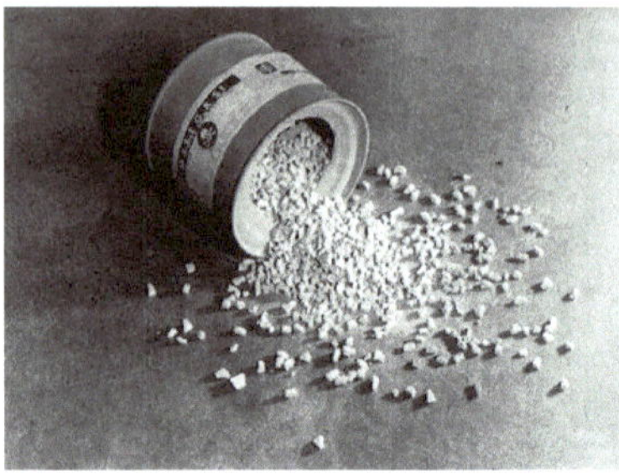

**Figure 12.30: A can of Zyklon B granules.**

The deliberate manufacture of Zyklon B without the odorant resulted in the material that was used on a group of 250 gypsy children at the Buchenwald concentration camp in early 1940. Subsequently in September 1940, a similar number of sick Polish prisoners of war and 600 Soviet prisoners of war were killed at Auschwitz (Figure 12.31). Once these horrific tests were performed, the systematic murder of millions of people in what was to be known as the 'Holocaust', including Jews, gypsies, and homosexuals, was accomplished using Zyklon B at Auschwitz, Majdanek, Sachsenhausen, and one of the Operation Reinhard camps. It is only fitting that many of the architects of the Holocaust themselves died from cyanide, including, Adolf Hitler (in addition to a bullet), Joseph Goebbles, Hermann Göring, and Heinrich Himmler.

**Figure 12.31: The main gate at Auschwitz, through which an estimated two million victims of Nazi genocide passed.**

Despite the horror associated with the use of hydrogen cyanide for the Holocaust, it was used by 11 US states for the death penalty (Figure 12.32). Arizona, Maryland, and Missouri retain the gas chamber as a secondary method of execution though they have lethal injection as the primary method. Potassium cyanide (KCN) pellets are placed into a compartment directly below the chair in the gas chamber (Figure 12.32). The condemned person is then strapped into the chair, and the airtight chamber sealed. Concentrated sulfuric acid ($H_2SO_4$) is then poured down a tube onto the cyanide pellets to generate hydrogen cyanide. Execution by gas chamber is especially unpleasant for the witnesses to the execution due to the physical responses exhibited during the process of dying, including convulsions and excessive drooling.

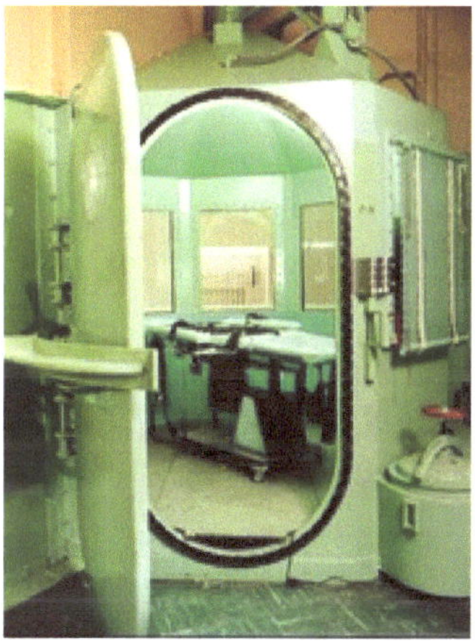

**Figure 12.32: The former gas chamber in San Quentin State Prison, now an execution chamber for lethal injection, in which the State of California executed 192 men and 4 women.**

## Oxides of carbon

### Carbon monoxide

Carbon monoxide (CO) is iso-electronic with nitrogen ($N_2$) and formed through the incomplete combustion of carbon or hydrocarbon compounds.

$$2\ C + O_2 \rightarrow 2\ CO$$

Carbon monoxide may also be made from steam and coal as part of synthesis gas,

$$C + H_2O \rightarrow CO + H_2$$

A convenient laboratory preparation of CO is the dehydration of formic acid by sulfuric acid,

$$HCO_2H \rightarrow CO + H_2O$$

Carbon monoxide is flammable, and has an explosive limit of 12.5 - 74% with an auto ignition temperature of 609 °C.

$$2\,CO + O_2 \rightarrow 2\,CO_2$$

Carbon monoxide is also very toxic; however, it is colorless, odorless, tasteless and non-irritating all of which increase its danger. The incomplete combustion of hydrocarbon (natural gas or heating oil) or carbon sources (coal or charcoal) is a common hazard in the home. In a closed environment (e.g., charcoal grill in room with no, or poor, ventilation) as carbon uses up oxygen in room the formation of toxic carbon monoxide results instead of $CO_2$. Typical sources of CO in the home, are shown in Figure 12.33.

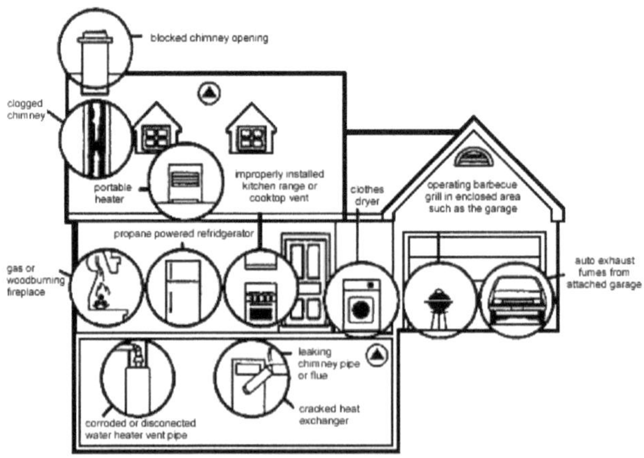


**Figure 12.33: Potential sources of carbon in the home. Copyright: Fire Safety Council.**

The toxicity of CO is due to its competition with oxygen at the heme binding site in hemoglobin. The binding affinity for CO is 200 times greater than that for oxygen, meaning that just small amounts of CO dramatically reduces hemoglobin's ability to transport oxygen around the body. The bright red color of the CO-heme complex is the reason that chronic exposure results in the skin adopting a bright red color. The symptoms of CO poisoning include headache, nausea, weakness, and eventually death. When air contains CO levels as low as 0.02%, headache and nausea occur; if the CO concentration is increased to 0.1%, unconsciousness will follow.

Cigarette smoke containing large amounts of carbon monoxide and as a result heavy-smokers can have up to 20% of the oxygen-active sites in their blood blocked by CO. However, despite this hazard, cigarettes would be even more deadly if they were not burnt. A key (and often engineered) ingredient in cigarettes is the addictive drug nicotine. Nicotine is an alkaloid (Figure 12.34a) as is caffeine (Figure 12.34b) that is in coffee and tea. The lethal dose of caffeine is approximately 10 g, which relates to approximately 70 - 100 cups of coffee (assuming a concentration of 100 - 150 mg per cup). Alternatively, a lethal dose of caffeine from cola would require approximately 180 - 280 12 oz bottles, each containing 35 - 44 mg. In comparison nicotine has a lethal dose of 50 mg. This means that 12 cigarettes can provide a lethal does if consumed. The only reason smoking 12 cigarettes do not kill immediately is that the majority of the nicotine is burnt in the smoking of the cigarette. If this were not true smokers would be killed before they could develop a habit.

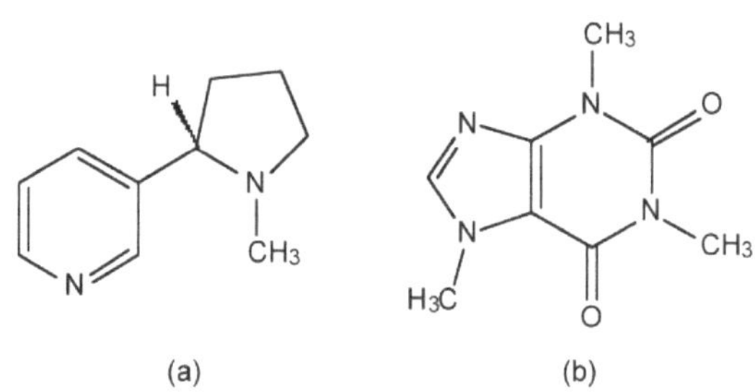


**Figure 12.34: The molecular structures of (a) nicotine and (b) caffeine.**

The bonding in CO involves 1 σ-bond and 2 sets of π-bonds (Figure 12.35). The C-O bond distance in carbon monoxide is 1.1 Å and therefore consistent with a triple bond. In comparison a typical C-O single bond is ca. 1.43 Å, and an average C-O double bond is ca. 1.23 Å. The major absorption band in the infra-red spectrum for CO is 2143 $cm^{-1}$, while for $^{13}CO$ it is 2099.2 $cm^{-1}$.

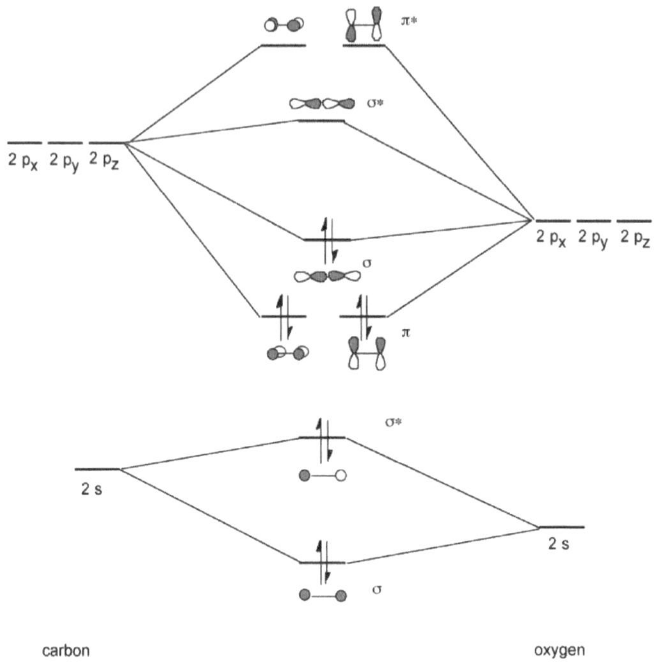


**Figure 12.35: The carbon monoxide molecular orbital diagram.**

# Carbon dioxide

Carbon dioxide ($CO_2$) is the most stable oxide of carbon and is formed from the burning of carbon or carbon containing compounds in air or an excess of oxygen,

$$C + O_2 \rightarrow CO_2$$

For industrial applications it is usually prepared from the decomposition of calcium carbonate (limestone) rather than separation from combustion products.

$$CaCO_3 \rightarrow CaO + CO_2$$

*Phase chemistry of carbon dioxide*

Carbon dioxide does not exist as a liquid under normal atmospheric pressure, but solid $CO_2$ (also known as dry ice) sublimes at -78.5 °C (Figure 12.36).

Dry ice is commonly used as a refrigerant for food or biological sample preservation.

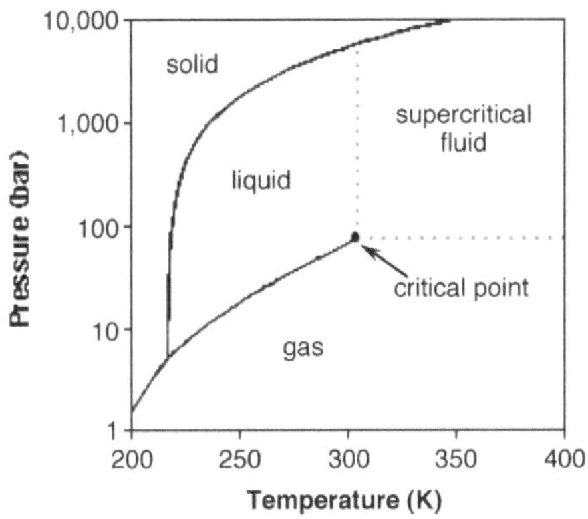


**Figure 12.36: The phase diagram for carbon dioxide.**

When dry ice is placed in water (espccially heated) sublimation is accelerated, and a low-sinking dense cloud of fog (smoke-like) is created. This is used in fog machines, at theaters, concerts, haunted houses, and nightclubs for dramatic effects. Fog from dry ice hovers above the ground unlike other artificial fog machines (that use partial combustion of oil) where the fog rises like smoke.

*Supercritical carbon dioxide*

As noted above carbon dioxide usually behaves as a gas in air at standard temperature and pressure (STP = 25 °C and 1 atm) or as a solid when frozen. However, if the temperature and pressure are both increased from STP to be at or above the critical point (Figure 12.36), carbon dioxide adopts properties midway between a gas and a liquid ($T_c$ = 31.1 °C and $P_c$ = 72.9 atm).

Supercritical $CO_2$ has become an important industrial solvent due to its role in chemical extraction in addition to its low toxicity and environmental impact. In this regard it is seen as a promising green solvent. One of the biggest applications is the decaffeination of coffee and tea without leaving any residue and allowing the caffeine to be separated and used in other beverage products.

*Structure and bonding*

Carbon dioxide is a linear molecule due to π-localization. The bonding in $CO_2$ involves 2 σ-bond and 2 sets of 3 center π-bonds (Figure 12.37). The C-O bond length of 1.2 Å should be compared to the value observed for organic carbonyls (e.g., ketones, esters, aldehydes) of 1.2 - 1.3 Å.

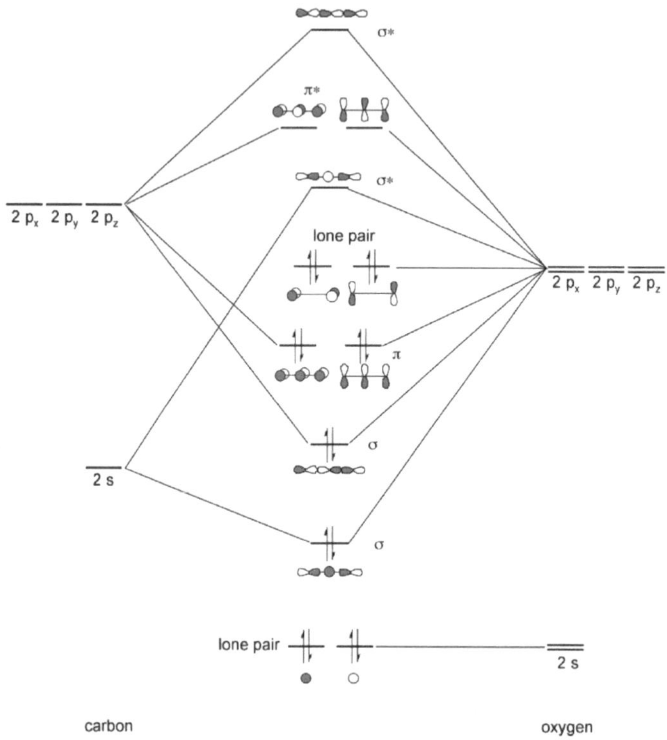


**Figure 12.37: The carbon dioxide molecular orbital diagram.**

*Dissolution and reaction with water*

Although $CO_2$ has no dipole moment it is very polar (dielectric constant = 1.60 at 0 °C, 50 atm) and consequently dissolves in polar solvents such as water up to a concentration of 0.18% (0.04 M). Most of it (+99%) is present as solvated $CO_2$ (Figure 7.38), and only ca. 0.2% is reacted to form carbonic acid, with subsequent equilibria resulting in the formation of bicarbonate ($HCO_3^-$) and carbonate $CO_3^{2-}$).

$$H_2O + CO_2 \rightarrow H_2CO_3$$

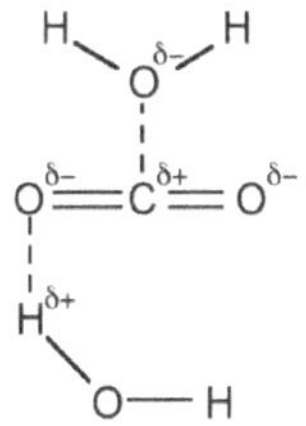

**Figure 12.38: Typical water solvation of carbon dioxide.**

The overall reaction involves a series of equilibria. The first equilibrium is the formation of carbonic acid,

$$CO_{2(solv)} + H_2O \underset{k_{H_2CO_3}}{\overset{k_{CO_2}}{\rightleftharpoons}} H_2CO_3$$

The reaction rates are on the magnitude of 1 second (i.e., slow), and as a consequence when carbon dioxide is carried in the body an enzyme is present to speed up the reaction.

$$K = \frac{k_{H_2CO_3}}{k_{CO_2}} = \frac{25}{0.04} = 600$$

The 2nd equilibrium is as a consequence of first ionization of carbonic acid to form bicarbonate ($HCO_3^-$),

$$H_2CO_3 \rightleftharpoons HCO_3^- + H^+$$

In contrast to the first reaction, this reaction is very fast with an equilibrium constant, $K_{eq} = 1.6\times10^{-4}$ @ 25 °C.

The 3rd equilibrium involves the formation of the carbonate ion and has a $K_{eq} = 4.84\times10^{-11}$.

$$HCO_3^- \rightleftharpoons CO_3^{2-} + H^+$$

Carbonate ($CO_3^{2-}$) is a delocalized ligand, which can act as a mono or bidentate or bridging group. The formation of carbonic acid is the reason that even in the absence of pollutants (such as $SO_2$) natural rainwater is slightly acidic due to dissolved $CO_2$. The equilibrium associated with carbonic acid is also responsible for the buffering of the pH in blood.

*Reaction chemistry*

Photosynthesis in plants reduces $CO_2$ to organic matter but similar reactions have yet to be developed in non-living systems.

Grignards react readily with carbon dioxide to form the carboxylate, which yields the associated carboxylic acid upon hydrolysis,

$$RMgX + CO_2 \rightarrow RCO_2MgX \xrightarrow{H_2O} RCO_2H + HOMgX$$

Similar reactions occur with other organometallic compounds. In addition, $CO_2$ reacts with alkali metal salts of phenols (phenolates) to yield the hydroxy-carboxylate,

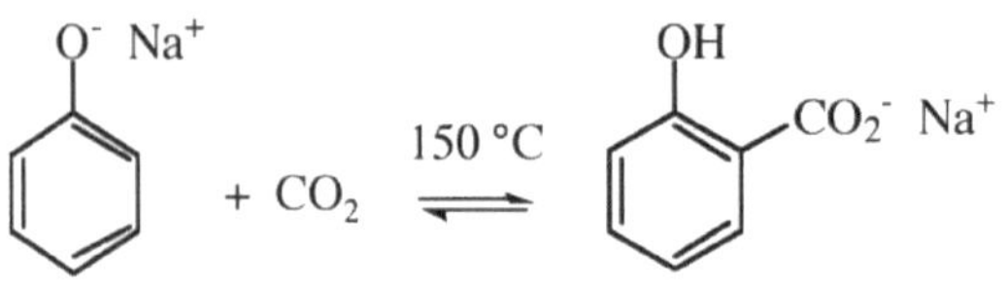


A number of complexes of $CO_2$ with transition metals are known in which the coordination can occur via the central carbon (Figure 12.39a) or the C=O bond (Figure 12.39b). Alternatively, $CO_2$ can bridge two metal centers.

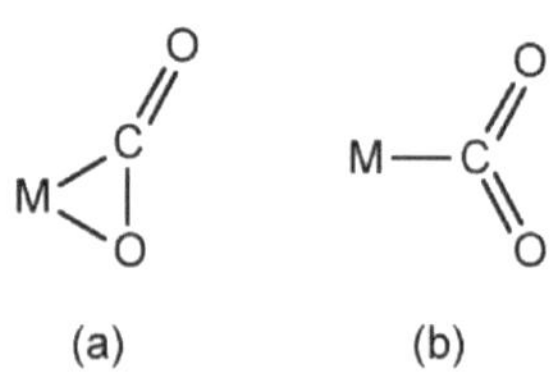


**Figure 12.39: Bonding modes of $CO_2$ to transition metals.**

*Global warming and carbon dioxide*

Global warming is the process of the observed increase in the average temperature of the Earth's near- surface air and oceans since the mid-20$^{th}$ century. Global surface temperature increased 0.74 °C (1.33 °F) between the start and the end of the 20$^{th}$ century (Figure 12.40). It is generally agreed that the majority of this temperature increase has occurred since the middle of the 20$^{th}$ century and was caused by increasing concentrations of greenhouse gases resulting from burning fossil fuels (the generation of additional $CO_2$) and deforestation (the loss of a mechanism for the consumption of $CO_2$), see Figure 12.40. While it is appreciated that natural phenomena (including solar radiation and volcanoes) produced most of the warming from pre- industrial times, the magnitude of the changes brought on by global industrialization is more significant.

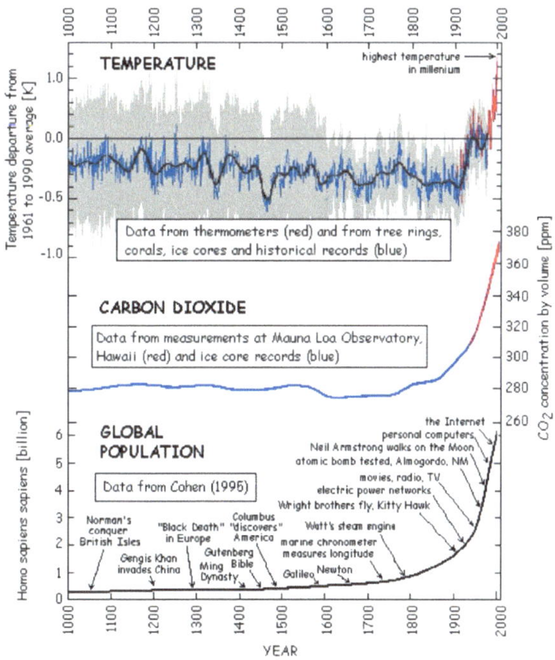


**Figure 12.40: Correlation of Earth average temperature with carbon dioxide and global population.**

The Earth's atmosphere has two functions. First, the ozone ($O_3$) in the upper atmosphere screens harmful UV from reaching the surface of the Earth.

Second, as solar radiation penetrates the atmosphere a portion of the heat is then retained as a consequence of the $CO_2$ in the atmosphere. It is this process that modulates the surface temperature and provides a stable environment for life. The failure or alteration of either of these processes can have a dramatic effect on the livability of a planet.

Consider the relative position of Venus, Earth, and Mars to the Sun (Figure 12.41). The closer a planet is to the sun the greater the UV radiation and the greater the heating of the planet; however, the temperature is also greatly modulated by the atmosphere. Venus has an atmosphere comprising 95% $CO_2$ and has a surface temperature of approximately 450 °C. In contrast, while Mars' atmosphere is also 95% $CO_2$, it is only 1% as dense as that of Earth's, and thus the surface temperatures range from 40 °C during the day (due to radiative heating) to 80 °C at night (due to the lack of retained heat because of the thin atmosphere). These should be compared to Earth's atmosphere which is 0.038% $CO_2$, which allows for the correct amount of heat to be retained to sustain life. Clearly any significant change in the $CO_2$ content of the atmosphere will change the global temperatures of a planet.

**Figure 12.41: The relative size and position of the planets from the sun.**

## Suboxides of carbon

*Carbon suboxide*

Carbon suboxide is the third oxide of carbon, $C_3O_2$. It is made from the dehydration of malonic acid with $P_4O_{10}$ above 140 °C

$$CH_2(CO_2H)_2 \rightarrow O{=}C{=}C{=}C{=}O + 2\ H_2O$$

Like carbon dioxide, the $C_3O_2$ molecule is linear, with pπ-pπ bonding. Gaseous carbon suboxide has an evil smell and while stable at 78 °C it polymerizes at 25 °C. Photolysis of $C_3O_2$ yields the unstable $C_2O$. As expected from its synthesis, carbon suboxide reacts slowly with water to form malonic acid, i.e., the reverse of; however, the reaction with stronger nuceophiles such as amines is rapid,

$$C_3O_2 + 2\ HNR_2 \rightarrow R_2NC(O)CH_2C(O)NR_2$$

*Mellitic acid anhydride*

The anhydride of mellitic acid (Figure 12.42a) may be considered as an oxide of carbon since its chemical formula contains only carbon and oxygen, i.e., C12O9 (Figure 12.42b).

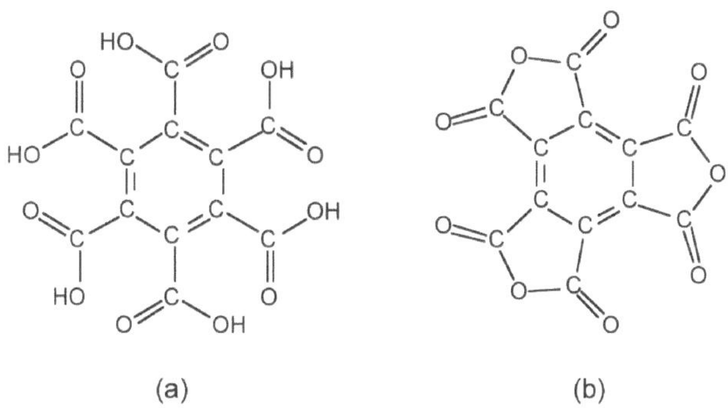

**Figure 12.42: Structures of (a) mellitic acid and (b) its anhydride.**

## Carbon halides

There are two general classes of carbon halides.

1. Homoleptic halides, e.g., $CCl_4$, $CCl_2F_2$, $C_6Cl_6$, etc.
2. Carbonyl halides, e.g., $Cl_2C{=}O$.

A summary of some simple carbon halides is given in Table 12.8.

| Compound | Melting point (°C) | Boiling point (°C) | Remarks |
|---|---|---|---|
| $CF_4$ | -185 | -128 | Very stable gas |
| $CCl_4$ | -23 | 76 | Colorless liquid, stable |
| $CBr_4$ | 93 | 190 | Pale yellow solid, decomposes upon boiling |
| $CI_4$ | 171 | - | Bright red solid, decom- poses prior to boiling, sublimed at low pressure |
| $F_2C=O$ | -114 | -83 | Decomposed by $H_2O$ |
| $Cl_2C=O$ | -118 | 8 | Phosgene, highly toxic |
| $Br_2C=O$ | - | 65 | Fumes in air |

**Table 12.8: Physical properties of simple halogen compounds of carbon.**

*Carbon tetrahalides*

The carbon tetrahalides are generally prepared by the direct (thermal) reaction of carbon with the appropriate halogen; however, specific syntheses are possible for each derivative.

$$C + 2\,X_2 \rightarrow CX_4$$

In addition to the direct reaction of fluorine with carbon, $CF_4$ can be prepared from SiC,

$$SiC + F_2 \rightarrow SiF_4 + CF_4$$

The $SiF_4$ side product is removed by passing the reaction mixture through NaOH solution, in which $SiF_4$ reacts to form silicate. The difference in reactivity of $SiF_4$ and $CF_4$ is attributable to the lack of an energetically accessible five-coordinate intermediate required for the associative mechanism.

Carbon tetrabromide can be obtained by bromination of $CH_4$ with HBr or $Br_2$, or by the reaction of $CCl_4$ with $AlBr_3$,

$$3\,CCl_4 + 4\,AlBr_3 \rightarrow 3\,CBr_4 + 4\,AlCl_3$$

Carbon tetraiodide ($CI_4$) can be made by the Lewis acid catalyzed halogen exchange reaction,

$$CCl_4 + 4\,C_2H_5I \rightarrow CI_4 + 4\,C_2H_5Cl$$

CF4 is very stable. In fact, it is so stable that it does not even react with molten sodium. In contrast to $CF_4$, carbon tetrachloride ($CCl_4$) reacts readily with alkali metals (K and Na) or other strong reducing agents (e.g., $F_2$, Al, Ba, Be, and Zn). While $CCl_4$ is thermodynamically unstable with respect to hydrolysis, it is kinetically stable, and thus finds extensive use as a solvent. Photolysis can result in the transfer of a chloride radical to various substrates. It is also used in the conversion of metal oxides to the chlorides. Carbon tetrabromide ($CBr_4$) is insoluble in water and other polar solvents, but soluble in benzene. Carbon tetraiodide ($CI_4$) decomposes thermally,

$$2\,CI_4 \rightarrow 2\,I_2 + I_2C{=}CI_2$$

The decreasing stability of $CX_4$, from fluorine to iodine, is directly related to the C-X bond energy (Table 12.9).

| C-X | Bond energy (kJ/mol) |
|---|---|
| C-F | 485 |
| C-Cl | 327 |
| C-Br | 285 |
| C-I | 213 |

**Table 12.9: Bond energies for carbon-halide bonds.**

*Hazards*

Despite its use as a solvent $CCl_4$ has significant hazardous effects. Inhalation of carbon tetrachloride vapor can cause headaches, mental confusion, depression, fatigue, loss of appetite, nausea, vomiting, and coma. The symptoms can take many hours to appear. The vapor and liquid irritate the eyes, and internal irritation, nausea, and vomiting are caused when taken orally. Chronic effects from prolonged inhalation include bronchitis and jaundice, while skin exposure can cause dermatitis.

Carbon tetrabromide is toxic by inhalation, and the vapor is narcotic if taken in high concentrations. As with $CCl_4$, $CBr_4$ can react explosively with alkali metals.

*Higher homoleptic halides*

Organic compounds that contain only carbon and a halogen are called halocarbons, and these include fluorocarbons and chlorocarbons. The easiest route to fluorocarbons involves the reaction of a hydrocarbon with a high valent fluoride (e.g., $CoF_3$) or the reaction of a chlorocarbon with $SbF_3$. In general, chlorocarbons with $sp^2$ carbon atoms are more stable than those with $sp^2$ carbon centers. The exception to this is aromatic compounds such as $C_6Cl_6$.

The physical properties of fluorocarbons range from inert to toxic. Thus, poly(tetrafluoroethylene), $(C_2F_4)_n$, known by either its acronym (PTFE) or its trade name (Teflon®), is chemically inert and has a low coefficient of friction (Table 12.10). As a consequence, its uses include coatings on armor-piercing bullets (to stop the wear on the gun barrel), laboratory containers and magnetic stirrers, tubing for corrosive chemicals, and thread seal tape in plumbing applications (Plumbers tape). A summary of the physical properties of PTFE is given in Table 12.10. PTFE is synthesized by the emulsion polymerization of tetra fluoroethylene monomer under pressure through free radical catalyst,

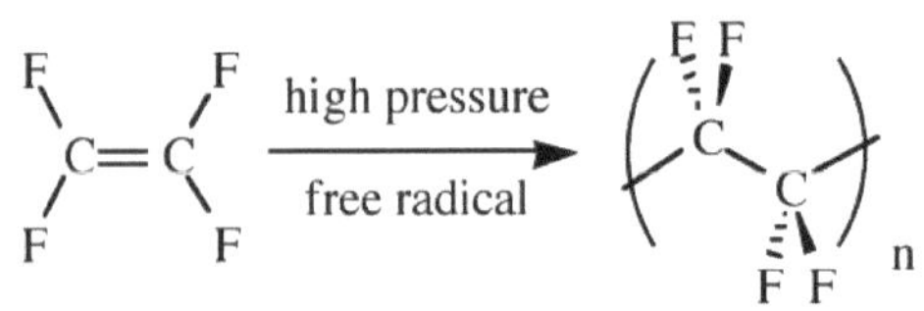


| Property | Value |
|---|---|
| Density | 2.2 g/cm$^3$ |
| Melting point | 327 °C |
| Young's modulus | 0.5 GPa |
| Yield strength | 23 MPa |
| Coe cient of friction | 0.05 - 0.10 |
| Dielectric constant | 2.1 |
| Dielectric strength (1 MHz) | 60 MV/m |

**Table 12.10: Physical properties of poly(tetrafluoroethylene) (PTFE).**

In contrast with PTFE, octafluoroisobutylene, $(CF_3)_2C=CF_2$, is highly toxic, while perfluorodecahydronapthalene ($C_{10}F_8$, Figure 12.43) is used as a blood substitute component.

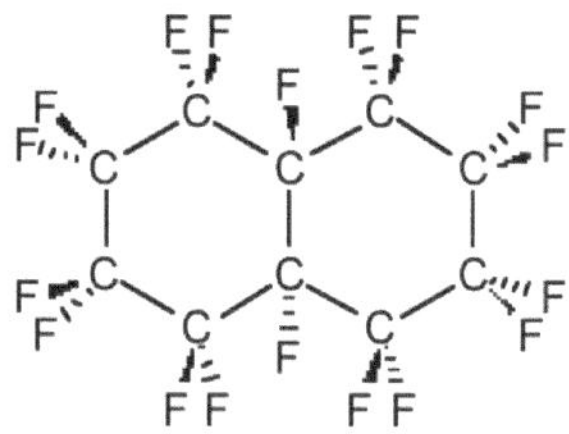

**Figure 12.43: Structure of perfluorodecahydronapthalene.**

*Mixed halides*

Mixed halides are an important class of halocarbon compound. They are synthesized by halide exchange,

$$3\ CCl_4 + 2\ SbF_3 \rightarrow 2\ CCl_2F_2 + 2\ SbCl_3$$

The high cost of $SbF_3$ means that the reaction is generally run with an excess of thc chloridc.

The ordinary name for mixed carbon halide is halon or Freon, although Freon is actually a Du Pont trade- mark. A list of selected Freon compounds is given in Table 12.11. Halons are non-toxic, non-flammable, and have no odor. However, it is their very lack of reactivity that has caused a problem.

| Freon | Formula | Uses |
|---|---|---|
| 12 | $CCl_2F_2$ | Refrigerant |
| 11 | $CCl_3F$ | Refrigerant |
| 114 | $ClF_2C\text{-}CClF_2$ | Refrigerant |
| 113 | $Cl_3C\text{-}CF_3$ | Solvent |
| 13B1 | $CBrF_3$ | Fire extinguisher |
| 1211 | $CBrClF_2$ | Fire extinguisher |

**Table 12.11: Selected Freons and their applications.**

*Environmental impact of chlorofluorcarbon compounds (CFCs)*

Chlorofluorcarbon compounds (CFCs) are very stable and are not degraded in the environment. As a consequence, they are transported to the stratosphere where they decomposed upon photolysis,

$$CCl_2F_2 + h\nu \rightarrow CClF_2 + Cl\cdot$$

The resulting chloride radical is a catalyst for the decomposition of ozone, as well as a catalyst for the reaction of ozone with molecular oxygen,

$$2\,O_3 \xrightarrow{Cl\cdot} 3\,O_2$$

$$O_3 + O \xrightarrow{Cl\cdot} 2\,O_2$$

The widespread use of CFCs as refrigerants and propellants meant that by 1986 there were 2.5 billion pounds of CFC being liberated to the atmosphere. This was equivalent to ½ lb per person on the planet. Since the ozone layer provides the vital protection to life on the Earth's surface from high energy UV radiation the release of CFC (along with other chemicals) caused a dramatic change in the ozone layer, including the increase in the polar hole in the ozone layer. As a result of the EU called for a complete ban of CFCs (which was followed by other countries). In their place new chemicals with similar refrigerant properties were developed. These compounds contained C-H bonds (e.g., $C_2HCl_2F_3$ and $C_2H_3Cl_2F$) that are readily broken in the lower atmosphere, thus limiting the transport to the stratosphere.

*Carbonyl halides*

All the carbonyl halides ($X_2C{=}O$, X = F, Cl, Br, I) are known (Table 7.8). Phosgene ($Cl_2C{=}O$) was first synthesized by John Davy (Figure 12.44) in 1812 by exposing a mixture of carbon monoxide and chlorine to sunlight, (7.50).

$$CO + Cl_2 \rightarrow Cl_2C{=}O$$

He named it phosgene from the Greek, *phos* (light) and *gene* (born), in reference to use of light to promote the reaction. The fluoride is also prepared by the reaction of carbon monoxide with the halogen, while the bromide is prepared by the partial hydrolysis of $CBr_4$ with sulfuric acid.

**Figure 12.44: British doctor and chemist John Davy FRS (1790 - 1868) was also the brother of the noted chemist Sir Humphry Davy (Figure 8.5).**

The synthesis of isocyanates from alkyl or aryl amines illustrates the electrophilic character of phosgene and its ability to introduce the equivalent of '$CO^{2+}$',

$$RNH_2 + Cl_2C{=}O \rightarrow RN{=}C{=}O + 2\ HCl$$

This reaction is conducted in the presence of a base such as pyridine that absorbs the hydrogen chloride. Phosgene may also be used to produce acyl chlorides from carboxylic acids; however, thionyl chloride is more commonly and more safely used in this reaction.

$$RCO_2H + Cl_2C{=}O \rightarrow RC(O)Cl + HCl + CO_2$$

*Phosgene as a weapon of war*

Phosgene is a toxic gas with the smell of new-mown hay and was used in chemical warfare during the First World War (Figure 12.45) where it was a more potent weapon than chlorine. A summary of the casualties inflicted by chemical warfare agents during the Great War is shown in Table 12.12.

While chlorine was potentially deadly it caused the victim to violently cough and choke (the body's natural defense to limiting inhalation), in contrast, phosgene caused much less coughing with the result that more of it was inhaled. Phosgene often had a delayed effect; apparently healthy soldiers were taken down with phosgene gas poisoning up to 48 hours after inhalation. A fatal dose of phosgene eventually led to shallow breathing and retching, pulse up to 120, an ashen face and the discharge of four pints of yellow liquid from the lungs each hour for the 48 of the drowning spasms.

**Figure 12.45: Australian infantry from the 45 Battalion, Australian 4$^{th}$ Division wearing Small Box Respirators (SBR) as protection of chemical warfare agents. Photograph taken at Zonnebeke, on the Ypres sector, 27$^{th}$ September 1917. Copyright: Australian War Memorial.**

| Country | Total casualties | Deaths |
|---|---|---|
| Russia | 419,340 | 56,000 |
| Germany | 200,000 | 9,000 |
| France | 190,000 | 8,000 |
| British Empire | 188,706 | 8,109 |
| Austria-Hungary | 100,000 | 3,000 |
| USA | 72,807 | 1,462 |
| Italy | 60,000 | 4,627 |
| Others | 10,000 | 1,000 |

**Table 12.12: Casualties from gas attacks during the First World War (including chlorine, phosgene, and mustard gas). British Empire includes troops from United Kingdom, Australia, Canada, India, New Zealand, and South Africa.**

Although phosgene's boiling point (7.6 °C) meant that is was a vapor, the so-called "white star" mixture of phosgene and chlorine was commonly used on the Somme, because the chlorine supplied the necessary vapor with which to carry the phosgene.

# Silicon

## Comparison between silicon and carbon

An understanding of the differences between carbon and silicon is important in understanding the relative chemistry of these Group 14 elements.

### *Size*

As expected, silicon is larger than carbon due to the presence of a second shell: i.e., C = $1s^2 2s^2 2p^2$ while Si = $1s^2 2s^2 2p^6 3s^2 3p^2$. A comparison of the relative sizes of carbon and silicon are given in Table 12.13.

| Element | Atomic radius (Å) | Covalent radius sp$^3$ (Å) | van der Waal radius (Å) |
|---|---|---|---|
| C | 0.91 | 1.11 | 0.77 |
| Si | 1.46 | 0.76 | 1.17 |

**Table 12.13: Atomic, covalent, and van der Waals radii of carbon and silicon.**

### *Coordination number*

Carbon is known to have a coordination number of 2, 3, and 4 in its compounds depending on the hybridization. A coordination number of 1 can also be considered for CO and $CN^-$. Four-coordinate carbon may be considered to be coordinatively saturated. In contrast, in the absence of overwhelming steric bulk, silicon is observed to have coordination numbers of 3, 4, 5, and 6. Examples of five and six-coordinate silicon include $Si(acac)_2Cl$ and $SiF_6^{2-}$, respectively. Coordination numbers of higher than 4 have been ascribed to the use of low-lying d orbitals; however, calculations show these are not significant. Instead, hypervalent silicon is better described by the formation of 3-center molecular orbitals, e.g., Figure 12.46. A hypervalent molecule is a molecule that contains one or more typical elements (Group 1, 2, 13-18) formally bearing more than eight electrons in their valence shells.

### *Electronegativity*

The electronegativities of silicon and carbon are given in Table 12.14, along with hydrogen. Since carbon is more electronegative than hydrogen the C-H bond is polarized towards carbon resulting in a more protic hydrogen (Figure 12.47a). In contrast, the lower electronegativity of silicon results in a more hydridic hydrogen (Figure 12.47b). This difference is reflected in the reaction chemistry of $SiH_4$ versus $CH_4$.

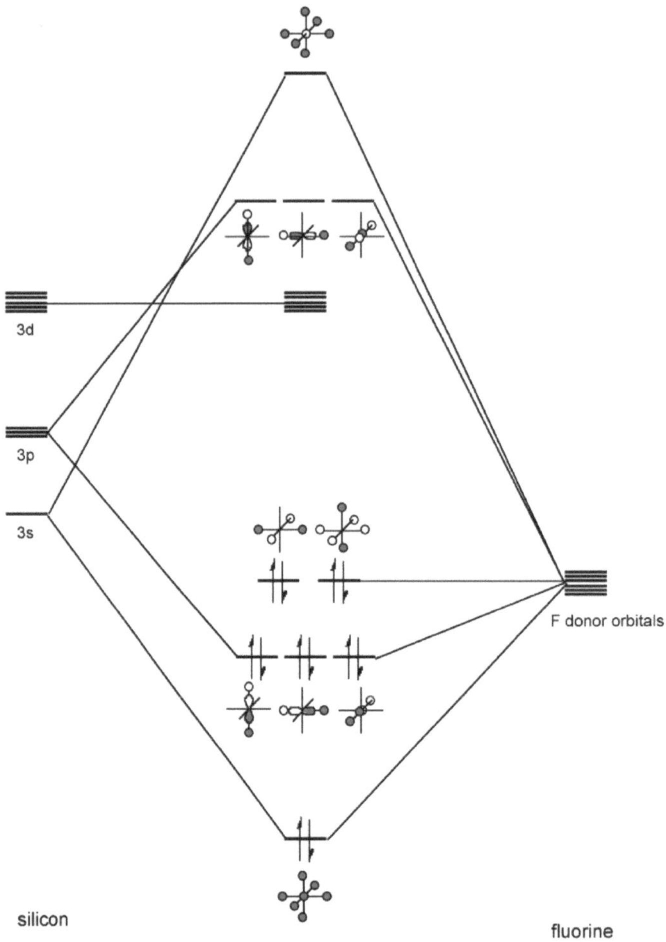


**Figure 12.46: Molecular orbital diagram for $SiF_6^{2-}$.**

| Element | Pauling scale |
|---|---|
| C | 2.5 |
| H | 2.1 |
| Si | 1.8 |

**Table 12.14: Selected Pauling electronegativity values.**

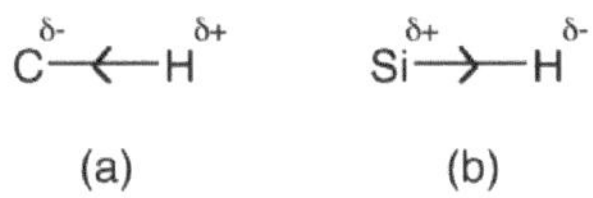

Figure 12.47: Relative polarization of C-H and Si-H bonds.

*Bond energies*

The E-E and E-O bond energies for carbon and silicon are given in Table 12.15. The bond energy for a C-C bond is slightly greater than for a C-O bond, while the Si-O bond is significantly stronger than the Si-Si bond. This difference is reflected in the chemistry of silicon versus carbon compounds. The chemistry of carbon is dominated by catenation: the ability of a chemical element to form a long chain-like structure via a series of covalent bonds. Although silicon does form Si-Si bonds, they are far more reactive than their C-C analogs, and polymers of silicon are predominantly comprised of Si-O chains (as a result of the very strong bond).

| Element | E-E bond energy (kJ/mol) | E-O bond energy (kJ/mol) |
|---|---|---|
| C | 356 | 335 |
| Si | 230 | 370 |

**Table 12.15: Selected bond energies for carbon and silicon.**

*Multiple bonds*

While unsaturated compounds for carbon (i.e., alkenes and alkynes) are common, the analogous silicon compounds (disilenes) were only reported in 1981, and disilynes in 2004. The Si=Si double bond lengths are 2.14 - 2.29 Å which is 5 - 10% shorter than the Si-Si single bond lengths. This bond shortening is less than ca. 13% in carbon compounds.

The traditional lack of multiple bonds for the Period 3 elements and lower led to the formulation of the double bond rule which states that chemical elements with a principal quantum number greater than 2 do not form multiple bonds (e.g., double bonds and triple bonds) with themselves or with other elements. This rule was made obsolete starting from 1981 with the discovery of silicon and phosphorus double bonds. Double bonds that would ordinarily not form can be stabilized with proper functional groups through kinetic stabilization, i.e., either electronically or sterically.

## Silicon and silica as electronic materials

*Semiconductor grade silicon*

The synthesis and purification of bulk polycrystalline semiconductor material represents the first step towards the commercial fabrication of an electronic device. This polycrystalline material is then used as the raw material for the formation of single crystal material that is processed to semiconductor wafers. The strong in fluence on the electric characteristics of a semiconductors exhibited by small amounts of some impurities requires that the bulk raw material be of very high purity (>99.9999%). Although some level of purification is possible during the crystallization process it is important to use as high a purity starting material as possible.

While 98% elemental silicon, known as metallurgical-grade silicon (MGS), is produced on a large scale, the requirements of extreme purity for electronic device fabrication require additional purification steps in order to produce electronic-grade silicon (EGS). In order for the purity levels to be acceptable for subsequent crystal growth and device fabrication, EGS must have carbon and oxygen impurity levels less than a few parts per million (ppm), and metal impurities at the parts per billion (ppb) range or lower. Table 12.16 and Table 12.17 give typical impurity concentrations in MGS and EGS, respectively. Besides the purity, the production

| Element | Concentration (ppm) | Element | Concentration (ppm) |
|---|---|---|---|
| aluminum | 1000-4350 | manganese | 50-120 |
| boron | 40-60 | molybdenum | < 20 |
| calcium | 245-500 | nickel | 10-105 |
| chromium | 50-200 | phosphorus | 20-50 |
| copper | 15-45 | titanium | 140-300 |
| iron | 1550-6500 | vanadium | 50-250 |
| magnesium | 10-50 | zirconium | 20 |

**Table 12.16: Typical impurity concentrations found in metallurgical-grade silicon (MGS).**

| Element | Concentration (ppb) | Element | Concentration (ppb) |
|---|---|---|---|
| arsenic | < 0.001 | gold | < 0.00001 |
| antimony | < 0.001 | iron | 0.1-1.0 |
| boron | ≤ 0.1 | nickel | 0.1-0.5 |
| carbon | 100-1000 | oxygen | 100-400 |
| chromium | < 0.01 | phosphorus | ≤ 0.3 |
| cobalt | 0.001 | silver | 0.001 |
| copper | 0.1 | zinc | < 0.1 |

**Table 12.17: Typical impurity concentrations found in electronic-grade silicon.**

*Metallurgical-grade silicon (MGS)*

The typical source material for commercial production of elemental silicon is quartzite gravel; a relatively pure form of sand ($SiO_2$). The first step in the synthesis of silicon is the melting and reduction of the silica in a submerged-electrode arc furnace. An example of which is shown schematically in Figure 7.48, along with the appropriate chemical reactions. A mixture of quartzite gravel and carbon are heated to high temperatures (ca. 1800 °C) in the furnace. The carbon bed consists of a mixture of coal, coke, and wood chips. The latter providing the necessary porosity such that the gases created during the reaction (SiO and CO) are able to ow through the bed.

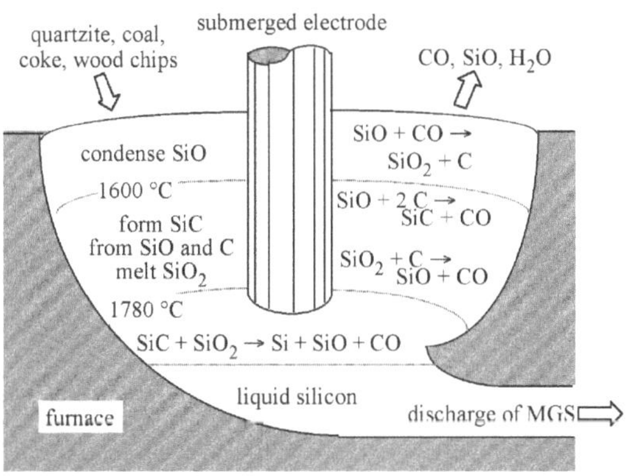


**Figure 12.48: Schematic of submerged-electrode arc furnace for the production of metallurgical-grade silicon (MGS).**

The overall reduction reaction of $SiO_2$ is expressed,

$$\mathrm{SiO_2(liquid) + 2\ C(solid) \rightarrow Si(liquid) + 2\ CO\ (gas)}$$

however, the reaction sequence is more complex than this overall reaction implies and involves the formation of SiC and SiO intermediates. The initial reaction between molten $SiO_2$ and C takes place in the arc between adjacent electrodes, where the local temperature can exceed 2000 °C.

$$\mathrm{SiO_2 + 2\ C \underset{<1600\ ^\circ C}{\overset{>1700\ ^\circ C}{\rightleftharpoons}} SiO + CO}$$

The SiO and CO thus generated ow to cooler zones in the furnace where SiC is formed, or higher in the bed where they reform $SiO_2$ and C.

$$\mathrm{SiO + 2\ C \rightarrow SiC + CO\ \ (1600 - 1700\ ^\circ C)}$$

The SiC reacts with molten $SiO_2$, producing the desired silicon along with SiO and CO.

$$\mathrm{SiC + SiO_2 \rightarrow Si + SiO + CO}$$

The molten silicon formed is drawn-off from the furnace and solidified.

The as-produced MGS is approximately 98-99% pure, with the major impurities being aluminum and iron (Table 12.16), however, obtaining low levels of boron impurities is of particular importance, because it is di cult to remove and serves as a dopant for silicon. The drawbacks of the above process are that it is energy and raw material intensive. It is estimated that the production of one metric ton (1,000 kg) of MGS requires 2500 - 2700 kg quartzite, 600 kg charcoal, 600 - 700 kg coal or coke, 300 - 500 kg wood chips, and 500,000 kWh of electric power. Currently, approximately 500,000 metric tons of MGS are produced per year, worldwide. Most of the production (ca. 70%) is used for metallurgical applications (e.g., aluminum-silicon alloys are commonly used for automotive engine blocks) from whence its name is derived. Applications in a variety of chemical products such as silicone resins account for about 30%, and only 1% or less of the total production of MGS is used in the manufacturing of high-purity EGS for the electronics industry. The current worldwide consumption of EGS is approximately $5\times10^6$ kg per year.

*Electronic-grade silicon (EGS)*

Electronic-grade silicon (EGS) is a polycrystalline material of exceptionally high purity and is the raw material for the growth of single-crystal silicon. EGS is one of the purest materials commonly available, see Table 12.17. The formation of EGS from MGS is accomplished through chemical purification processes. The basic concept of which involves the conversion of MGS to a volatile silicon compound, which is purified by distillation, and subsequently decomposed to re-form elemental silicon of higher purity (i.e., EGS). Irrespective of the purification route employed, the first step is physical pulverization of MGS followed by its conversion to the volatile silicon compounds.

A number of compounds, such as monosilane ($SiH_4$), dichlorosilane ($SiH_2Cl_2$), trichlorosilane ($SiHCl_3$), and silicon tetrachloride ($SiCl_4$), have been considered as chemical intermediates. Among these, $SiHCl_3$ has been used predominantly as the intermediate compound for subsequent EGS formation, although $SiH_4$ is used to a lesser extent. Silicon tetrachloride and its lower chlorinated derivatives are used for the chemical vapor deposition (CVD) growth of Si and $SiO_2$. The boiling points of silane and its chlorinated products (Table 12.18) are such that they are conveniently separated from each other by fractional distillation.

| Compound | Boiling point (°C) |
|---|---|
| $SiH_4$ | -112.3 |
| $SiH_3Cl$ | -30.4 |
| $SiH_2Cl_2$ | 8.3 |
| $SiHCl_3$ | 31.5 |
| $SiCl_4$ | 57.6 |

**Table 12.18: Boiling points of silane and chlorosilanes at 760 mmHg (1 atmosphere).**

The reasons for the predominant use of $SiHCl_3$ in the synthesis of EGS are as follows:

- $SiHCl_3$ can be easily formed by the reaction of anhydrous hydrogen chloride with MGS at reasonably low temperatures (200 - 400 °C);

- it is liquid at room temperature so that purification can be accomplished using standard distillation techniques;
- it is easily handled and if dry can be stored in carbon steel tanks;
- its liquid is easily vaporized and, when mixed with hydrogen it can be transported in steel lines without corrosion;
- it can be reduced at atmospheric pressure in the presence of hydrogen;
- its deposition can take place on heated silicon, thus eliminating contact with any foreign surfaces that
- may contaminate the resulting silicon; and
- it reacts at lower temperatures (1000 – 1200 °C) and at faster rates than does $SiCl_4$.

*Chlorosilane (Seimens) process*

Trichlorosilane is synthesized by heating powdered MGS with anhydrous hydrogen chloride (HCl) at around 300 °C in a fluidized-bed reactor,

$$\text{Si(solid)} + 3\ \text{HCl(gas)} \underset{>900\ ^\circ\text{C}}{\overset{\text{ca. }300\ ^\circ\text{C}}{\rightleftharpoons}} \text{SiHCl}_3\text{(vapor)} + \text{H}_2\ \text{(gas)}$$

Since the reaction is actually an equilibrium and the formation of $SiHCl_3$ highly exothermic, efficient removal of generated heat is essential to assure a maximum yield of $SiHCl_3$. While the stoichiometric reaction is that shown above, a mixture of chlorinated silanes is actually prepared which must be separated by fractional distillation, along with the chlorides of any impurities. In particular iron, aluminum, and boron are removed as $FeCl_3$ (boiling point = 316 °C), $AlCl_3$ (melting point = 190 °C, sublimes), and $BCl_3$ (boiling point = 12.65 °C), respectively. Fractional distillation of $SiHCl_3$ from these impurity halides result in greatly increased purity with a concentration of electrically active impurities of less than 1 ppb.

EGS is prepared from purified $SiHCl_3$ in a chemical vapor deposition (CVD) process similar to the epitaxial growth of Si. The high-purity $SiHCl_3$ is vaporized, diluted with high-purity hydrogen, and introduced into the Seimens deposition reactor, shown schematically in Figure 12.49. Within the reactor, thin silicon rods called slim rods (ca. 4 mm diameter) are supported by graphite electrodes. Resistance heating of the slim rods causes the decomposition of the $SiHCl_3$ to yield silicon.

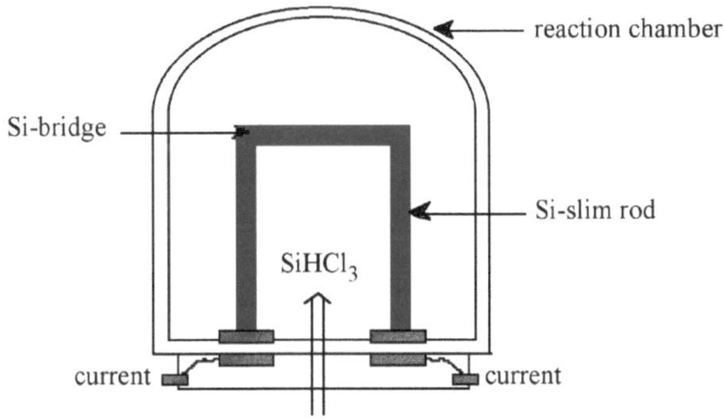


**Figure 12.49: Schematic representation of a Seimens deposition reactor.**

The shift in the equilibrium from forming $SiHCl_3$ from Si at low temperature, to forming Si from $SiHCl_3$ at high temperature is as a consequence of the temperature dependence,

$$\ln K_p = \frac{-\Delta H}{RT}$$

of the equilibrium constant, where ρ = partial pressure.

$$K_p = \frac{\rho_{SiHCl_3}\ \rho_{H_2}}{\rho_{HCl}}$$

Since the formation of $SiHCl_3$ is exothermic, i.e., $\Delta H < 0$, an increase in the temperature causes the partial pressure of $SiHCl_3$ to decrease. Thus, the Siemens process is typically run at ca. 1100 °C, while the reverse fluidized bed process is carried out at 300 °C.

The slim rods act as a nucleation point for the deposition of silicon, and the resulting polycrystalline rod consists of columnar grains of silicon (polysilicon) grown perpendicular to the rod axis. Growth occurs at less than 1 mm per hour, and after deposition for 200 to 300 hours high-purity (EGS) polysilicon rods of 150 - 200 mm in diameter are produced. For subsequent oat-zone refining the polysilicon EGS rods are cut into long cylindrical rods. Alternatively, the as-formed polysilicon rods are broken into chunks for single crystal growth processes, for example Czochralski melt growth.

In addition to the formation of silicon, the HCl coproduct reacts with the $SiHCl_3$ reactant to form silicon tetrachloride ($SiCl_4$) and hydrogen as major byproducts of the process,

$$HCl + SiHCl_3 \rightarrow SiCl_4 + H_2$$

This reaction represents a major disadvantage with the Seimens process: poor efficiency of silicon and chlorine consumption. Typically, only 30% of the silicon introduced into CVD reactor is converted into high-purity polysilicon.

In order to improve efficiency, the HCl, $SiCl_4$, $H_2$, and unreacted $SiHCl_3$ are separated and recovered for recycling. Figure 12.50 illustrates the entire chlorosilane process starting with MGS and including the recycling of the reaction byproducts to achieve high overall process efficiency. As a consequence, the production cost of high-purity EGS depends on the commercial usefulness of the byproduct, $SiCl_4$. Additional disadvantages of the Seimens process are derived from its relatively small batch size, slow growth rate, and high-power consumption. These issues lead to the investigation of alternative cost-efficient routes to EGS.

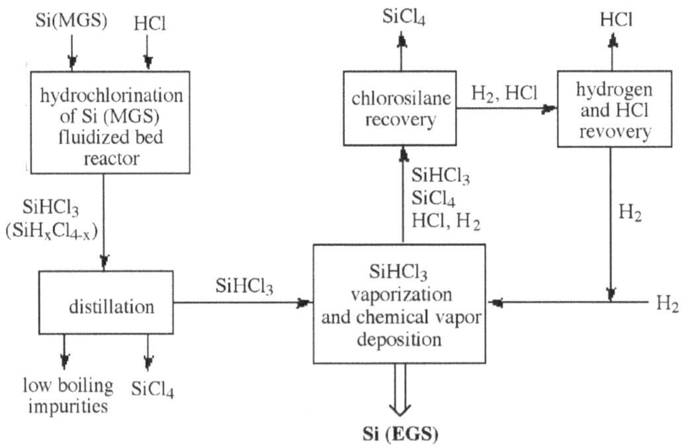


**Figure 12.50: Schematic representation of the reaction pathways for the formation of EGS using the chlorosilane process.**

*Silane process*

An alternative process for the production of EGS that has begun to receive commercial attention is the pyrolysis of silane ($SiH_4$). The advantages of producing EGS from $SiH_4$ instead of $SiHCl_3$ are potentially lower costs associated with lower reaction temperatures, and less harmful byproducts. Silane decomposes < 900 °C to give silicon and hydrogen,

$$SiH_4(vapor) \rightarrow Si(solid) + 2\ H_2(gas)$$

Silane may be prepared by a number of routes, each having advantages with respect to purity and production cost. The simplest process involves the direct reaction of MGS powders with magnesium at 500 °C in a hydrogen atmosphere, to form magnesium silicide ($Mg_2Si$). The magnesium silicide is then reacted with ammonium chloride in liquid ammonia below 0 °C,

$$Mg_2Si + 4\ NH_4Cl \rightarrow SiH_4 + 2\ MgCl_2 + 5\ NH_3$$

This process is ideally suited to the removal of boron impurities (a p-type dopant in Si), because the diborane ($B_2H_6$) produced during the reaction forms the Lewis acid-base complex, $H_3B(NH_3)$, whose volatility is sufficiently lower than $SiH_4$, allowing for the purification of the latter. It is possible to prepare EGS with a boron content of ≤20 ppt using $SiH_4$ synthesized in this manner. However, phosphorus (another dopant) in the form of $PH_3$ may be present as a contaminant requiring subsequent purification of the $SiH_4$.

Alternative routes to $SiH_4$ involve the chemical reduction of $SiCl_4$ by either lithium hydride,

$$SiCl_4 + 4\ LiH \rightarrow SiH_4 + 4\ LiCl$$

lithium aluminum hydride,

$$SiCl_4 + 4\ LiAlH_4 \rightarrow SiH_4 + LiCl + AlCl_3$$

or via hydrogenation in the presence of elemental silicon.

$$SiCl_4 + 2\ H_2 + Si(98\%) \rightarrow 4\ SiHCl_3$$

$$2\,SiHCl_3 \rightarrow SiH_2Cl_2 + SiCl_4$$

$$3\,SiH_2Cl_2 \rightarrow SiH_3Cl + 2\,SiHCl_3$$

$$2\,SiH_3Cl \rightarrow SiH_4 + SiH_2Cl_2$$

The hydride reduction reactions may be carried-out on relatively large scales (ca. 50 kg), but only batch processes. In contrast, Union Carbide has adapted the hydrogenation to a continuous process, involving disproportionation reactions of chlorosilanes and the fractional distillation of silane, Table 12.18.

Pyrolysis of silane on resistively heated polysilicon laments at 700 – 800 °C yields polycrystalline EGS. As noted above, the EGS formed has remarkably low boron impurities compared with material prepared from trichlorosilane. Moreover, the resulting EGS is less contaminated with transition metals from the reactor container because $SiH_4$ decomposition does not cause as much of a corrosion problem as is observed with halide precursor compounds.

*Granular polysilicon deposition*

Both the chlorosilane (Seimens) and silane processes result in the formation of rods of EGS. However, there has been increased interest in the formation of granular polycrystalline EGS. This process was developed in 1980's and relies on the decomposition of $SiH_4$ in a fluidized-bed deposition reactor to produce free- owing granular polysilicon.

Tiny silicon particles are fluidized in a $SiH_4/H_2$ flow, and act as seed crystal onto which polysilicon deposits to form free- owing spherical particles. The size distribution of the particles thus formed is over the range from 0.1 to 1.5 mm in diameter with an average particle size of 0.7 mm. The fluidized-bed seed particles are originally made by grinding EGS in a ball (or hammer) mill and leaching the product with acid, hydrogen peroxide, and water. This process is time-consuming and costly, and tended to introduce undesirable impurities from the metal grinders. In a new method, large EGS particles are fired at each other by a high-speed stream of inert gas and the collision breaks them down into particles of suitable size for a fluidized bed. This process has the main advantage that it introduces no foreign materials and requires no leaching or other post purification.

The fluidized-bed reactors are much more efficient than traditional rod reactors as a consequence of the greater surface area available during CVD growth of silicon. It has been suggested that fluidized-bed reactors require $^1/_5$ to $^1/_{10}$ the energy, and half the capital cost of the traditional process. The quality of fluidized-bed polysilicon has proven to be equivalent to polysilicon produced by the conventional methods. Moreover, granular EGS in a free- owing form, and with high bulk density, enables crystal growers to obtain the high, reproducible production yields out of each crystal growth run. For example, in the Czochralski crystal growth process, crucibles can be quickly and easily filled to uniform loading with granular EGS, which typically exceed those of randomly stacked polysilicon chunks produced by the Siemens silane process.

*Zone refining*

The technique of zone refining is used to purify solid materials and is commonly employed in metallurgical refining. In the case of silicon may be used to obtain the desired ultimate purity of EGS, which has already been purified by chemical processes. Zone refining was invented by Pfann (Figure 12.51) and makes use of the fact that the equilibrium solubility of any impurity (e.g., Al) is different in the solid and liquid phases of a material (e.g., Si). For the dilute solutions, as is observed in EGS silicon, an equilibrium segregation coefficient ($k_0$) is defined by $k_0 = C_s/C_l$, where $C_s$ and $C_l$ are the equilibrium concentrations of the impurity in the solid and liquid near the interface, respectively.

**Figure 12.51. American materials scientist William Gardner Pfann (on left) inventor of zone refining (1917 - 1982).**

If $k_0$ is less than 1 then the impurities are left in the melt as the molten zone is moved along the material. In a practical sense a molten zone is established in a solid rod. The zone is then moved along the rod, from left to right. If $k_0$ $< 1$ then the frozen part left on the trailing edge of the moving molten zone will be purer than the material that melts in on the right-side leading edge of the moving molten zone. Consequently, the solid to the left of the molten zone is purer than the solid on the right. At the completion of the first pass the impurities become concentrated to the right of the solid sample. Repetition of the process allows for purification to exceptionally high levels. Table 12.19. lists the equilibrium segregation coefficients for common impurity and dopant elements in silicon; it should be noted that they are all less than 1.

| Element | $k_0$ | Element | $k_0$ |
|---|---|---|---|
| aluminum | 0.002 | iron | $8 \times 10^{-6}$ |
| boron | 0.8 | oxygen | 0.25 |
| carbon | 0.07 | phosphorus | 0.35 |
| copper | $4 \times 10^{-6}$ | antimony | 0.023 |

**Table 12.19: Segregation coefficients for common impurity and dopant elements in silicon.**

*Oxidation of silicon*

In the fabrication of integrated circuits (ICs), the oxidation of silicon is essential, and the production of superior ICs requires an understanding of the oxidation process and the ability to form oxides of high quality. Silicon dioxide has several uses:

1. Serves as a mask against implant or diffusion of dopant into silicon.
2. Provides surface passivation.
3. Isolates one device from another (dielectric isolation).
4. Acts as a component in MOS structures.
5. Provides electrical isolation of multi-level metallization systems.

Methods for forming oxide layers on silicon have been developed, including thermal oxidation, wet anodization, chemical vapor deposition (CVD), and plasma anodization or oxidation. Generally, CVD is used when putting the oxide layer on top of a metal surface, and thermal oxidation is used when a low-charge density level is required for the interface between the oxide and the silicon surface.

Silicon's surface has a high affinity for oxygen and thus an oxide layer rapidly forms upon exposure to the atmosphere. The chemical reactions which describe this formation are:

$$Si_{(s)} + O_{2(g)} \rightarrow SiO_{2(s)}$$

$$Si_{(s)} + 2\,H_2O_{(g)} \rightarrow SiO_{2(s)} + 2\,H_{2(g)}$$

In the first reaction a dry process is utilized involving oxygen gas as the oxygen source and the second reaction describes a wet process which uses steam. The dry process provides a "good" silicon dioxide but is slow and mostly used at the beginning of processing. The wet procedure is problematic in that the purity of the water used cannot be guaranteed to a suitable degree. This problem can be easily solved using a pyrogenic technique which combines hydrogen and oxygen gases to form water vapor of very high purity. Maintaining reagents of high quality is essential to the manufacturing of integrated circuits and is a concern which plagues each step of this process.

The formation of the oxide layer involves shared valence electrons between silicon and oxygen, which allows the silicon surface to rid itself of "dangling" bonds, such as lone pairs and vacant orbitals, Figure 12.52. These vacancies create mid-gap states between the valence and conduction bands, which prevents the desired band gap of the semiconductor. The Si-O bond strength is covalent (strong), and so can be used to achieve the loss of mid-gap states and passivate the surface of the silicon.

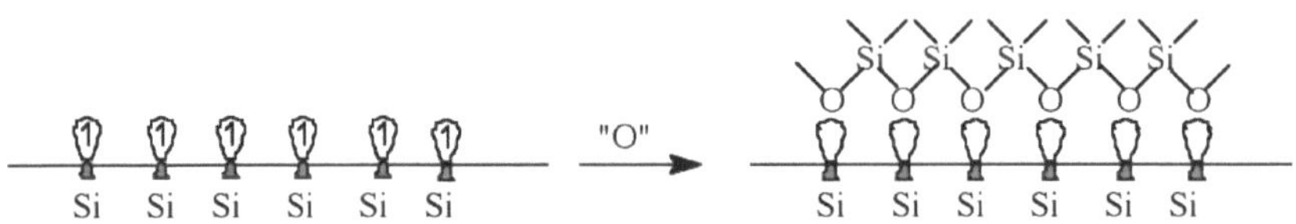


**Figure 12.52: Removal of dangling bonds by oxidation of surface.**

The oxidation of silicon occurs at the silicon-oxide interface and consists of four steps:

1. Diffusive transport of oxygen across the diffusion layer in the vapor phase adjacent to the silicon oxide-vapor interface.
2. Incorporation of oxygen at the outer surface into the silicon oxide film.
3. Diffusive transport across the silicon oxide film to its interface with the silicon lattice.
4. Reaction of oxygen with silicon at this inner interface.

As the Si-$SiO_2$ interface moves into the silicon its volume expands and based upon the densities and molecular weights of Si and $SiO_2$, 0.44 Å Si is used to obtain 1.0 Å $SiO_2$.

*Pre-oxidation cleaning*

The first step in oxidizing a surface of silicon is the removal of the native oxide which forms due to exposure to open air. This may seem redundant to remove an oxide only to put on another, but this is necessary since uncertainty exists as to the purity of the oxide which is present. The contamination of the native oxide by both organic and inorganic materials (arising from previous processing steps and handling) must be removed to prevent the degradation of the essential electrical characteristics of the device. A common procedure uses a $H_2O$-$H_2O_2$-$NH_4OH$ mixture which removes the organics present, as well as some group I and II metals. Removal of heavy metals can be achieved using a $H_2O$-$H_2O_2$-HCl mixture, which complexes with the ions which are formed. After removal of the native oxide, the desired oxide can be grown. This growth is useful because it provides: chemical protection, conditions suitable for lithography, and passivation. The protection prevents unwanted reactions from occurring and the passivation fills vacancies of bonds on the surface not present within the interior of the crystal. Thus, the oxidation of the surface of silicon fulfils several functions in one step.

*Thermal oxidation*

The growth of oxides on a silicon surface can be a particularly tedious process, since the growth must be uniform and pure. The thickness wanted usually falls in the range 50 - 500 Å, which can take a long time and must be done on a large scale. This is done by stacking the silicon wafers in a horizontal quartz tube while the oxygen source flows over the wafers, which are situated vertically in a slotted paddle (boat), see Figure 12.53. This procedure is performed at 1 atm pressure, and the temperature ranges from 700 to 1200 °C, being held to within ±1 °C to ensure uniformity. The choice of oxidation technique depends on the thickness and oxide properties required. Oxides that are relatively thin and those that require low charge at the interface are typically grown in dry oxygen. When thick oxides are required (>0.5 mm) are desired, steam is the source of choice. Steam can be used at wide range of pressures (1 atm to 25 atm), and the higher pressures allow thick oxide growth to be achieved at moderate temperatures in reasonable amounts of time.

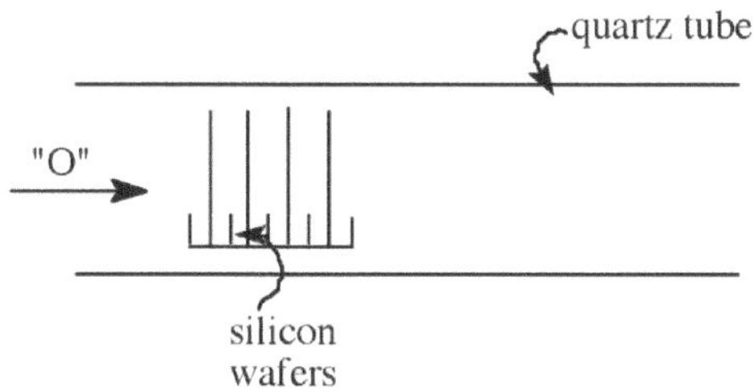


**Figure 12.53: Horizontal diffusion tube showing the oxidation of silicon wafers at 1 atm pressure.**

The thickness of $SiO_2$ layers on a Si substrate is readily determined by the color of the film. Table 12.20 provides a guideline for thermal grown oxides.

*High pressure oxidation*

High pressure oxidation is another method of oxidizing the silicon surface which controls the rate of oxidation. This is possible because the rate is proportional to the concentration of the oxide, which in turn is proportional to the partial pressure of the oxidizing species, according to Henry's law,

$$C = H_{(pO)}$$

where C is the equilibrium concentration of the oxide, H is Henry's law constant, and $p_O$ is the partial pressure of the oxidizing species. This approach is fast, with a rate of oxidation ranging from 100 to 1000 mm/h, and also occurs at a relatively low temperature. It is a useful process, preventing dopants from being displaced and also forms a low number of defects, which is most useful at the end of processing.

*Plasma oxidation*

Plasma oxidation and anodization of silicon is readily accomplished by the use of activated oxygen formed within an electrical discharge or plasma. The oxidation is carried out in a low pressure (0.05 - 0.5 Torr) chamber, and the plasma is produced either by a DC electron source or a high-frequency discharge. In simple plasma oxidation the silicon wafer is held at ground potential. In contrast, aniodization systems usually have a DC bias between the sample and an electrode with the sample biased positively with respect to the cathode. Platinum electrodes are commonly used as the cathodes.

| Film thickness (μm) | Color | Film thickness (μm) | Color |
|---|---|---|---|
| 0.05 | tan | 0.63 | violet-red |
| 0.07 | brown | 0.68 | "bluish" |
| 0.10 | dark to red-violet | 0.72 | blue-green to green |
| 0.12 | royal blue | 0.77 | "yellowish" |
| 0.15 | light to metallic blue | 0.80 | orange |
| 0.17 | light yellow- green | 0.82 | salmon |
| 0.20 | light gold | 0.85 | light red-violet |
| 0.22 | gold | 0.86 | violet |
| 0.25 | orange to melon | 0.87 | blue violet |
| 0.27 | red-violet | 0.89 | blue |
| 0.30 | blue to violet blue | 0.92 | blue-green |
| 0.31 | blue | 0.95 | yellow-green |
| 0.32 | blue to blue-green | 0.97 | yellow |
| 0.34 | light green | 0.99 | orange |
| 0.35 | green to yellow-green | 1.00 | carnation pink |
| 0.36 | yellow-green | 1.02 | violet red |
| 0.37 | green-yellow | 1.05 | red-violet |
| 0.39 | yellow | 1.06 | violet |
| 0.41 | light orange | 1.07 | blue-violet |
| 0.42 | carnation pink | 1.10 | green |
| 0.44 | violet-red | 1.11 | yellow-green |
| 0.46 | red-violet | 1.12 | green |
| 0.47 | violet | 1.18 | violet |
| 0.48 | blue-violet | 1.19 | red-violet |
| 0.49 | blue | 1.21 | violet-red |
| 0.50 | blue green | 1.24 | carnation pink |
| 0.52 | green | 1.25 | orange |
| 0.54 | yellow-green | 1.28 | "yellowish" |
| 0.56 | green-yellow | 1.32 | sky to green-blue |
| 0.57 | "yellowish" | 1.40 | orange |
| 0.58 | light orange to pink | 1.46 | blue-violet |
| 0.60 | carnation pink | 1.50 | blue |

**Table 12.20: Color chart for thermally grown $SiO_2$ films, observed under day-light fluorescent lighting.**

There have been at least 34 different reactions reported to occur in an oxygen plasma, however, the vast majority of these are inconsequential with respect to the formation of active species. Furthermore, many of the potentially active species are sufficiently short lived that it is unlikely that they make a significant contribution. The primary active species within the oxygen plasma are undoubtedly $O^-$ and $O^{2+}$. Both being produced in near equal quantities, although only the former is relevant to plasma aniodization. While these species may be active with respect to surface oxidation, it is more likely that an electron transfer occurs from the semiconductor surface yields activated oxygen species, which are the actual reactants in the oxidation of the silicon.

The significant advantage of plasma processes is that while the electron temperature of the ionized oxygen gas is in excess of 10,000 K, the thermal temperatures required are significantly lower than required for the high-pressure method, i.e., < 600 °C. The advantages of the lower reaction temperatures include: the minimization of dopant diffusion and the impediment of the generation of defects. Despite these advantages there are two primary disadvantages of any plasma-based process. First, the high electric fields present during the processes cause damage to the resultant oxide, in particular, a high density of interface traps often results. However, post annealing may improve film quality. Second, the growth rates of plasma oxidation are low, typically 1000 Å/h. This growth rate is increased by about a factor of 10 for plasma aniodization, and further improvements are observed if 1 - 3% chlorine is added to the oxygen source.

*Masking*

A selective mask against the diffusion of dopant atoms at high temperatures can be found in a silicon dioxide layer, which can prove to be very useful in integrated circuit processing. A pre-deposition of dopant by ion implantation, chemical diffusion, or spin-on techniques typically results in a dopant source at or near the surface of the oxide. During the initial high-temperature step, diffusion in the oxide must be slow enough with respect to diffusion in the silicon that the dopants do not di use through the oxide in the masked region and reach the silicon surface. The required thickness may be determined by experimentally measuring, at a particular temperature and time, the oxide thickness necessary to prevent the inversion of a lightly doped silicon substrate of opposite conductivity. To this is then added a safety factor, with typical total values ranging from 0.5 to 0.7 mm. The impurity masking properties result when the oxide is partially converted into a silica impurity oxide "glass" phase and prevents the impurities from reaching the $SiO_2$-Si interface.

*Applications for Silica Thin Films*

While the physical properties of silica make it suitable for use in protective and optical coating applications, the biggest application of insulating $SiO_2$ thin films is undoubtedly in semiconductor devices, in which the insulator performs a number of specific tasks, including, surface passivation, field effect transistor (FET) gate layer, isolation layers, planarization and packaging.

The term insulator generally refers to a material that exhibits low thermal or electrical conductivity; electrically insulating materials are also called dielectrics. It is in regard to the high resistance to the ow of an electric current that $SiO_2$ thin films are of the greatest commercial importance. The dielectric constant (ε) is a measure of a dielectric materials ability to store charge and is characterized by the electrostatic energy stored per unit volume across a unit potential gradient. The magnitude of ε is an indication of the degree of polarization or charge displacement within a material. The dielectric constant for air is 1, and for ionic solids is generally in the range of 5 - 10. Dielectric constants are defined as the ratio of the material's capacitance to that of air, i.e.,

$$\varepsilon = C_{material}/C_{air}$$

The dielectric constant for silicon dioxide ranges from 3.9 to 4.9, for thermally and plasma CVD grown films, respectively.

An insulating layer is a film or deposited layer of dielectric material separating or covering conductive layers. Ideally, in these applications an insulating material should have a surface resistivity of greater than $10^{13}$ $\Omega/cm^2$ or a volume resistivity of greater than $10^{11}$ Ω.cm. However, for some applications, lower values are acceptable; an electrical insulator is generally accepted to have a resistivity greater than $10^5$ Ω.cm. CVD $SiO_2$ thin films have a resistivity of $10^6$ - $10^{16}$ Ω.cm, depending on the film growth method.

As a consequence of its dielectric properties $SiO_2$ and related silicas, are used for isolating conducting layers, to facilitate the diffusion of dopants from doped oxides, as diffusion and ion implantation masks, capping doped films to prevent loss of dopant, for gettering impurities, for protection against moisture and oxidation, and for electronic passivation. Of the many methods used for the deposition of thin films, chemical vapor deposition (CVD) is most often used for semiconductor processing. In order to appreciate the unique problems associated with the CVD of insulating $SiO_2$ thin films it is worth first reviewing some of their applications. Summarized below are three areas of greatest importance to the fabrication of contemporary semiconductor devices: isolation and gate insulation, passivation, and planarization.

*Device isolation and gate insulation*

A microcircuit may be described as a collection of devices each consisting of "an assembly of active and passive components, interconnected within a monolithic block of semiconducting material". Each device is required to be isolated from adjacent devices in order to allow for maximum efficiency of the overall circuit. Furthermore, within a device, contacts must also be electrically isolated. While there are a number of methods for isolating individual devices within a circuit (reverse-biased junctions, mesa isolation, use of semi-insulating substrates, and oxide isolation), the isolation of the active components in a single device is almost exclusively accomplished by the deposition of an insulator.

In Figure 12.54 is shown a schematic representation of a silicon MOSFET (metal-oxide-semiconductor field effect transistor). The MOSFET is the basic component of silicon-CMOS (complimentary metal-oxide-semiconductor) circuits which, in turn, form the basis for logic circuits, such as those used in the CPU (central processing unit) of a modern personal computer. It can be seen that the MOSFET is isolated from adjacent devices by a reverse-biased junction ($p^+$-channel stop) and a thick oxide layer. The gate, source and drain contact are electrically isolated from each other by a thin insulating oxide. A similar scheme is used for the isolation of the collector from both the base and the emitter in bipolar transistor devices.

As a transistor, a MOSFET has many advantages over alternate designs. The key advantage is low power dissipation resulting from the high impedance of the device. This is a result of the thin insulation layer between the channel (region between source and drain) and the gate contact, see Figure 12.54. The presence of an insulating gate is characteristic of a general class of devices called MISFETs (metal-insulator-semiconductor field effect transistor). MOSFETs are a subset of MISFETs where the insulator is specifically an oxide, e.g., in the case of a silicon MISFET device the insulator is $SiO_2$, hence MOSFET. It is the fabrication of MOSFET circuits that has allowed silicon technology to dominate digital electronics (logic circuits). However, increases in computing power and speed require a constant reduction in device size and increased complexity in device architecture.

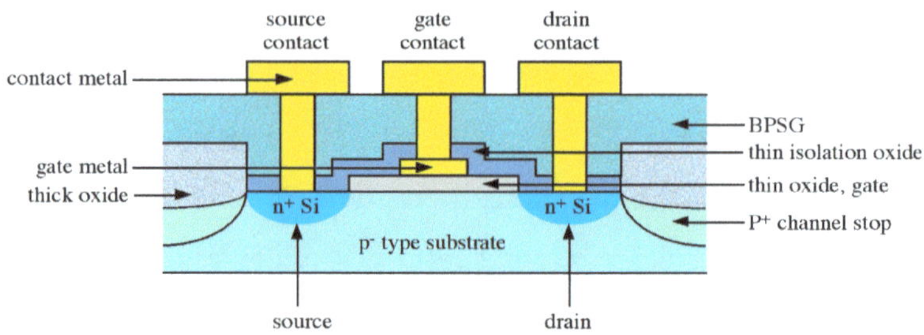


**Figure 12.54: Schematic diagrams of a Si-MOSFET (metal-oxide-semiconductor field effect transistor).**

*Passivation*

Passivation is often defined as a process whereby a film is grown on the surface of a semiconductor to:

- chemically protect it from the environment,
- provide electronic stabilization of the surface.

From the earliest days of solid-state electronics, it has been recognized that the presence or absence of surface states plays a decisive role in the usefulness of any semiconducting material. On the surface of any solid-state material there are sites in which the coordination environment of the atoms is incomplete. These sites, commonly termed "dangling bonds", are the cause of the electronically active states which allow for the recombination of holes and electrons. This recombination occurs at energies below the bulk value and interferes with the inherent properties of the semiconductor. In order to optimize the properties of a semiconductor device it is desirable to covalently satisfy all these surface bonds, thereby shifting the surface states out of the band gap and into the valence or conduction bands. Electronic passivation may therefore be described as a process which reduces the density of available electronic states present at the surface of a semiconductor, thereby limiting hole and electron recombination possibilities. In the case of silicon both the native oxide and other oxides admirably fulfil these requirements.

Chemical passivation requires a material that inhibits the diffusion of oxygen, water, or other species to the surface of the underlying semiconductor. In addition, the material is ideally hard and resistant to chemical attack. A perfect passivation material would satisfy both electronic and chemical passivation requirements.

*Planarization*

For the vast majority of electronic devices, the starting point is a substrate consisting of a flat single crystal wafer of semiconducting material. During processing, which includes the growth of both insulating and conducting films, the surface becomes increasingly non-planar. For example, a gate oxide in a typical MOSFET (see Figure 12.54) may be typically 100 – 250 Å thick, while the isolation or field oxide may be 10,000 Å. In order for the successful subsequent deposition of conducting layers (metallization) to occur without breaking metal lines (often due to the difficulty in maintaining step coverage), the surface must be at and smooth. This process is called planarization and can be carried out by a technique known as sacrificial etchback. The steps for the etchback process are outlined in Figure 12.55.

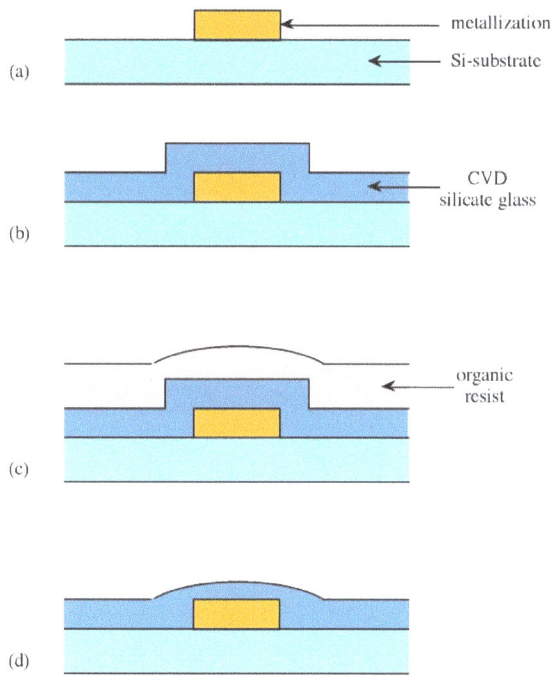


**Figure 12.55: Schematic representation of the planarization process. A metallization feature (a) is CVD covered with silicate glass (b), and subsequently coated with an organic resin (c). After etching the resist, a smooth silicate surface is produced (d).**

An abrupt step (Figure 12.55a) is coated with a conformal layer of a low melting dielectric, e.g., borophosphorosilicate glass, BPSG (Figure 12.55b), and subsequently a sacrificial organic resin (Figure 12.55c). The sample is then plasma etched such that the resin and dielectric are removed at the same rate. Since the plasma etch follows the contour of the organic resin, a smooth surface is left behind (Figure 12.55d). The planarization process thus reduces step height differentials significantly. In addition, regions or valleys between individual metallization elements (vias) can be completely filled allowing for a route to producing uniformly at surfaces, e.g., the BPSG film shown in Figure 12.54.

The process of planarization is vital for the development of multilevel structures in VLSI circuits. To minimize interconnection resistance and conserve chip area, multilevel metallization schemes are being developed in which the interconnects run in 3-dimensions.

# Chapter 13: Group 15 The Pnictogens

## The elements

The Group 15 elements have a particular name *pnictogens*. Despite the modern IUPAC notation, the Group 15 elements are still referred to as Group V elements in particular by the semiconductor industry. Table 13.1 lists the derivation of the names of the Group 15 elements.

| Element | Symbol | Name |
|---|---|---|
| Nitrogen | N | Latin *nitrogenium*, where *nitrum* (derived from Greek *nitron*) means saltpetre |
| Phosphorus | P | From the Greek *phosphoros* meaning bringer of light |
| Arsenic | As | Derived from Syriac *zarniqa* and Persian *zarnikh*, meaning yellow orpiment |
| Antimony | Sb | Greek *anti* and *monos* meaning not alone. The symbol Sb from Latin *stibium* |
| Bismuth | Bi | New Latin *bisemutum* from German w*ismuth*, meaning white mass |

**Table 13.1: Derivation of the names of each of the Group 15 (V) elements.**

According to the Oxford English Dictionary, the correct spelling of the element is phosphorus. The word phosphorous is the adjectival form of the $P^{3+}$ valence. In the same way that sulfur forms sulfurous and sulfuric compounds, phosphorus forms phosphorous compounds (e.g., phosphorous acid) and $P^{5+}$ valency phosphoric compounds (e.g., phosphoric acids and phosphates).

## Discovery

### *Nitrogen*

Nitrogen was discovered by Rutherford (Figure 13.1) in 1772. He called it noxious air or fixed air because there it had been known since the late 18$^{th}$ century that there was a fraction of air that did not support combustion. Nitrogen was also studied by Scheele (Figure 9.5), Cavendish (Figure 7.3), and Priestley (Figure 13.2), who referred to it as burnt air or *phlogisticated* air.

**Figure 13.1: Scottish chemist and physician Daniel Rutherford (1749 - 1819).**

**Figure 13.2: Portrait (by Ellen Sharples) of British clergyman natural philosopher, educator, and political theorist Joseph Priestley (1733 - 1804).**

*Phosphorus*

German alchemist Hennig Brand (Figure 13.3) was experimenting with urine (which contains dissolved phosphates) in 1669. While attempting to create the fabled philosopher's stone (the legendary alchemical substance capable of turning base metals, such as lead, into gold) by the distillation of salts from urine, he produced a white material that glowed in the dark and burned with a brilliant light. He gave the substance the name phosphorus mirabilis (miraculous bearer of light). His process involved letting the urine stand for days then boiling it down to a paste which led to a white waxy substance, white phosphorus.

**Figure 13.3: The discovery of phosphorus by German merchant and alchemist Hennig Brand (1630 - 1710) as depicted by Joseph Wright (with significant artistic license with regard to the brightness of the chemiluminescence).**

Brand sold the recipe for 200 thaler (a silver coin from whose name the dollar is derived) to D Krafft who toured much of Europe showing it. During his journeys he met Robert Boyle (Figure 7.2) who without learning the details of the synthesis recreated and improved it by using sand in the reduction of the phosphate,

$$4\ NaPO_3 + 2\ SiO_2 + 10\ C \rightarrow 2\ Na_2SiO_3 + 10\ CO + P_4$$

*Arsenic*

Arsenic sulfides and oxides were known since ancient times. Zosimos (ca. 300 AD) describes roasting *sandarach* (realgar, $\alpha$-$As_4S_4$) to obtain cloud of arsenious oxide ($As_2O_3$) that he reduced to metallic arsenic (Figure 13.4).

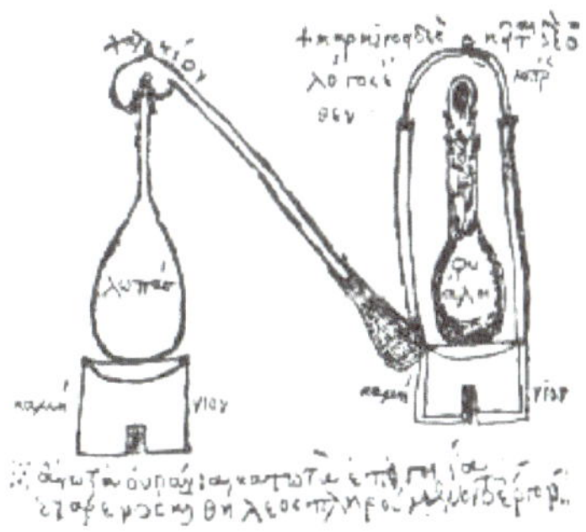

**Figure 13.4: The distillation apparatus of Zosimos.**

*Antimony*

Antimony(III) sulfide, $Sb_2S_3$ was known as early as 3000 BC. Pastes of $Sb_2S_3$ powder in fat were used as eye cosmetics in the Middle East. An artifact made of antimony dating to about 3000 BC was found at Tello (part of present-day Iraq), and copper objects plated with antimony from 2500 - 2200 BC have been found in Egypt. The first European description of a procedure for isolating antimony is in the book *De la pirotechnia* by Vannoccio Biringuccio (Figure 13.5).

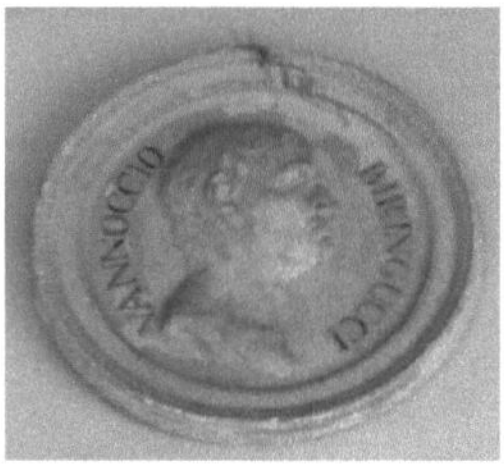


**Figure 13.5: Italian metallurgist Vannoccio Biringuccio (1480 - 1539) as depicted in the Specola Museum in Florence.**

*Bismuth*

Since bismuth was known in ancient times, no one person is credited with its discovery. However, the French chemist Claude François Geoffroy (Figure 13.6) demonstrated in 1753 that this metal is distinct from lead and tin.

**Figure 13.6. French chemist Claude François Geoffroy (1729 - 1753). Before his work time, bismuth-containing minerals were frequently misidentified as either lead, tin or antimony ores.**

## Abundance

The abundance of the Group 15 elements is given in Table 13.2.

| Element | Terrestrial abundance (ppm) |
|---|---|
| N | 25 (Earth's crust), 5 (soil), 0.5 (sea water), $78 \times 10^4$ (atmosphere) |
| P | 1000 (Earth's crust), 0.65 (soil), $60 \times 10^{-3}$ (sea water), trace (atmosphere) |
| As | 1.5 (Earth's crust), 10 (soil), $16 \times 10^{-3}$ (sea water), trace (atmosphere) |
| Sb | 0.2 (Earth's crust), 1 (soil), $0.3 \times 10^{-3}$ (sea water) |
| Bi | $48 \times 10^{-3}$ (Earth's crust), 0.25 (soil), $400 \times 10^{-6}$ (sea water) |

**Table 13.2: Abundance of the Group 15 elements.**

## Isotopes

The naturally abundant isotopes of the Group 15 elements are listed in Table 13.3.

| Isotope | Natural abundance (%) |
|---|---|
| Nitrogen-14 | 99.634 |
| Nitrogen-15 | 0.0366 |
| Phosphorus-31 | 100 |
| Arsenic-75 | 100 |
| Antimony-121 | 57.36 |
| Antimony-123 | 42.64 |
| Bismuth-209 | 100% |

**Table 13.3: Abundance of the non-synthetic isotopes of the Group 15 elements.**

Two radioactive isotopes of phosphorus ($^{32}P$ and $^{33}P$) have half-lives that make them useful for scientific experiments (14.262 and 25.34 days, respectively). $^{32}P$ is a β-emitter (1.71 MeV) and is used to produce radiolabeled DNA and RNA probes. Due to the high energy of the β particles which can penetrate skin and corneas, and because any $^{32}P$ ingested, inhaled, or absorbed is incorporated into bone and nucleic acids extreme care needs to be taken in handling. The lower energy β particles emitted from $^{33}P$ (0.25 MeV) make it useful for applications such as DNA sequencing.

While bismuth is traditionally regarded as the element with the heaviest stable isotope, $^{209}Bi$, it had long been suspected to be unstable on theoretical grounds. In 2003 researchers at the Institut d'Astrophysique Spatiale in Orsay,

France, measured the alpha emission half-life of $^{209}Bi$ to be $1.9 \times 10^{19}$ years, over a billion times longer than the current estimated age of the universe!

## Industrial production of the elements

Nitrogen is the largest constituent of the Earth's atmosphere (78.082% by volume, 75.3% by weight). It is created by fusion processes in stars and is estimated to be the 7th most abundant element by mass in the universe. Industrial gas produced is by the fractional distillation of liquid air, or by mechanical means using gaseous air (i.e., pressurized reverse osmosis membrane or pressure swing adsorption). Commercial nitrogen is often a by-product of air processing for industrial concentration of oxygen for steelmaking, etc.

White phosphorus was originally made commercially for the match industry in the 19th century, by distilling o phosphorus vapor from precipitated phosphates, mixed with ground coal or charcoal,

$$Ca_{10}(PO_4)_6F_2 + 15\,C + 9\,SiO_2 \rightarrow 6\,P_{n(g)} + 9\,|(CaO.SiO_2)| + CaF_2 + 15\,CO_{(g)}$$

The precipitated phosphates were made from ground-up bones that had been de-greased and treated with strong acids. This process is, however, obsolete due to the submerged-arc furnace for phosphorus production was introduced to reduce phosphate rock. Calcium phosphate (phosphate rock) is heated to 1200 - 1500 °C with $SiO_2$ and coke (impure carbon) to produce vaporized tetraphosphorus, $P_4$.

## Physical properties

The physical properties of the Group 15 elements (Table 13.4) encompasses a gas ($N_2$), a non-metallic solid ($P_4$), metalloids (As and Sb), and a metal (Bi).

*Vapor phase*

Nitrogen forms a dimer in the vapor phase with a triple bond (Figure 8.8). In the vapor phase above 800 °C tetraphosphorus ($P_4$) is partially dissociated to $P_2$.

| Element | Melting point (°C) | Boiling point (°C) | Density (g/cm³) |
|---|---|---|---|
| N | -210.00 | -195.79 | 1.251 g/L (0 °C @ 101.325 kPa) |
| P | 44.2 (white), 610 (black) | 280.5 (white), 416 - 590 (sub., red), 620 (sub., violet) | 1.823 (white), 2.2 - 2.34 (red), 2.36 (violet), 2.69 (black) |
| As | 817 | 615 (sub.) | 5.727 |
| Sb | 630.63 | 1587 | 6.697 (solid), 6.53 (liquid) |
| Bi | 271.5 | 1564 | 9.78 (solid), 10.05 (liquid) |

**Table 13.4: Selected physical properties of the Group 15 elements.**

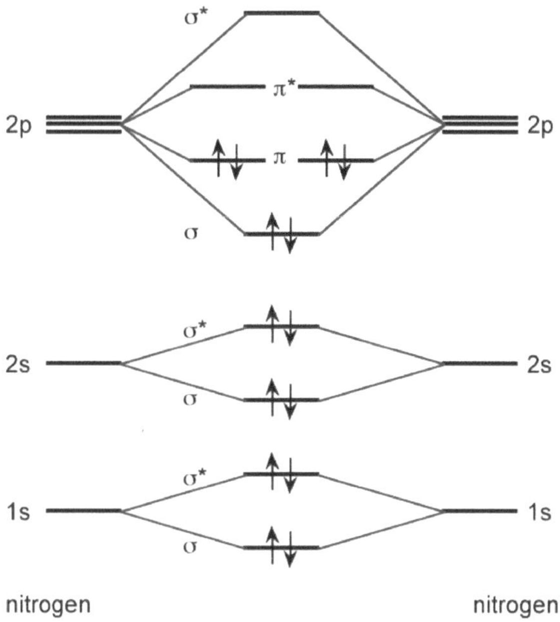


**Figure 13.7: Molecular orbital diagram for the formation of $N_2$.**

*Solid state*

Phosphorus forms a number of allotropes with very different properties (Figure 13.8). Red phosphorus is an intermediate phase between the white and violet forms. Scarlet phosphorus is obtained by allowing a solution of white phosphorus in carbon disulfide to evaporate in sunlight. Black phosphorus is formed by heating white phosphorus under high pressures (ca. 12,000 atmospheres).

**Figure 13.8: Images of (left to right) white, red, violet and black allotropes of phosphorus.**

White phosphorus has two forms, low-temperature β form and high-temperature α form; both of which contain the $P_4$ tetrahedron (Figure 13.9). White phosphorus is the least stable, the most reactive, most volatile, less dense, and most toxic of the allotropes.

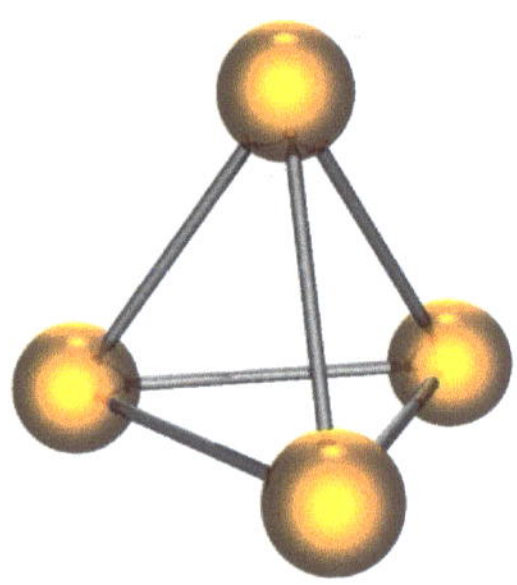

**Figure 13.9: The structure of the $P_4$ molecule in white phosphorus.**

The structural relationship between white and red phosphorus involves breaking one of the P-P bonds in the $P_4$ unit and forming a bond with a neighboring tetrahedron to give a chain structure (Figure 13.10). Red phosphorus is formed by heating white phosphorus to 250 °C or by exposing white phosphorus to sunlight. Actually, red phosphorus is not a single allotrope, but rather an intermediate phase between the white and violet phosphorus, and most of its properties have a range of values (Table 13.4).

Violet phosphorus (Figure 13.11) is the thermodynamic stable form of phosphorus that is produced by heating red phosphorus above 550 °C. Due to the synthesis being developed by Johann Hittorf (Figure 13.12) it is sometimes known as Hittorf's phosphorus.

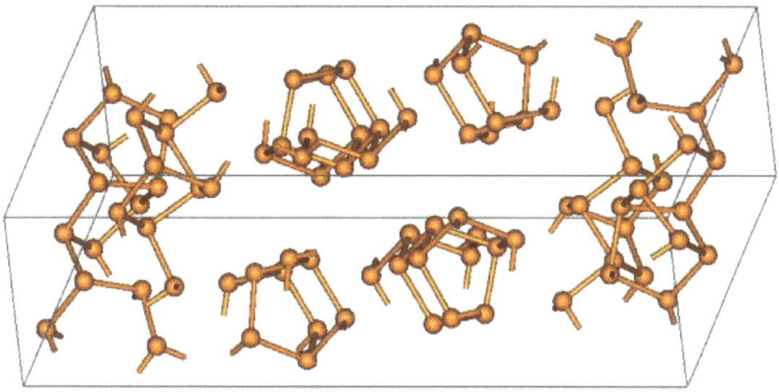

**Figure 13.10: Crystal unit cell of red phosphorus.**

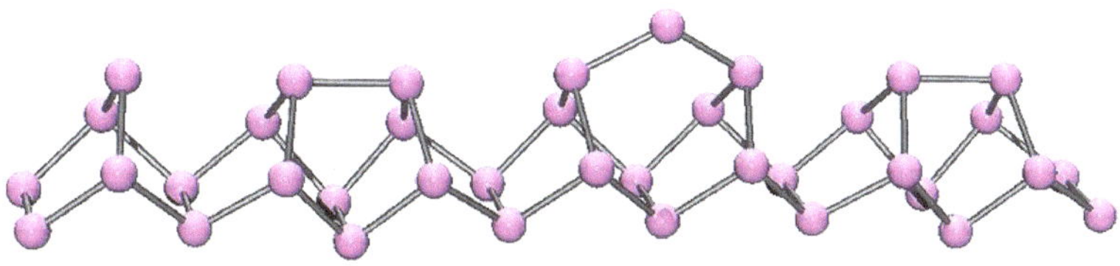

**Figure 13.11: Structure of violet (Hittorf's) phosphorus.**

**Figure 13.12: German physicist Johann Wilhelm Hittorf (1824 - 1914).**

Black phosphorus is the least reactive allotrope and the thermodynamic stable form below 550 °C. It is also known as β-metallic phosphorus and has a structure somewhat resembling that of graphite (Figure 13.13).

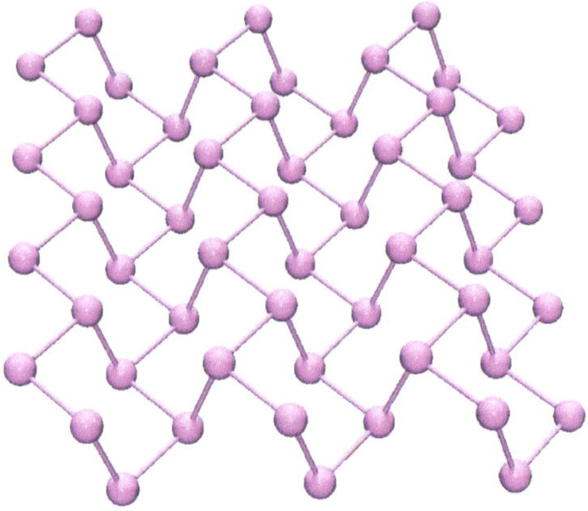

**Figure 8.14: Crystal structure of black phosphorus.**

In a similar manner to phosphorus, arsenic has several allotropes some of which a structurally related to those of phosphorus. Grey arsenic has a structure similar to black phosphorus (Figure 13.13). Yellow arsenic ($As_4$) is soft and waxy with a structure similar to too $P_4$ (Figure 13.9). Finally, black arsenic is similar in structure to red phosphorus (Figure 13.10). Antimony and bismuth are both traditional metals and have trigonal hexagonal structures ($a$ = 4.299, $c$ = 11.25 Å, and $a$ = 4.537, $c$ = 11.838 Å, respectively).

## Reaction chemistry of nitrogen

Despite nitrogen being the inert component of the Earth's atmosphere, dinitrogen undergoes a range of reactions, although it only reacts with a few reagents under standard temperature and pressure. Nitrogen reacts with oxygen in an electric arc both in the laboratory and within lightening-strikes.

$$N_2 + O_2 \rightarrow 2\,NO$$

The synthesis of ammonia is accomplished by the Haber process, using an iron oxide ($Fe_3O_4$) catalyst,

$$N_2 + 3\,H_2 \rightarrow 2\,NH_3$$

at about 500 °C and 200 atmospheres pressure. Nitrogen reacts with lithium metal at room temperature to form the nitride,

$$6\,Li + N_2 \rightarrow 2\,Li_3N$$

Magnesium also burns in nitrogen, forming magnesium nitride,

$$3\,Mg + N_2 \rightarrow Mg_3N_2$$

Nitrogen forms complexes with transition metals yielding nitrogeno complexes,

$$[(NH_3)_5Ru^{II}(H_2O)]^{2+} + N_2 \rightarrow [(NH_3)_5Ru^{II}(N_2)]^{2+}$$

Under some conditions these complexes react to give ammonia and as such may give a hint to the action of nitrogenase in which molybdenum in the active site.

$$[W(N_2)_2(PR_3)_4 \xrightarrow{H_2SO_4/MeOH} 2\,NH_3 + N_2\uparrow + W^{VI}\text{ compounds}$$

# Hydrides

## Trihydrides of the Group 15 elements

All five of the Group 15 elements form hydrides of the formula $EH_3$. Table 13.5 lists the IUPAC names along with those in more common usage.

| Compound | Traditional name | IUPAC name |
|---|---|---|
| $NH_3$ | Ammonia | Azane |
| $PH_3$ | Phosphine | Phosphane |
| $AsH_3$ | Arsine | Arsane |
| $SbH_3$ | Stibine | Stibane |
| $BiH_3$ | Bismuthine | Bismuthane |

**Table 13.5: Traditional and IUPAC (International Union of Pure and Applied Chemistry) names for the Group 15 hydrides.**

The boiling point and melting point increase increases going down the Group (Table 13.6) with increased molecular mass, with the exception of $NH_3$ whose anomalously high melting and boiling points (Figure 13.14) are a consequence of strong N-H···H hydrogen bonding. A similar (and stronger) effect is observed for the Group 16 hydrides ($H_2E$).

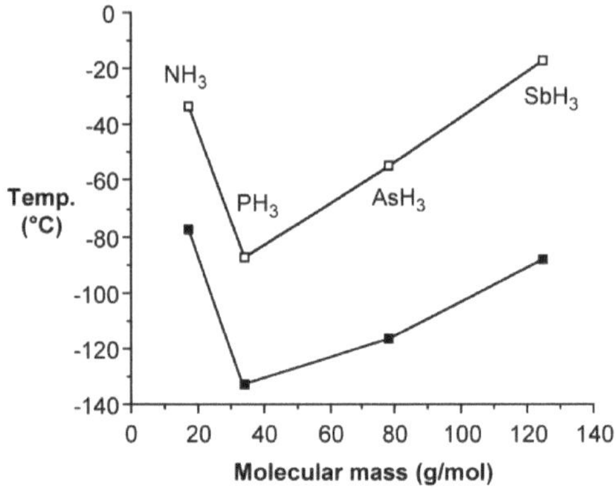


**Figure 13.14: Plot of melting and boiling points of $EH_3$ (E = N, P, As, and Sb) as a function of molecular mass.**

| Compound | Melting point (°C) | Boiling point (°C) | $\Delta H_f$ (kJ/mol) | E-H bond energy (kJ/mol) | H-E-H bond angle (°) |
|---|---|---|---|---|---|
| $NH_3$ | -77.7 | -33.35 | -46.2 | 391 | 107 |
| $PH_3$ | -133 | -87.7 | 9.3 | 322 | 93.5 |
| $AsH_3$ | -116.3 | -55 | 172.2 | 247 | 92 |
| $SbH_3$ | -88 | -17.1 | 142.8 | 255 | 91.5 |

**Table 8.6: Selected physical properties of Group 15 hydrides.**

The E-H bond strengths decrease down the group and this correlates with the overall stability of each compound (Table 13.6). The H-E-H bond angles (Table 13.6) also decrease down the Group. The H-E-H bond angle is expected to be a tetrahedral ideal of 109.5°, but since lone pairs repel more than bonding pairs, the actual angle would be expected to be slightly smaller. Two possible explanations are possible for the difference between $NH_3$ and the other hydrides.

The N-H bond is short (1.015 Å) compared to the heavier analogs, and nitrogen is more electronegative than hydrogen, so the bonding pair will reside closer to the central atom and the bonding pairs will repel each other opening the H-N-H angle more than observed for $PH_3$, etc.

The accessibility of the 2s and 2p orbitals on nitrogen allows for hybridization and the orbitals associated with N-H bonding in $NH_3$ are therefore close to $sp^3$ in character, resulting in a close to tetrahedral geometry. In contrast, hybridization of the ns and np orbitals for P, As, etc., is less accessible, and as a consequence the orbitals associated with P-H bonding in $PH_3$ are closer to p in character resulting in almost 90° H-P-H angle. The lower down the Group the central atom the less hybridization that occurs and the closer to pure p-character the orbitals on E associated with the E-H bond.

# Ammonia

Ammonia ($NH_3$) is a colorless, pungent gas (Boiling point = -33.5 °C) whose odor can be detected at concentrations as low 20 - 50 ppm. Its high boiling point relative to its heavier congeners is indicative of the formation of strong hydrogen bonding. The strong hydrogen bonding also results in a high heat of vaporization (23.35 kJ/mol) and thus ammonia can be conveniently used as a liquid at room temperature despite its low boiling point.

Ammonia solution causes burns and irritation to the eyes and skin. The vapor causes severe irritation to the respiratory system. If swallowed the solution causes severe internal damage.

*Synthesis*

Ammonia is manufactured on the industrial scale by the Haber process using the direct reaction of nitrogen with hydrogen at high pressure ($10^2$ - $10^2$ atm) and high temperature (400 550 °C) over a catalyst (e.g., α-iron),

$$N_2 + 3\,H_2 \rightarrow 2\,NH_3$$

On the smaller scale ammonia is prepared by the reaction of an ammonium salt with a base,

$$NH_4X + OH^- \rightarrow NH_3 + H_2O + X^-$$

or hydrolysis of a nitride.

$$Mg_3N_2 + H_2O \rightarrow 3\ Mg(OH)_2 + 2\ NH_3$$

The latter is a convenient route to $ND_3$ by the use of $D_2O$.

*Structure*

The nitrogen in ammonia adopts $sp^3$ hybridization, and ammonia has an umbrella structure (Figure 13.15) due to the stereochemically active lone pair.

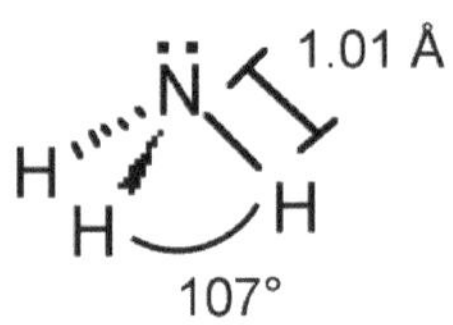


**Figure 13.15: The structure of ammonia.**

The barrier to inversion of the umbrella is very low ($E_a$ = 24 kJ/mol) and the inversion occurs 100's of times a second,

As a consequence, it is not possible to isolate chiral amines in the same manner that is possible for phosphines. In a similar manner to water,

$$2\ H_2O \rightleftharpoons H_3O^+ + OH^-$$

ammonia is a self-ionizing,

$$2\ NH_3 \rightleftharpoons \underset{\text{ammonium}}{NH_4^+} + \underset{\text{amide}}{NH_2^-}$$

The equilibrium constant ($K = 10^{-33}$) is much lower than water ($K = 10^{-14}$). The lower dielectric constant of ammonia (16.5 @ 20 °C) as compared to water (80.4 @ 20 °C) means that ammonia is not as good as water as a solvent for ionic compounds but is better for covalent organic compounds.

*Reactions*

The similarity of ammonia and water means that the two compounds are miscible. In fact, ammonia forms a series of solid hydrates, analogous to ice in which hydrogen bonding defines the structures (Figure 13.16). Several hydrates of ammonia are known, including: $NH_3.2H_2O$ (ammonia dihydrate, ADH), $NH_3.H_2O$ (ammonia monohydrate, AMH), and $2NH_3.2H_2O$ (ammonia hemihydrate, AHH).

It should be noted that these hydrates do not contain discrete NH $^{+}$ or $OH^{-}$ ions, indicating that ammonium hydroxide does not exist as a discrete species despite the common usage of the name. In aqueous solution, ammonia is a weak base ($pK_b = 4.75$),

$$NH_{3(aq)} + H_2O_{(l)} \rightleftharpoons NH_{4(aq)}^{+} + OH^{-}_{(aq)}$$

Ammonia solutions commonly used in the laboratory is a 35% solution in water. In warm weather the solution develops pressure and the cap must be released with care. The 25% solution sold commercially (for home use) is free from this problem.

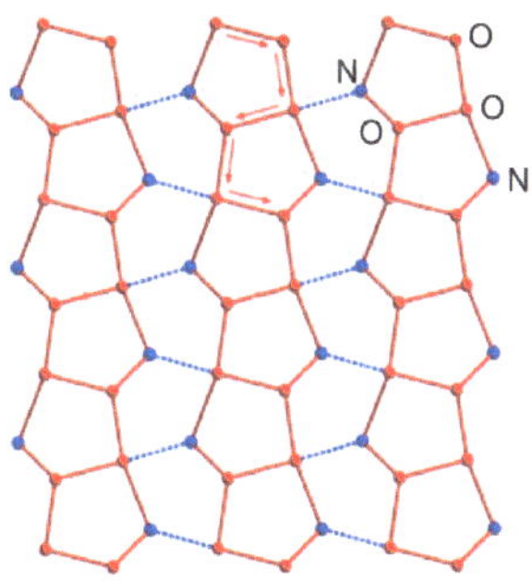


**Figure 13.16: The crystal structure of ammonia monohydrate (ANH-II) with hydrogen atoms omitted. Adapted from A. D. Fortes, E. Suard, M. -H. Lemée-Cailleau, C. J. Pickard, and R. J. Needs, Crystal structure of ammonia monohydrate phase II. *J. Am. Chem. Soc.*, 2009, 131, 13508. Copyright: American Chemical Society (2009).**

Ammonia is a Lewis base and readily forms Lewis acid-base complexes with both transition metals,

$$[Ni(H_2O)_6]^{2+} + 6\,NH_3 \rightleftharpoons [Ni(NH_3)_6]^{2+} + 6\,H_2O$$

and main group metals (Figure 13.17).

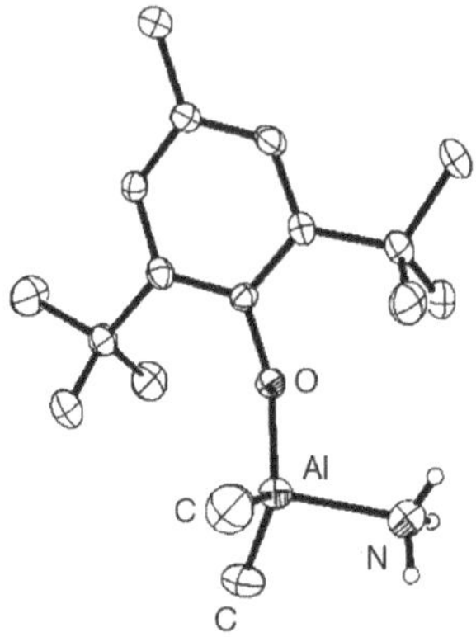


**Figure 13.17: The molecular structure of $AlMe_2(BHT)NH_3$. Hydrogen atoms, except those attached to nitrogen, are omitted for clarity. Adapted from M. D. Healy, J. T. Leman, and A. R. Barron, $AlMe_2(BHT)NH_3$: an unusually stable organoaluminum-ammonia complex and its extended coordination sphere solvate. *J. Am. Chem. Soc.*, 1991, 113, 2776. Copyright: American Chemical Society (1991).**

The formation of stable ammonia complexes is the basis of a simple but effective method of detection: Nessler's reagent,

$$NH_4^+ + 2\,[HgI_4]^{2-} + 4\,OH^- \rightarrow HgO{\cdot}Hg(NH_2)I + 7\,I^- + 3\,H_2O$$

Using a 0.09 mol/L solution of potassium tetraiodomercurate(II), $K_2[HgI_4]$, in 2.5 mol/L potassium hydroxide. A yellow coloration indicates the presence of ammonia: at higher concentrations, a brown precipitate may form. The sensitivity as a spot test is about 0.3 μg $NH_3$ in 2 μL.

Ammonia forms a blue solution with Group 1 metals. As an example, the dissolution of sodium in liquid ammonia results in the formation of solvated $Na^+$ cations and electrons,

$$Na_{(s)} \rightarrow Na_{(solv)} \rightleftharpoons Na^+_{(solv)} + e^-_{(solv)}$$

where 'solv' = $NH_3$. The solvated electrons are stable in liquid ammonia and form a complex: $[e^-(NH_3)_6]$.

It is this solvated electron that gives the strong reducing properties of the solution as well as the characteristic signal in the ESR spectrum associated with a single unpaired electron. The blue color of the solution is often ascribed to these solvated electrons; however, their absorption is in the far infra-red region of the spectrum. A second species, $Na^-_{(solv)}$, is actually responsible for the blue color of the solution.

$$2\ Na_{(solv)} \rightleftharpoons Na^+_{(solv)} + Na^-_{(solv)}$$

The reaction of ammonia with oxygen is highly favored,

$$4\ NH_3 + 3\ O_2 \rightarrow 2\ N_2 + 6\ H_2O$$

$$4\ NH_3 + 5\ O_2 \rightarrow 4\ NO + 6\ H_2O$$

and the flammability limit of ammonia is 16 25 vol%. If the reaction is carried out in the presence of a catalyst (Pt or Pd) the reaction can be limited

*Ammonium salts*

The ammonium cation ($NH_4^+$) behaves in a similar manner to the Group 1 metal ions. The solubility and structure of ammonium salts particularly resembles those of potassium and rubidium because of their relative size (Table 13.7). One difference is that ammonium salts often decompose upon heating,

$$NH_4Cl_{(s)} \rightarrow NH_{3(g)} + HCl_{(g)}$$

| Cation | Ionic radius (Å) |
|---|---|
| $K^+$ | 1.33 |
| $NH_4^+$ | 1.43 |
| $Rb^+$ | 1.47 |

**Table 13.7: Ionic radius of the ammonium ion compared to those of potassium and rubidium.**

The decomposition of ammonium salts of oxidizing acids can often be violent to highly explosive, and they should be treated with care. For example, while ammonium dichromate, $(NH_4)_2Cr_2O_7$, decomposes to give a volcano (Figure 13.18), ammonium permanganate, $NH_4[MnO_4]$, is friction sensitive and explodes at 60 °C. Ammonium nitrate, $NH_4[NO_3]$, can cause re if contacted with a combustible material and is a common ingredient in explosives since it acts as the oxygen source due to its positive oxygen balance, i.e., the compound liberates oxygen surplus to its own needs upon decomposition,

$$NH_4[NO_3] \rightarrow H_2O + N_2 + O$$

**Figure 13.18: A demonstration of an ammonium dichromate volcano.**

*Liquid ammonia as a solvent*

Ammonia has a reasonable liquid range (-77 to 33 °C), and as such it can be readily liquefied with dry ice (solid $CO_2$, $T_{sub}$ = -78.5 °C), and handled in a thermos flask. Ammonia's high boiling point relative to its heavier congeners is indicative of the formation of strong hydrogen bonding, which also results in a high heat of vaporization (23.35 kJ/mol). As a consequence, ammonia can be conveniently used as a liquid at room temperature despite its low boiling point.

Liquid ammonia is a good solvent for organic molecules (e.g., esters, amines, benzene, and alcohols). It is a better solvent for organic compounds than water, but a worse solvent for inorganic compounds. The solubility of inorganic salts is highly dependent on the identity of the counter ion (Table 13.8).

| Soluble in liquid $NH_3$ | Generally insoluble in liquid $NH_3$ |
|---|---|
| $SCN^-$, $I^-$, $NH_4^-$, $NO_3^-$, $NO_2^-$, $ClO_4^-$ | $F^-$, $Cl^-$, $Br^-$, $CO_3^{2-}$, $SO_4^{2-}$, $O^{2-}$, $OH^-$, $S^{2-}$ |

**Table 13.8: General solubility of inorganic salts in liquid ammonia as a function of the counter ion.**

The difference in solubility of inorganic salts in ammonia as compared to water, as well as the lower temperature of liquid ammonia, can be used to good advantage in the isolation of unstable compounds. For example, the attempted synthesis of ammonium nitrate by the reaction of sodium nitrate and ammonium chloride in water results in the formation of nitrogen and water due to the decomposition of the nitrate,

$$NaNO_2 + NH_4Cl \xrightarrow{H_2O} NaCl + NH_4(NO_2) \rightarrow N_2 + 2\,H_2O$$

By contrast, if the reaction is carried out in liquid ammonia, the sodium chloride side product is insoluble and the ammonium nitrate may be isolated as a white solid after filtration and evaporation below its decomposition temperature of 0 °C,

$$NaNO_2 + NH_4Cl \xrightarrow{NH_3} NaCl\downarrow + NH_4(NO_2)$$

Ammonation is defined as a reaction in which ammonia is added to other molecules or ions by covalent bond formation utilizing the unshared pair of electrons on the nitrogen atom, or through ion-dipole electrostatic interactions. In simple terms the resulting ammine complex is formed when the ammonia is acting as a

*Ammonation*

Lewis base to a Lewis acid,

$$SiF_4 + 2\,NH_3 \rightarrow SiF_4(NH_3)_2$$

$$BF_3 + NH_3 \rightarrow BF_3(NH_3)$$

or as a ligand to a cation, e.g., $[Pt(NH_3)_4]^{2+}$, $[Ni(NH_3)_6]^{2+}$, $[Cr(NH_3)_6]^{3+}$, and $[Co(NH_3)_6]^{3+}$.

*Ammonolysis*

Ammonolysis with ammonia is an analogous reaction to hydrolysis with water, i.e., a dissociation reaction of the ammonia molecule producing $H^+$ and an $NH_2^-$ species. Ammonolysis reactions occur with inorganic halides,

$$P(O)Cl_3 + 6\,NH_3 \rightarrow P(O)(NH_2)_3 + 3\,NH_4Cl$$

$$BCl_3 + 6\,NH_3 \rightarrow B(NH_2)_3 + 3\,NH_4Cl$$

and organometallic compounds,

$$Al(CH_3)_3 + NH_3 \rightarrow {}^1/_n\,[(H_3C)_2Al(NH_2)]_n + CH_4$$

In both case the $NH^-$ moiety forms a substituent or ligand.

The reaction of esters and aryl halides are also examples of ammonolysis reactions.

$$RC(O)OR' + NH_3 \rightarrow RC(O)NH_2 + R'OH$$

$$C_6H_5Cl + 2\,NH_3 \rightarrow C_6H_5NH_2 + NH_4Cl$$

*Homoleptic amides*

A homoleptic compound is a compound with all the ligands being identical, e.g., $M(NH_2)_n$. A general route to homoleptic amide compounds is accomplished by the reaction of a salt of the desired metal that is soluble in liquid ammonia (Table 13.8) with a soluble Group 1 amide. The solubility of the Group 1 amides is given in Table 13.9. Since all amides are insoluble (except those of the Group 1 metals) are insoluble in liquid ammonia, the resulting amide may be readily isolated, e.g.,

$$Mn(SCN)_2 + 2\ KNH_2 \rightarrow Mn(NH_2)_2 \downarrow + 2\ KSCN$$

$$Cr(NO_3)_3 + 3\ KNH_2 \rightarrow Cr(NH_2)_3 \downarrow + 3\ KNO_3$$

| Amide | Solubility in liquid ammonia |
|---|---|
| $LiNH_2$ | Sparingly soluble |
| $NaNH_2$ | Sparingly soluble |
| $KNH_2$ | Soluble |
| $RbNH_2$ | Soluble |
| $CsNH_2$ | Soluble |

**Table 13.9: Solubility of Group 1 amides in liquid ammonia.**

*Redox reactions*

Ammonia is poor as an oxidant since it is relatively easily oxidized, e.g.,

$$4\ NH_3 + 5\ O_2 \rightarrow 4\ NO + 6\ H_2O$$

$$2\ NH_3 + 3\ CuO \rightarrow N_2 + 3\ H_2O + 3\ Cu$$

Thus, if it is necessary to perform an oxidation reaction ammonia is not a suitable solvent; however, it is a good solvent for reduction reactions.

Liquid ammonia will dissolve Group 1 (alkali) metals and other electropositive metals such as calcium, strontium, barium, magnesium, aluminum, europium, and ytterbium. At low concentrations (ca. 0.06 mol/L), deep blue solutions are formed: these contain metal cations and solvated electrons,

$$Na_{(s)} \rightarrow Na_{(solv)} \rightleftharpoons Na^+_{(solv)} + e^-_{(solv)}$$

The solvated electrons are stable in liquid ammonia and form a complex: $[e^-(NH_3)_6]$.

The solvated electrons provide a suitable and powerful reducing agent for a range of reactions that are not ordinarily accomplished, e.g.,

$$[Ni(CN)_4]^{2-} + 2\,e^-_{(solv)} \rightarrow [Ni(CN)_4]^{4-}$$

$$Co_2(CO)_8 + 2\,e^-_{(solv)} \rightarrow 2\,[Co(CO)_4]^-$$

*Ammonia: from uses in agriculture and beer production to prolonging the First World War*

One of the cornerstones of the industrial expansion in Germany between 1870 and 1910 (when the population grew from 41 million to 66 million) was the chemicals industry. In particular companies such as Badische Anilin und Soda Fabrik (Baden Aniline and Soda Factory or BASF) had developed their business on the formation of synthetic dyes derived from the chemicals in coal tar.

Coal tar was formed as a by-product of the gasification of coal. A key ingredient of coal tar is naphtha (Table 13.10) that was used by the rubber industry; however, the remainder was disposed of. It was not long before chemists started to investigate the constituents present in coal tar. One of the first discoveries was by a chemistry student at the Royal College of Chemistry (now Imperial College) in London, William Perkin (Figure 8.20), who isolated the first synthetic dye: mauveine (Figure 13.20). This discovery led to an explosion in the chemical industry based on the extraction of compounds from coal tar, which was an essentially free waste product from every gasworks. Nowhere was this research more commercially driven that in Germany.

| Property | Typical value |
|---|---|
| Molecular weight | 100 - 215 g/mol |
| Specific gravity | 0.75 - 0.85 g/cm$^3$ |
| Boiling point | 160 – 220 °C |
| Vapor pressure | 5 mmHg |
| Solubility in water | Insoluble |

**Table 13.10: The typical physical properties of naphtha.**

The rapid increase in the German population also put a strain on the country's food resources. What compounded this issue was that the aristocratic Junkers families of East Prussia who owned much of the land in what was known as Germany's breadbasket. Junkers grew rye on their estates because the soil was too light for wheat, and since rye was fertilized with potash (potassium oxide,

$K_2O$) of which Germany had vast resources. However, in 1870 grain from the US was becoming cheaper and thus competitive with German rye. To protect their profits the Junkers demanded both subsidies for the export of rye and tariffs for the import of wheat. The result of this was that all the German rye was leaving the country and there was not enough wheat being produced to satisfy the needs of the local population. Sufficient wheat could be grown in Germany if a suitable fertilizer was available.

**Figure 8.19: A portrait of English chemist Sir William Perkin (1838 - 1907) holding a sample of cloth dyed with his discovery, mauveine, which is often called Perkin's purple.**

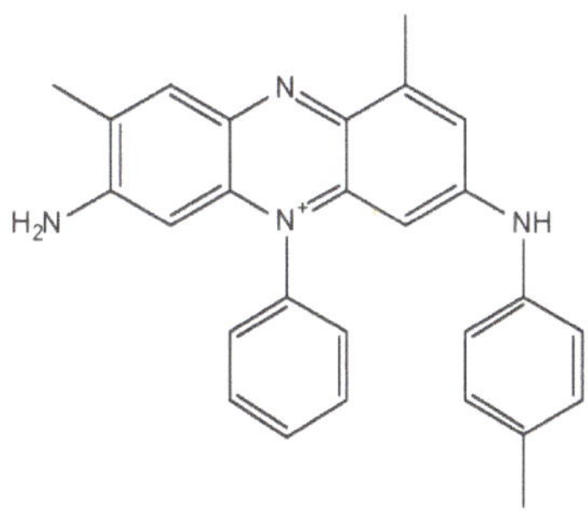


**Figure 13.20. The structure of mauveine-B.**

Sodium nitrate ($NaNO_3$), also known as Chile saltpeter, was the most common fertilizer. Unfortunately, by 1900 the deposits looked to be depleted and an alternative was needed. The alternative was found as a component in coal tar. It was known that one of the chemicals that caused the stink associated with coal gas and coal tar was ammonia ($NH_3$). Chemist George Fownes

(Figure 8.22) had suggested that ammonia be turned into a salt and used as a fertilizer. Unfortunately, the amount of ammonia that could be separated from coal tar was still insufficient, so if ammonia could be made on a large enough scale, then large-scale manufacture of a fertilizer could be realized.

**Figure 13.21: British chemist George Fownes, FRS (1815 - 1849).**

In 1909 Fritz Haber (Figure 13.22) presented a method of ammonia synthesis to BASF. His work in collaboration with Carl Bosch (Figure 13.23) resulted in the process known as the Haber-Bosch process in which nitrogen and hydrogen are mixed at high temperature (600 °C) under pressure (200 atm) over an osmium catalyst,

$$N_2 + 3\,H_2 \xrightarrow{\text{Os cat}} 6\,NH_3$$

It is interesting to note that the realization of the Haber-Bosch process required not only high-pressure vessels to be constructed by the steel industry, but also, the liquid forms of nitrogen and hydrogen. As it turned out ammonia was a necessary component for enabling the production of liquid nitrogen and hydrogen, and involved a false hypothesis of what caused malaria, which led to a desire to keep drinks cold.

Long before it was understood the real cause of malaria, John Gorrie (Figure 13.24), a doctor working in Apalachicola on the Gulf Coast of Florida, was obsessed with finding a cure for malaria. The term malaria originated from Medieval Italian: mala aria (bad air), and it was associated with swamps and marshlands. Gorrie noticed that malaria was connected to hot humid weather, so he began hanging bowl of ice in wards and circulating the air with a fan. However, ice was cut from frozen lakes and rivers in the North East of the US, stored and then shipped all over the world, and Apalachicola was so small that ice was seldom delivered.

**Figure 13.22: German chemists Fritz Haber (1868 - 1934) who received the Nobel Prize in Chemistry in 1918 for his development for synthesizing ammonia.**

**Figure 13.23: German chemist Carl Bosch (1874 - 1940) who received the Nobel Prize in Chemistry in 1931 for his work in high-pressure chemistry.**

**Figure 13.24: Portrait of American physician and scientist John Gorrie (1803 - 1855) is considered the father of refrigeration and air conditioning.**

Gorrie started looking into methods of making ice. It was well known that when a compressed gas expands it takes heat from its surroundings. Gorrie made a steam engine that compressed air in a piston, which when the piston retracted the air cooled. On the next compression stroke the cold air was pushed out across a brine solution (saturated aqueous NaCl) cooling it. When he brought water in contact with the cold brine, it froze creating the first man made ice. On 14 July 1850 Gorrie produced ice for the French Consul to cool the champagne for the celebration of Bastille Day. Just before he died, Gorrie suggested that his (by then) Patented process could be used for cooling food for transport, and it was this application that was used extensively by British merchants to bring meat from Australia to Britain. However, in Germany, Gorrie's invention was more useful for beer.

Whereas the British traditionally brewed beer in which the yeast ferments on the surface (top fermentation) at a temperature of 60 70 °F, in Germany beer was made using bottom fermentation. This style of fermentation requires a temperature just above freezing. Traditionally, cold cellars were used to store the fermenting beer, and it is from here the name lager is derived from the German verb *largern*: to store. There had been a law in Germany preventing brewing in the summer, but with Gorrie's process the possibility was to be able to brew beer all year. Carl von Linde (Figure 13.25) was asked to develop a refrigeration system. He used ammonia instead of air in Gorrie's system, and in 1879 he set up a company to commercialize his ideas. The success of his refrigerator was such that by 1891 there were over 12,000 fridges being used, and more importantly there was now a convenient method of liquefying gases such as hydrogen and nitrogen; both of which were needed for the Haber-Bosch process.

**Figure 13.25: German engineer Carl Paul Gottfried von Linde (1842 - 1934).**

As a consequence of the use of ammonia as a refrigerant, it was possible to prepare ammonia on a large industrial scale. Ammonia prepared by the Haber-Bosch process can be converted to nitric acid by the Ostwald process developed by Wilhelm Ostwald (Figure 13.26). Treatment of ammonia with air over a platinum catalyst yields initially nitric oxide,

$$4\,NH_3 + 5\,O_2 \xrightarrow{\text{Pt cat}} 4\,NO + 6\,H_2O$$

and subsequently to nitrogen dioxide,

$$2\,NO + O_2 \xrightarrow{\text{Pt cat}} 2\,NO_2$$

which dissolves in water to give nitric acid,

$$3\,NO_2 + H_2O \rightarrow 2\,HNO_3 + NO$$

Addition of soda (sodium hydroxide, NaOH) to nitric acid results in the formation of sodium nitrate,

$HNO_3 + NaOH \rightarrow NaNO_3 + H_2O$

which was the same fertilizer produced from the deposits in Chile.

**Figure 13.26: German chemist Friedrich Wilhelm Ostwald (1853 - 1932) received the Nobel Prize in 1909 for his work on catalysis.**

Unfortunately, for the Haber-Bosch and Ostwald processes, an even cheaper form of fertilizer was synthesized around the same time using calcium carbide to prepare calcium cyanamide ($CaCN_2$),

$CaC_2 + N_2 \rightarrow CaCN_2 + C$

As a consequence, the Haber-Bosch process was forgotten until the outbreak of the First World War in 1914. Within weeks of the outbreak Germany realized it had only enough explosives for about a year of conflict. This was because the main source of explosives, sodium nitrate was the same source that gave fertilizer, i.e., Chile. Realizing this, the Royal Navy effectively blockaded the supply lines. If Germany did not find another source of Great War would have been over early in 1916, however, it was remembered that the Haber-Bosch process in combination with the Ostwald processes would allow the synthesis of nitric acid, which when mixed with cotton, made nitrocellulose (Figure 13.27), also known as gun cotton, an explosive,

$3\ HNO_3 + C_6H_{10}O_5 \rightarrow C_6H_7(NO_2)_3O_5 + 3\ H_2O$

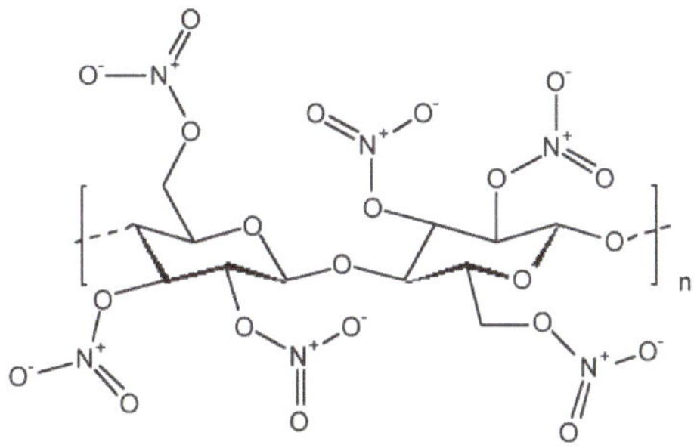

**Figure 13.27: The structure of nitrocellulose.**

As a result of the industrial synthesis of ammonia Germany was able to manufacture sufficient explosives to fight until 11 November 1918, by which time almost 10 million were dead, almost 7 million missing, and over 21 million were wounded (Figure 8.28).

**Figure 13.28: The Douamont Ossuary cemetery and World War I memorial in Verdun, France.**

## Hydrazine

Hydrazine ($H_2H_4$) is a colorless liquid with an odor similar to ammonia. Hydrazine has physical properties very close to water, with a melting point of 2 °C and a boiling point of 114 °C. The similarity in its chemistry to water is as a result of strong intermolecular hydrogen bonding.

warning: Hydrazine is highly toxic and dangerously unstable and is usually handled as aqueous solution for safety reasons. Even so, hydrazine hydrate causes severe burns on the skin and eyes. Contact with transition metals, their oxides (e.g., rust), or salts cause catalytic decomposition and possible ignition of evolved hydrogen. Reactions with oxidants are violent.

*Synthesis*

Hydrazine is manufactured on the industrial scale by the Olin Raschig process using the reaction of a sodium hypochlorite solution with ammonia at 5 °C to form chloramine ($NH_2Cl$) and sodium hydroxide,

$$NH_3 + OCl^- \rightarrow OH^- + NH_2Cl$$

The chloramines solution is the reacted with ammonia under pressure at 130 °C,

$$NH_2Cl + OH^- + NH_3 \rightarrow N_2H_4 + Cl^- + H_2O$$

Ammonia is used in a 33-fold excess. If transition metals are present then decomposition occurs,

$$2\ NH_2Cl + N_2H_4 \rightarrow 2\ NH_4^+ + 2\ Cl^- + N_2$$

and therefore, ethylenediaminetetraacetic acid (EDTA, Figure 13.29) is added to complex the transition metal ions. The as produced hydrazine solution can be concentrated by distillation to give a 65% solution. Anhydrous hydrazine is formed by the distillation from NaOH.

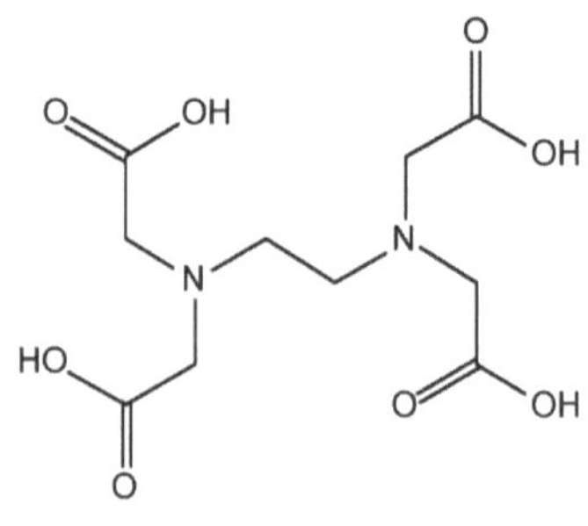


**Figure 13.29: The structure of ethylenediaminetetraacetic acid (EDTA).**

Alternative routes to hydrazine include the oxidation of urea with sodium hypochlorite,

$$(H_2N)_2C{=}O + NaOCl + 2\,NaOH \rightarrow N_2H_4 + H_2O + NaCl + Na_2CO_3$$

and the reaction of ammonia and hydrogen peroxide,

$$2\,NH_3 + H_2O_2 \rightarrow N_2H_4 + 2\,H_2O$$

*Structure*

The nitrogen atoms in hydrazine adopt sp$^3$ hybridization (Figure 13.30a), and molecule adopts a gauche conformation in the vapor, liquid and solid states (Figure 13.30b).

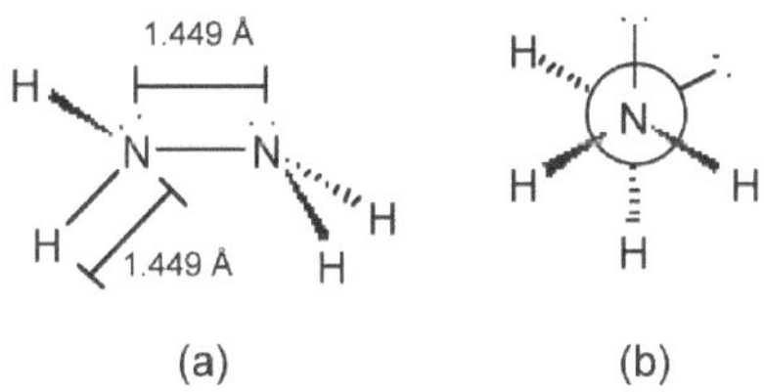


**Figure 13.30: The structure of hydrazine.**

In a similar manner to ammonia,

$$2\,NH_3 \rightleftharpoons \underset{\text{ammonium}}{NH_4^+} + \underset{\text{amide}}{NH_2^-}$$

hydrazine is a self-ionizing,

$$2\,N_2H_4 \rightleftharpoons N_2H_5^+ + N_2H_3^-$$

While there is a wide range of salts of the N H$^{+}$ cation, only the sodium and potassium salts of $N_2H_3^-$ are stable.

*Reaction chemistry and uses*

Hydrazine is polar, highly ionizing, and forms stable hydrogen bonds, and its resemblance to water is reflected in the formation of aqueous solutions and hydrates. In the solid state the monohydrate is formed, i.e., $N_2H_4.H_2O$. In solution hydrazine acts as a base to form the hydrazinium ion,

$$2\,NH_3 \rightleftharpoons NH_4^+ + NH_2^-$$

where $K_b = 8.5{\times}10^{-7}$. The presence of a second Lewis base site means that hydrazine can be protonated twice to form the hydrazonium ion,

$$N_2H_5^+ + H_2O \rightleftharpoons N_2H_5^{2+} + OH^-$$

where $K = 8.9{\times}10^{-16}$. Salts of the cation $N_2H_5^+$ are stable in water; however, the salts of the dication are less stable. The reaction of hydrazine with oxygen is highly favored,

$$N_2H_4 + O_2 \rightarrow N_2 + 2\,H_2O$$

and the explosive limit is 1.8 100 vol%.

Hydrazine is useful in a number of organic reactions for the synthesis of a wide range of compounds used in pharmaceuticals, textile dyes, and in photography, including:

- Hydrazone formation

$$2\,(CH_3)_2C{=}O + N_2H_4 \rightarrow 2\,H_2O + [(CH_3)_2C{=}N]_2$$

$$[(CH_3)_2C{=}N]_2 + N_2H_4 \rightarrow 2\,(CH_3)_2C{=}NNH_2$$

- Alkyl-substituted hydrazine synthesis via direct alkylation with alkyl halides.
- Reaction with 2-cyanopyridines to form amide hydrazides, which can be converted using 1,2-diketones into triazines.
- Use in the Wolff-Kishner reduction that transforms the carbonyl group of a ketone or aldehyde into a methylene (or methyl) group via a hydrazone intermediate,

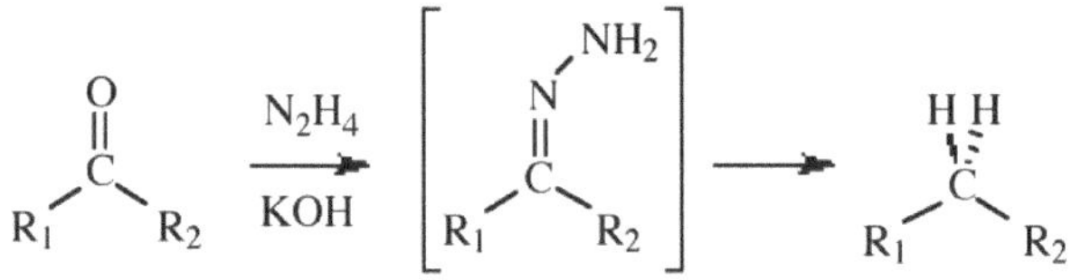


- As a building block for the preparation of many heterocyclic compounds via condensation with a range of difunctional electrophiles.
- Cleavage N-alkylated phthalimide derivatives.
- As a convenient reductant because the by-products are typically nitrogen gas and water.

*Messerschmitt Me 163 Komet*

Designed by Alexander Lippisch (Figure 13.31), the Messerschmitt Me 163B Komet (Figure 13.32) was the first rocket-powered fighter plane. With a top speed of around 596 mph (Mach 0.83) and a service ceiling of 40,000 ft, the Komet's performance of the Me 163B far exceeded that of contemporary piston engine fighters. However, despite its impressive performance, it was only produced in limited numbers (ca. 370 as compared to the 1,430 built of its jet powered compatriot the Me 262) and was not an effective combat airplane.

**Figure 13.31: German pioneer of aerodynamics Alexander Martin Lippisch (1894 - 1976).**

The Komet was powered by the HWK 109-509 hot engine (Figure 8.34) that used a mixture of a fuel and an oxidizer. The fuel was a mixture of hydrazine hydrate (30%), methanol (57%), and water (13%) that was designated by the code name, *C-Stoff*, that burned with the oxygen-rich exhaust from hydrogen peroxide (*T-Stoff*) used as the oxidizer. The *C-Stoff* was stored in a glass tank on the plane, while the *T-Stoff* was stored in an aluminum container. An oxidizing agent cocktail of $CaMnO_4$ and/or $K_2CrO_4$ was added to the *T-Stoff* generating steam and high temperatures, this intern reacted violently with the *C-Stoff*. The ow of reagents was controlled by two pumps, to regulate the rate of combustion and thereby the amount of thrust. The violent combustion process resulted in the formation of water, carbon dioxide and nitrogen, and a huge amount of heat sending out a superheated stream of steam, nitrogen and air that was drawn in through the hole in the mantle of the engine, thus providing a forward thrust of approximately 3,800 lbf. Because of the potential hazards of mixing the fuels, they were stored at least ½ mile apart, and the plane was washed with water between fueling steps and after missions.

**Figure 13.32: The Messerschmitt Me 163B-1.**

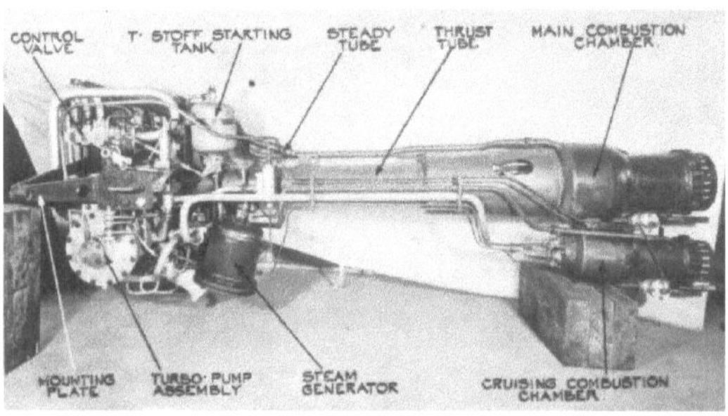


**Figure 13.33: A photograph taken by Professor A. D. Baxter of the HWK 109-509 hot engine.**

## Phosphine and arsine

Because of their use in metal organic chemical vapor deposition (MOCVD) of 13-15 (III-V) semiconductor compounds phosphine ($PH_3$) and arsine ($AsH_3$) are prepared on an industrial scale.

*Synthesis*

Phosphine ($PH_3$) is prepared by the reaction of elemental phosphorus ($P_4$) with water,

$$2\ P_4 + 12\ H_2O \rightarrow 5\ PH_3 + 3\ H_3PO_4$$

Ultra-pure phosphine that is used by the electronics industry is prepared by the thermal disproportionation of phosphorous acid,

$$4\ H_3PO_3 \rightarrow PH_3 + 3\ H_3PO_4$$

Arsine can be prepared by the reduction of the chloride,

$$4\ AsCl_3 + 3\ LiAlH_4 \rightarrow 4\ AsH_3 + 3\ LiAlCl_4$$

$$4\ AsCl_3 + 3\ NaBH_4 \rightarrow 4\ AsH_3 + 3\ NaCl + 3\ BCl_3$$

The corresponding syntheses can also be used for stibine and bismuthine. The hydrolysis of calcium phosphide or arsenide can also generate the trihydrides.

*Structure*

The phosphorus in phosphine adopts $sp^3$ hybridization, and thus phosphine has an umbrella structure (Figure 13.34a) due to the stereochemically active lone pair. The barrier to inversion of the umbrella ($E_a$ = 155 kJ/mol) is much higher than that in ammonia ($E_a$ = 24 kJ/mol). Putting this difference in context, ammonia's inversion rate is $10^{11}$ while that of phosphine is $10^3$. As a consequence, it is possible to isolate chiral organophosphines (PRR'R"). Arsine adopts the analogous structure (Figure 13.34b).

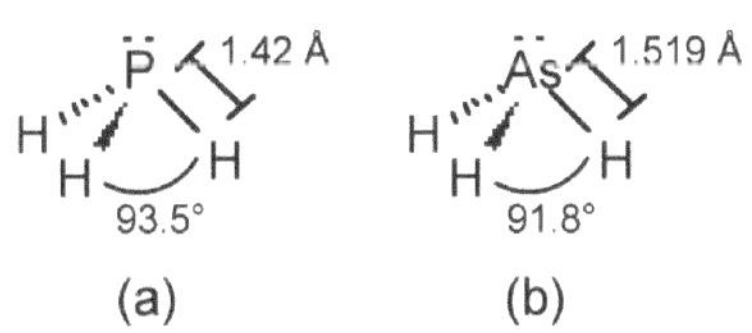


**Figure 13.34: The structures of (a) phosphine and (b) arsine.**

*Reactions*

Phosphine is only slightly soluble in water (31.2 mg/100 mL) but it is readily soluble in non-polar solvents. Phosphine acts as neither an acid nor a base in water; however, proton exchange proceeds via the phosphonium ion ($PH_4^+$) in acidic solutions and via $PH_2^-$ at high pH, with equilibrium constants $K_b = 4\times10^{-28}$ and $K_a = 41.6\times10^{-29}$, respectively.

Arsine has similar solubility in water to that of phosphine (i.e., 70 mg/100 mL), and $AsH_3$ is generally considered non-basic, but it can be protonated by superacids to give isolable salts of $AsH^+$. Arsine is readily oxidized in air,

$$2\,AsH_3 + 3\,O_2 \rightarrow As_2O_3 + 3\,H_2O$$

Arsine will react violently in presence of strong oxidizing agents, such as potassium permanganate, sodium hypochlorite or nitric acid. Arsine decomposes to its constituent elements upon heating to 250 - 300 °C.

*Gutzeit test*

The Gutzeit test is the characteristic test for arsenic and involves the reaction of arsine with $Ag^+$. Arsine is generated by reduction of aqueous arsenic compounds, typically arsenites, with Zn in the presence of $H_2SO_4$. The evolved gaseous $AsH_3$ is then exposed to silver nitrate either as powder or as a solution. With solid $AgNO_3$, $AsH_3$ reacts to produce yellow $Ag_4AsNO_3$, while with a solution of $AgNO_3$ black $Ag_3As$ is formed.

*Hazards*

Pure phosphine is odorless, but technical grade phosphine has a highly unpleasant odor like garlic or rotting fish, due to the presence of substituted phosphine and diphosphine ($P_2H_4$). The presence of $P_2H_4$ also causes spontaneous combustion in air. Phosphine is highly toxic; symptoms include pain in the chest, a sensation of coldness, vertigo, shortness of breath, and at higher concentrations lung damage, convulsions and death. The recommended limit (RL) is 0.3 ppm.

Arsine is a colorless odorless gas that is highly toxic by inhalation. Owing to oxidation by air it is possible to smell a slight, garlic-like scent when arsine is present at about 0.5 ppm. Arsine attacks hemoglobin in the red blood cells, causing them to be destroyed by the body. Further damage is caused to the kidney and liver. Exposure to arsine concentrations of 250 ppm is rapidly fatal: concentrations of 25 - 30 ppm are fatal for 30 min exposure, and concentrations of 10 ppm can be fatal at longer exposure times. Symptoms of poisoning appear after exposure to concentrations of 0.5 ppm and the recommended limit (RL) is as low as 0.05 ppm.

## Oxides and oxoacids

### Oxides of nitrogen

A summary of the physical properties of the oxides of nitrogen is given in Table 13.11.

| Oxide | Formula | Melting point (°C) | Boiling point (°C) |
|---|---|---|---|
| Nitrous oxide | $N_2O$ | -90.8 | -88.5 |
| Nitric oxide | NO | -163.6 | -161.8 |
| Dinitrogen trioxide | $N_2O_3$ | -100.6 | 3.5 (dec.) |
| Nitrogen dioxide (dinitrogen tetroxide) | $NO_2/N_2O_4$ | -11.2 ($NO_2$) | 21.2 ($N_2O_4$) |
| Nitrogen pentoxide | $N_2O_5$ | 30 | 47 |

**Table 13.11: Physical properties of the oxides of nitrogen.**

### Nitrous oxide

Gaseous nitrous oxide ($N_2O$) is prepared by the careful thermal decomposition of ammonium nitrate ($NH_4NO_2$),

$$NH_4NO_2 \xrightarrow{\Delta} N_2O + 2\,H_2O$$

Nitrous oxide is a linear molecule (Figure 13.35a) that is isoelectronic (and isostructural) with carbon dioxide. Despite its use as a power enhancement for automobiles, nitrous oxide is actually not very reactive, and a major use is as an aerosol propellant.

*Nitrous oxide as an anesthetic drug*

Nitrous oxide is known as "laughing gas" due to the euphoric effects of inhaling it, a property that has led to its recreational use as a hallucinogen. However, it is as an anesthetic that it has a legitimate application.

The first use of nitrous oxide as anesthetic drug was when dentist Horace Wells (Figure 13.36) with assistance by Gardner Quincy Colton (Figure 8.37) and John Mankey Riggs (Figure 13.38), demonstrated insensitivity to pain from a dental extraction in December 1844. Wells subsequently treated 12 - 15 patients, and according to his own record it only failed as an anesthetic in two cases. In spite of these results, the method was not immediately adopted, probably because during his first public demonstration was only partly successful.

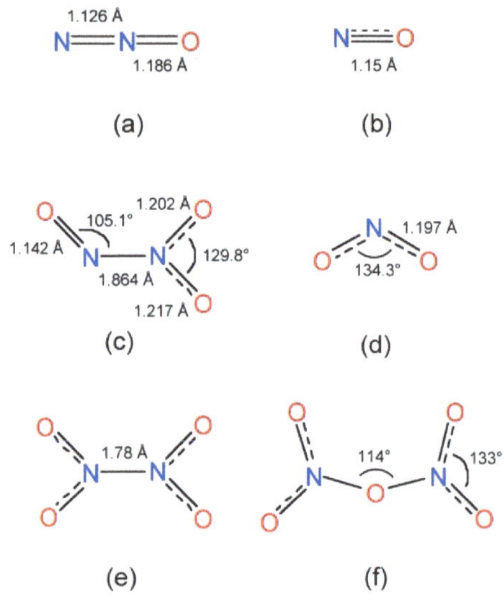


**Figure 13.35: Structures of (a) nitrous oxide, (b) nitric oxide, (c) dinitrogen trioxide, (d) nitrogen dioxide, (e) nitrogen tetroxide, and (f) nitrogen pentoxide.**

**Figure 8.37: American dentist Horace Wells (1815 - 1848).**

**Figure 13.37: American showman, lecturer, and former medical student Gardner Quincy Colton (1814 - 1898).**

**Figure 13.38: John Mankey Riggs (1811 - 1885) was the leading authority on periodontal disease in the United States, to the point that periodontal disease was known as Riggs disease.**

The method did not come into general use until 1863, when Colton successfully used it for more than 25,000 patients. As such, the usage of nitrous oxide rapidly became the preferred anesthetic method in dentistry. Because the gas is mild enough to keep a patient in a conscious and conversational state, and yet in most cases strong enough to suppress the pain caused by dental work, it remains the preferred gas anesthetic in today's dentistry.

*Nitrous: the secret to more power.*

In motorsports, nitrous oxide (often referred to as nitrous or NOS) allows the engine to burn more fuel, resulting in a more powerful combustion, and hence greater horsepower. The gas itself is not flammable, but it delivers more oxygen (33%) than atmospheric air (21%) by breaking down at elevated temperatures. When $N_2O$ breaks down in during fuel combustion, the decomposition of nitrous is exothermic, contributing to the overall power increase.

Nitrous oxide is stored as a compressed liquid (Figure 13.39); the evaporation and expansion of liquid nitrous oxide in the intake manifold causes a large drop in intake charge temperature, resulting in a denser charge, further allowing more air/fuel mixture to enter the cylinder. Nitrous oxide is sometimes injected into (or prior to) the intake manifold, whereas other systems directly inject right before the cylinder (direct port injection) to increase power.

**Figure 13.39: A typical NOS system for automotive use.**

One of the major problems of using nitrous oxide in a reciprocating engine is that it can produce enough power to damage or destroy the engine. Very large power increases are possible, and if the mechanical structure of the engine is not properly reinforced, the engine may be severely damaged or destroyed during this kind of operation. Automotive-grade liquid nitrous oxide differs slightly from medical-grade nitrous oxide in that a small amount of sulfur is added to prevent substance abuse.

## Nitric oxide

Nitric oxide (NO) is formed by the high temperature oxidation of nitrogen,

$$N_2 + O_2 \rightleftharpoons 2\,NO$$

or the platinum catalyzed oxidation of ammonia at 800 °C,

$$4\ NH_3 + 5\ O_2 \xrightarrow{Pt} 4\ NO + 6\ H_2O$$

Nitric oxide (Figure 13.35b) is electronically equivalent to dinitrogen ($N_2$) plus an electron, and as a consequence it is paramagnetic with one unpaired electron. The location of the unpaired electron in the π* orbital (Figure 13.40a) results in a bond order of 2.5 rather than the triple bond observed for $N_2$ (Figure 13.40b). The N-O distance of 1.15 Å is intermediate between the triple bond distance in $NO^+$ (1.06 Å) and the typical double bond distance (ca. 1.20 Å). Furthermore, because of the location of the electron it is easy to oxidize nitric oxide to the nitrosonium ion ($NO^+$),

$$NO \rightarrow NO^+ + e^-$$

Nitric oxide is unstable to heat,

$$3\ NO \rightarrow N_2O + NO_2$$

and oxidation,

$$2\ NO + O_2 \rightarrow 2\ NO_2$$

It will also react with halogens to form the nitrosyl halides, XNO.

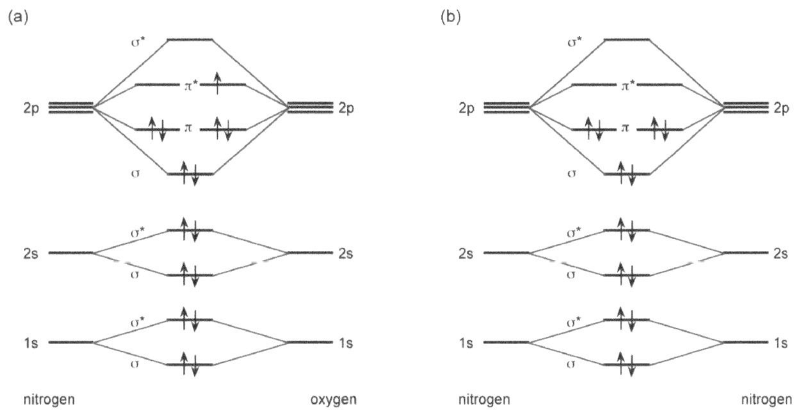


**Figure 13.40: Molecular orbital diagrams for (a) nitric oxide and (b) dinitrogen.**

## Dinitrogen trioxide

Dinitrogen trioxide ($N_2O_3$) is formed from the stoichiometric reaction between NO and $O_2$ or NO and $N_2O_4$. Dinitrogen trioxide has an intense blue color in the liquid phase and a pale blue color in the solid state. Thermal dissociation of $N_2O_3$,

$$N_2O_3 \rightleftharpoons NO + NO_2$$

occurs above -30 °C, and some self-ionization of the pure liquid is observed, (8.72).

$$N_2O_3 \rightleftharpoons NO^+ + NO_2^-$$

The asymmetric structure of $N_2O_3$ (Figure 8.36c) results in a polar molecule (Figure 8.42).

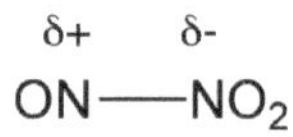


**Figure 13.41: The polarization of the N-N bond in dinitrogen trioxide.**

## Nitrogen dioxide (and tetroxide)

Formed from the oxidation of nitric oxide,

$$2\,NO + O_2 \rightleftharpoons 2\,NO_2$$

brown nitrogen dioxide is actually in equilibrium with its colorless dimeric form, nitrogen tetroxide ($N_2O_4$),

$$\underset{\text{brown, paramagnetic}}{2\,NO_2} \rightleftharpoons \underset{\text{colorless, diamagnetic}}{N_2O_4}$$

Nitrogen dioxide (Figure 13.35d) is electronically equivalent to the nitrate anion ($NO_2^-$) less one electron, and as such it is paramagnetic with one unpaired electron. The location of the unpaired electron in a nitrogen $sp^2$ orbital, and a consequently it forms a dimer through a N-N bond (Figure 13.35e).

Furthermore, it is easy to oxidize nitrogen dioxide to the nitronium ion ($NO_2^+$),

$$NO_2 \rightarrow NO_2^+ + e^-$$

Nitrogen dioxide dissolves in water to form a mixture of nitric and nitrous acids,

$$2\,NO_2 + H_2O \rightarrow NO_3^- + NO_2^- + 2\,H^+$$

Nitrogen dioxide acts as an oxidizing agent with the formation of nitrate anion,

$$2\,NO_2 + 2\,I^- \rightarrow I_2 + 2\,NO_2^-$$

The most common structural form of $N_2O_4$ (Figure 13.35e) is planar with a long N-N bond (1.78 Å) that is significantly longer than observed in hydrazine (1.47 Å). Rationalization of this structural effect is obtained from a consideration of the molecular orbitals, which show that the electrons in the σ-bond are actually delocalized over the whole molecule. The rotation about the N-N bond is 9.6 kJ/mol.

## Nitrogen pentoxide

The dehydration of nitric acid, with $P_2O_5$, yields nitrogen pentoxide,

$$2\,HNO_3 \rightleftharpoons N_2O_5 + H_2O$$

which is an unstable solid at room temperature (Table 13.11). In the solid-state nitrogen pentoxide is actually nitronium nitrate ($NO_2^+NO_3^-$); however, in the vapor phase it exists as a molecular species (Figure 13.35f) with a bent N-O-N unit. Nitrogen pentoxide is a very powerful nitrating and oxidation agent.

## Nitrogen oxides as precursors to smog and acid rain

Nitrogen oxides ($NO_x$) emissions are estimated to be in the range of 25 - 100 megatonnes of nitrogen per year. Natural sources are thought to make up approximately $^1/_3$ of the total. The generations of $NO_x$ (primarily a mixture of

$NO_2$ and NO) is the main source of smog and a significant contribution to atmospheric pollution; however, $NO_x$ is also responsible for much of the acidity in acid rain.

*Atmospheric reactions leading to acid rain*

In the dry atmosphere, nitric oxide reacts is oxidized rapidly in sunlight by ozone,

$$NO + O_3 \rightarrow NO_2 + O_2$$

The nitrogen dioxide reacts with the hydroxide radical, formed by the photochemical decomposition of ozone,

$$O_3 + hv \rightarrow O^* + O_2$$

$$O^* + H_2O \rightarrow 2\ HO\cdot$$

in the presence of a non-reactive gas molecule such as nitrogen to form nitric acid vapor,

$$NO_2 + \cdot OH + N_2 \rightarrow HNO_3 + N_2$$

The conversion rate for $NO_x$ to $HNO_3$ is approximately ten times faster than the equivalent reaction for sulfur dioxide. For example, conversion is essentially complete for a $NO_x$ plume, by the time it transverses the North Sea from the UK to Scandinavia.

At night, conversion takes place via the formation of a nitrate radical,

$$NO_2 + O_3 \rightarrow NO_3 + O_2$$

which subsequently photo- chemically unstable under sunlight forming nitrogen pentoxide,

$$NO_2 + NO_3 + N_2 \rightarrow N_2O_5 + N_2$$

that reacts with water on the surface of aerosol particles to form nitric acid,

$$N_2O_5 + H_2O \rightarrow 2\ HNO_3$$

Both NO and $NO_2$ are only slightly soluble in water and it is therefore more probable that the nitric acid content of rain is more likely due to the dissolution of nitric acid vapor into raindrops rather than a separate reaction.

$$N_2O_5 + H_2O_{(l)} \rightarrow 2\ HNO_{3(aq)}$$

## Oxoacids of Nitrogen

*Nitrous acid*

In the gas phase nitrous acid can be made by the following reaction:

$$NO + NO_2 + H_2O \rightleftharpoons 2\ HNO_2$$

The gas phase structure as determined by IR spectroscopy is shown in Figure 13.42a, in which the nitrogen is planar with $sp^2$ hybridization.

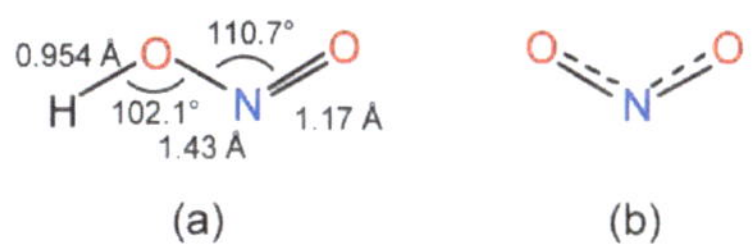


**Figure 13.42 The structures of (a) nitrous acid and the structurally related (b) nitrite ion.**

In basic aqueous solution the same reaction results in the formation of the nitrite ion,

$$NO + NO_2 + 2\ OH^- \rightleftharpoons 2\ NO_2^- + H_2O$$

which can be precipitated as the $Ba^{2+}$ salt. After separation the addition to sulfuric acid yields a solution of nitrous acid. However, it is not possible to concentrate by heating since decomposition occurs,

$$3\ HNO_2 \rightarrow H_3O^+ + NO_3^- + 2\ NO$$

Nitrous acid is a fairly weak acid in water (pK = 5.22); however, many salts are known of the nitrite ion, $NO_2^-$ (Figure 13.42b). Complexes of the nitrite ion can be monodentate with bonding via nitrogen (nitro) or an oxygen (nitrito). Both isomers can be isolated in the case of an inert metal, i.e., substitutionally inert $d^6$ octahedral complexes (Figure 13.43). Bidentate and bridging modes of coordination are also known for the nitrite anion (Figure 8.44).

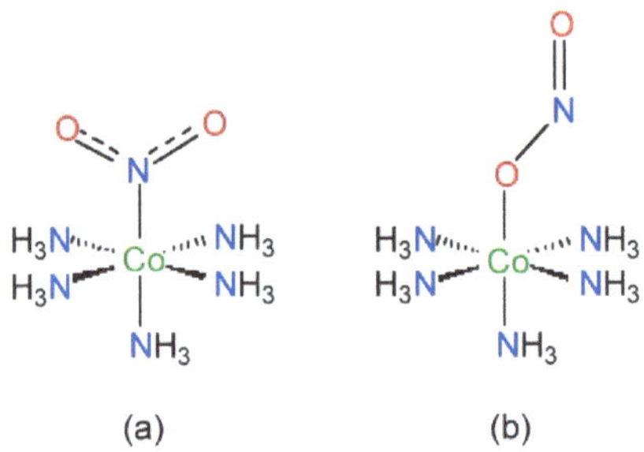


**Figure 13.43: Structures of (a) orange nitro isomer and (b) pink nitrito isomer of [Co $(NH_3)_5(NO_2)$]**

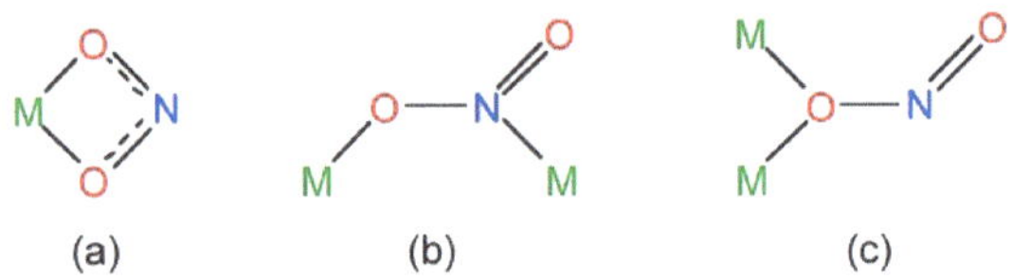


**Figure 13.44: Bidentate (a) and bridging (b and c) modes of coordination of the nitrite anion.**

Nitrite can act as either an oxidizing agent,

$$NO_2^- + 2\,H^+ + Fe^{2+} \rightarrow NO + H_2O + Fe^{3+}$$

or a reducing agent,

$$NO_2^- + H_2O + 2\,Ce^{4+} \rightarrow NO_3^- + 2\,H^+ + 2\,Ce^{3+}$$

The most important use of nitrous acid is in the diazotization reactions in which nitrous acid is generated by acidifying nitrite solution.

*Nitric acid*

Nitric acid ($HNO_3$), also known as *aqua fortis* and *spirit of nitre*, is made by the dissolution of nitrogen dioxide in water,

$$3\ NO_2 + H_2O \rightarrow 2\ HNO_3 + NO$$

Nitric acid can be concentrated by distillation from concentrated sulfuric acid. The structure of $HNO_3$ in the gas phase is planar at the sp$^2$ nitrogen (Figure 13.45a).

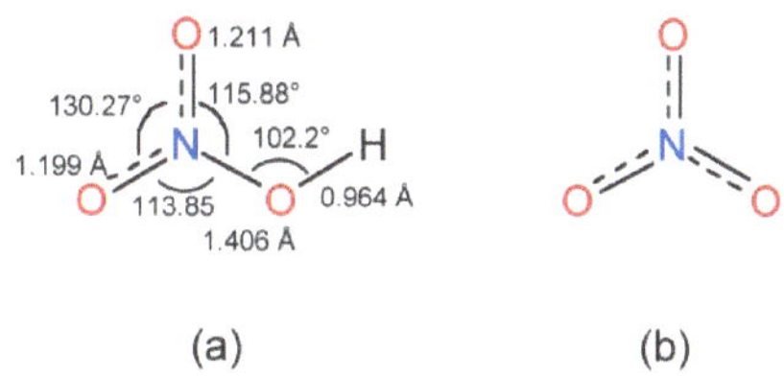


**Figure 13.45: The structures of (a) nitric acid and (b) the nitrate anion.**

Pure 100% nitric acid is a very corrosive liquid that is strongly acidic and protonates and dissolves organic species. In the liquid phase it is slightly dissociated,

$$3\ HNO_3 \rightleftharpoons H_3O^+ + NO_2^+ + 2\ NO_3^-$$

It is also a powerful oxidizing agent, converting non-metal elements to either the oxide or oxoacid. In contrast with metals it forms either salts or complexes in which the metal is in its highest oxidation state. It is unstable and decomposes upon heating or photolysis.

The pure acid has the highest self-ionization of pure liquid acids,

$$2\ HNO_3 \rightleftharpoons H_2NO_3^+ + NO_3^-$$

However, the loss of water results,

$$H_2NO_3^+ \rightleftharpoons H_2O + NO_2^+$$

such that the overall reaction can be described by

$$2\ HNO_3 \rightleftharpoons NO_3^- + NO_2^+ + H_2O$$

The common concentration of nitric acid is 70%. While the pure acid is colorless, samples often take on a yellow color due to the photochemical decomposition of nitric acid to give brown $NO_2$.

$$4\ HNO_3 \xrightarrow{h\nu} 4\ NO_2 + 2\ H_2O + O_2$$

The term the acid test is derived from the medieval practice of debasing of gold and silver currencies (often by the Monarchs who issued them) by debasing with copper. With ducats (Milan), livres (France), florins (Florence), maravedies (Spain) and bezants (Constantinople) in widespread use, and each with a nominal gold or silver content, it was important for a merchant to be able verify the worth of any particular coin. If a drop of dilute nitric acid was placed onto a silver coin adulterated with copper, it turned green, due to the formation of copper(II) nitrate. Conversely, if a gold coin reacted in any way with the nitric acid it was not pure. In both cases the coins failed the acid test.

*Aqua regia* (so called because it dissolves gold) is a mixture of 70% nitric acid and hydrochloric acid 1:3 ratio. Aqua regia is a very powerful oxidizing agent (it contains $Cl_2$) and stabilizes some metals as their chloro complexes (Table 13.12). If HF is added in place of HCl, tantalum may be dissolved with the formation of $[TaF_2]^-$.

| Metal | Chloride salt formed |
|---|---|
| Au | $[AuCl_4]^-$ |
| Pt | $[PtCl_6]^{2-}$ |

**Table 13.12: Metal chloride salts formed by the dissolution of metals in aqua regia.**

Fuming nitric acid (100% nitric) is exceedingly corrosive and should not be used. In 100% sulfuric acid, nitric acid acts as a base and gets protonated,

$$HNO_3 + H_2SO_4 \rightarrow ON(OH)_2^+ + HSO_4^-$$

$$ON(OH)_2^+ + HSO_4^- + H_2SO_4 \rightarrow H_3O^+ + NO_2^+ + 2\ HSO_4^-$$

and acts as a powerful nitrating agent. In water dilute nitric acid is fully ionized,

$$HNO_3 + H_2O \rightarrow H_3O^+ + NO_3^-$$

The nitrate ion is planar (Figure 13.45b) and forms many salts and complexes. The nitrate anion is most commonly a monodentate ligand but can also be a bidentate ligand (Figure 8.46a) or a bridging ligand (Figure 8.46b and c).

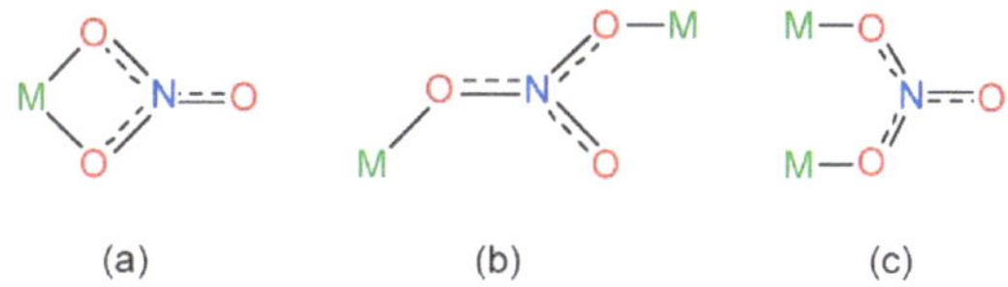


**Figure 13.46: Coordination modes of the nitrate anion.**

## Phosphorous Oxides

*Phosphorous trioxide*

The stoichiometric oxidation of white phosphorus yields phosphorous trioxide,

$$P_4 + 3\,O_2 \rightarrow 2\,P_2O_3$$

Several structural forms of phosphorous trioxide are known, but $P_4O_6$ is a stable molecule structure (Figure 13.47), while the rest are polymers.

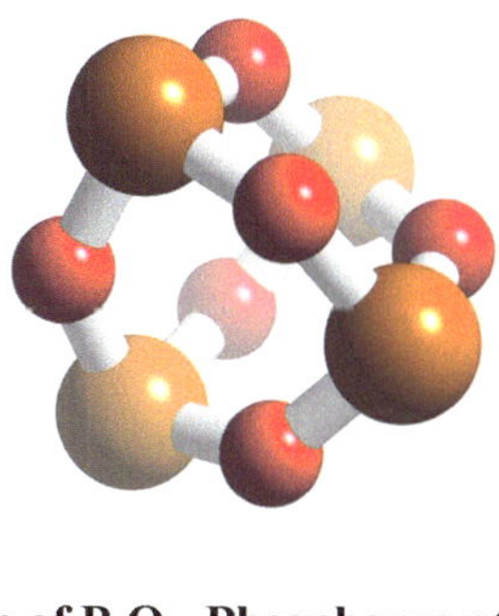

**Figure 13.47: The structure of $P_4O_6$. Phosphorus atoms are represented by the large spheres, and oxygen atoms by the small spheres (Copyright: Karl Harrison, 2005).**

*Phosphorous pentoxide*

The reaction of white phosphorus with excess oxygen (or the oxidation of phosphorous trioxide) yields phosphorous pentoxide,

$$P_4 + 5\ O_2 \rightarrow 2\ P_2O_5$$

The structure of hexagonal phosphorous pentoxide is actually that of the dimeric form, $P_4O_{10}$, and is based upon the structure of $P_4O_6$, but with P=O units instead of lone pairs (Figure 13.48). The structure is maintained in the vapor phase; however, other crystalline and glassy forms comprise of a sheet-like structure (Figure 13.49).

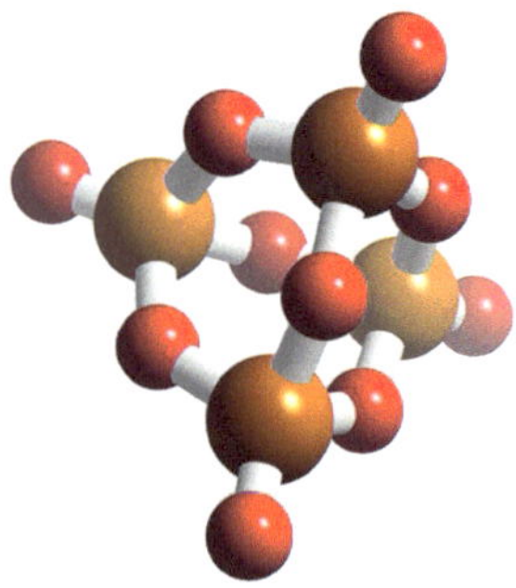

**Figure 8.47: The structure of hexagonal $P_4O_{10}$. Phosphorus atoms are represented by the large spheres, and oxygen atoms by the small spheres (Copyright: Karl Harrison, 2005).**

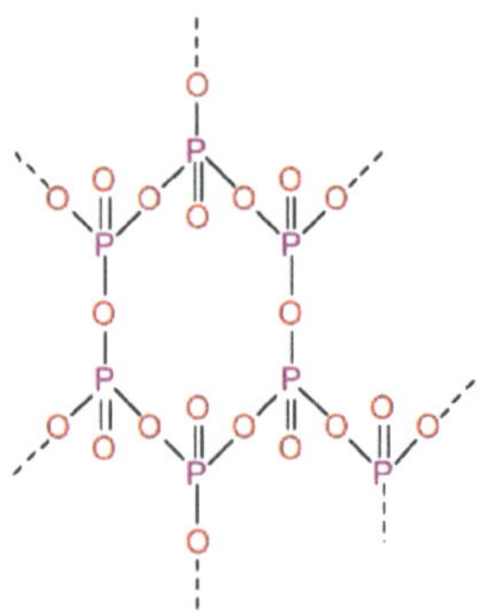

**Figure 13.49: The structure of a sheet in glassy $P_4O_{10}$.**

Phosphorous pentoxide is an excellent drying agent below 100 °C. It reacts with water to form various phosphoric acids, and it will extract water from other `drying agents', e.g.,

$$2\ HNO_3 \xrightarrow[-H_2O]{P_4O_{10}} N_2O_5$$

$$H_2SO_4 \xrightarrow[-H_2O]{P_4O_{10}} SO_3$$

Phosphorous pentoxide will dehydrate amides to give nitriles,

$$RC(O)NH_2 \xrightarrow[-H_2O]{P_4O_{10}} RC{\equiv}N$$

## Oxoacids of Phosphorus

Phosphorous pentoxide ($P_2O_5$) is an excellent drying agent, and its action is a result of the formation of a range of oxoacids.

*Hypophosphorous acid*

Hypophosphorous acid, $H_3PO_2$, is easily prepared pure buy the reaction of white phosphorous with base, followed by acidification,

$$P_4 + 4\ OH^- + 4\ H_2O \rightarrow 4\ H_2PO_2^- + 2\ H_2$$
$$\downarrow \text{acidify}$$
$$H_3PO_2$$

The pure acid is a solid (Melting point = 27 °C) and very soluble in water.

The structure of $H_3PO_2$ is determined by X-ray crystallography to be tetrahedral with two hydride ligands and a hydroxide (Figure 13.50a). The presence of two hydrides is confirmed by NMR spectroscopy. The $^1$H NMR resonance shows an OH line and a doublet from the P-H with a large one-bond coupling constant to the $^{31}$P nucleus. The non-decoupled $^{31}$P NMR spectrum shows a triplet [$\delta$ = 13 ppm, $J$(P-H) = 530 Hz] due to the two hydrides.

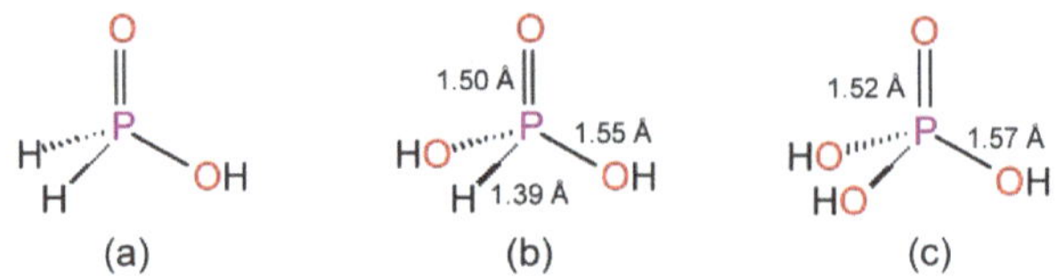


**Figure 13.50: Oxoacids of phosphorous: (a) hypophosphorous acid, (b), phosphorous acid, and (c) phosphoric acid.**

In water hypophosphorous acid is a monobasic acid (pK = 1.2),

$$H_2P(O)OH + H_2O \rightleftharpoons H_2P(O)_2^- + H_3O^+$$

and it forms a wide range of salts. It is also a powerful reducing agent, but its reaction kinetics is slow.

*Phosphorous acid*

The reaction of $P_4O_6$ or $PCl_3$ with water yields phosphorous acid, $H_3PO_3$; which like hypophosphorous acid is a solid (Melting point = 70.1 °C) and very soluble in water. The structure is shown by X-ray crystallography to be comprised of a tetrahedral phosphorus with one hydride and two hydroxides (Figure 13.50b). $^{31}$P NMR spectroscopy demonstrates the presence of a single hydride by the presence of a doublet as a consequence of the phosphorous center being split by a single hydride [δ = 4 ppm, *J*(P-H) = 700 Hz]. The $^{1}$H NMR spectrum shows a doublet for the hydride and a single resonance of twice the intensity for the hydroxide.

As expected, in water phosphorous acid is dibasic,

$$HP(O)(OH)_2 \underset{pK_1 = 1.8}{\rightleftharpoons} [HP(O)_2OH]^- \underset{pK_2 = 6.7}{\rightleftharpoons} [HP(O)_3]^{2-}$$

The acid (and the anions) are strong reducing agents, yielding phosphoric acid. They actually react very slowly, and it is thought that this may be due to the reaction being in the tautomeric form, $P(OH)_3$. Although this has not been isolated, the trialkyl derivatives exist in both forms, i.e., esters of phosphoric acids (Figure 13.51a) and trialkyl phosphates (Figure 13.51b).

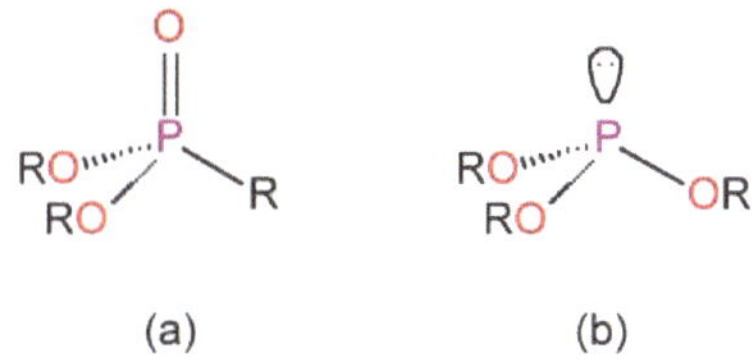

**Figure 13.51: Structures of two tautomeric forms; (a) an ester of phosphoric acid and (b) a trialkyl phosphate.**

*Ortho phosphoric acid*

Orthophosphoric acid ($H_3PO_4$) is the most common oxoacid of phosphorus (Figure 8.50c). The term acid phosphoric acid is commonly used. It is made from the reaction of phosphates with sulfuric acid ($H_2SO_4$), or from the hydrolysis of $P_4O_{10}$. The pure acid is a colorless crystalline solid (Melting point = 42.35 °C) with extensive hydrogen bonding. The inter-phosphoric acid hydrogen bonding is partially maintained in aqueous solutions above 50% solutions.

Phosphoric acid is very stable and shows no oxidation chemistry below 350 °C. As expected, phosphoric acid is tribasic,

$$PO(OH)_3 \underset{pK_1 = 2.15}{\rightleftharpoons} [PO_2(OH)_2]^- \underset{pK_2 = 7.1}{\rightleftharpoons} [PO_3(OH)]^{2-} \underset{pK_3 = 12.4}{\rightleftharpoons} [PO_4]^{3-}$$

The anions $H_2PO_4^-$ and $HPO_4^{2-}$ have particular names, dihydrogen phosphate and monohydrogen phosphate, respectively. Many salts of are known for all three anions; those with phosphate ($PO_4^{2-}$) are often insoluble in water. Many coordination complexes are known, especially with $M^{3+}$ and $M^{4+}$ ions.

*Phosphorous esters*

The alkyl and aryl homologs of phosphoric acid (i.e., the phosphate triester $O=P(OR)_3$) are prepared by the reaction of $P_4O_{10}$ with the appropriate alcohol. The triesters are good solvents and Lewis basic ligands with coordination via the oxide moiety. The diesters and monoesters can also be made, and they are important in biochemical processes.

Phosphite triesters, $P(OR)_3$, may be made by the reaction of $PCl_3$ with alcohols or phenols in the presence of an organic base as the acceptor for the HCl formed. Alternatively, they can be prepared directly from white phosphorous,

$$P_4 + 6\,OR^- + 6\,CCl_4 + 6\,ROH \rightarrow 4\,P(OR)_3 + 6\,CHCl_3 + 6\,Cl^-$$

Phosphite triesters are readily oxidized to the appropriate phosphate triester,

$$2\,P(OR)_3 + O_2 \rightarrow 2\,O{=}P(OR)_3$$

They also react with alkyl halides to form the dialkyl phosphonate via the Michaelis-Arbusov reaction,

$$P(OR)_3 + R'X \rightarrow O{=}P(OR)_2R' + RX$$

*Polyphosphates*

Polyphosphates contain the $PO_4$ unit, and the simplest example (pyrophosphate or diphosphate, Figure 8.52a) can be considered as condensation products of monohydrogen phosphate,

$$2\,HPO_4^{2-} \xrightarrow{\Delta} [O_3P\text{-}O\text{-}PO_3]^{4-} + H_2O$$

Longer chains can be formed, e.g., the triphosphate $P_3O_{10}^{5-}$ (Figure 8.52b). The formation and reverse (hydrolysis) reaction are slow, but are readily catalyzed, e.g., by enzymes.

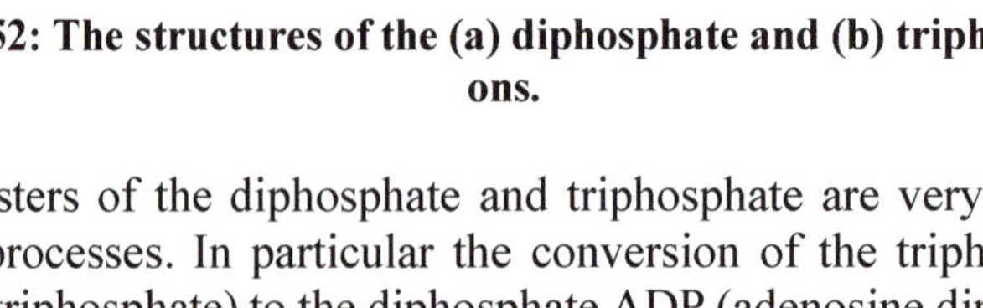

**Figure 13.52: The structures of the (a) diphosphate and (b) triphosphate anions.**

The monoesters of the diphosphate and triphosphate are very important in biological processes. In particular the conversion of the triphosphate ATP (adenosine triphosphate) to the diphosphate ADP (adenosine diphosphate) by the transfer of a phosphate group is important in the energetics of biological reactions.

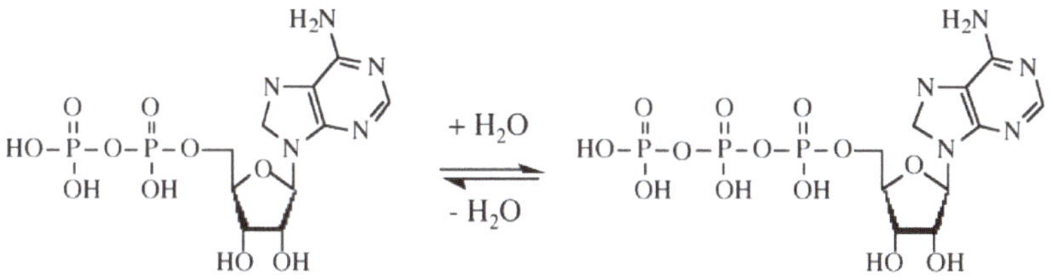

*Cyclic polyphosphates*

Also known as meta phosphates, cyclic phosphates are also made from fused $PO_4$ units. The simplest, $P_3O_9^{3-}$, is shown in Figure 13.53. Slow addition of water to P O results in the formation of the tetrameric polyphosphate, $P_4O_{12}^{4-}$ (Figure 8.54), which is known as *calgon* due to its ability to complex $Ca^{2+}$ as well as other metals.

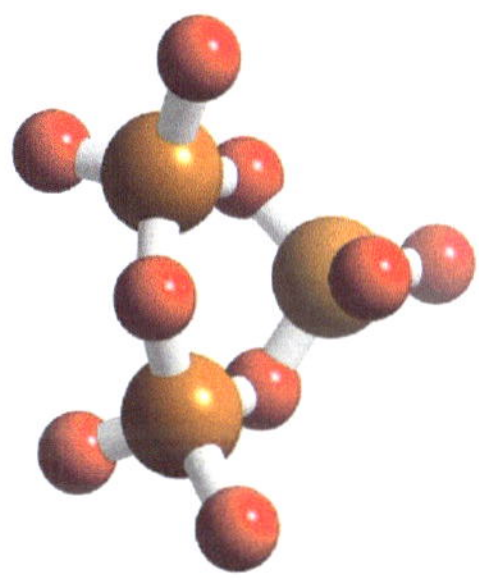

**Figure 13.53: The structure of $P_3O_9^{3-}$. Phosphorus atoms are represented by the large spheres, and oxygen atoms by the small spheres (Copyright: Karl Harrison, 2005).**

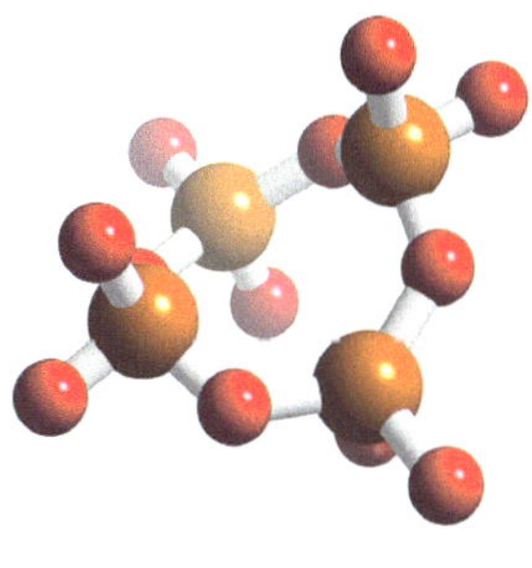

**Figure 13.54: The structure of $P_4O_{12}^{4-}$ calgon. Phosphorus atoms are represented by the large spheres, and oxygen atoms by the small spheres (Copyright: Karl Harrison, 2005).**

# Halides of phosphorous

## Phosphorous trihalides

Phosphorous trihalides, $PX_3$, are produced from the direct reaction of phosphorous with the appropriate halogen,

$$P_4 + 6\ Cl_2 \rightarrow 4\ PCl_3$$

The fluoride is readily made from the halide exchange reaction of $PCl_3$ with fluoride salts,

$$2\ PCl_3 + 3\ CaF_2 \rightarrow 2\ PF_3 + 3\ CaCl_2$$

Mixed trihalides are formed by halide exchange,

$$PCl_3 + PBr_3 \rightleftharpoons PCl_2Br + PClBr_2$$

A summary of the physical properties of the phosphorous trihalides is given in Table 13.13. All the compounds have a pyramidal structure in the vapor phase and in solution.

| Compound | Melting point (°C) | Boiling point (°C) | P-X (Å) | X-P-X (°) |
|---|---|---|---|---|
| $PF_3$ | -151.5 | -101.8 | 1.56 | 96.3 |
| $PCl_3$ | -93.6 | 76.1 | 2.04 | 100 |
| $PBr_3$ | -41.5 | 173.2 | 2.22 | 101 |
| $PI_3$ | 61.2 | 200 (dec.) | 2.43 | 102 |

**Table 13.13: Selected physical properties of the phosphorous trihalides.**

The phosphorous trihalides hydrolyze to phosphoric acid,

$$PCl_3 + 3\ H_2O \rightarrow H_3PO_3 + 3\ HCl$$

and undergo alcoholysis to form the trialkyl phosphite derivative,

$$PCl_3 + 3\ HOR \rightarrow P(OR)_3 + 3\ HCl$$

Phosphorous trifluoride is only slowly hydrolyzed by water but reacts readily with alkaline solutions. In contrast, phosphorus triiodide is an unstable red solid that reacts violently with water. Phosphorous trichloride in particular is an excellent synthon for most trialkyl phosphines,

$$PCl_3 + 3\ RMgBr \rightarrow PR_3 + 3\ MgBrCl$$

As with other P(III) compounds such as trialkyl phosphines, phosphorous trihalides can be oxidized to the analogous phosphene oxide, e.g.,

$$2\ PCl_3 + O_2 \rightarrow O{=}PCl_3$$

Phosphorous trihalides form Lewis acid-base complexes with main group metals, but the bonding to $d^n$ ($n \neq 0$) transition metals occurs in the same manner to that of trialkyl phosphine, i.e., with dπ-pπ back donation. In fact, one of the first examples of complexation of phosphorous to a low oxidation metal was the formation of $PF_3$ complexes with Fe-porphyrin.

*Phosphorous pentahalides*

The reaction of white phosphorous with excess halogen yields the pentahalides,

$$P_4 + 10\ Cl_2 \rightarrow 4\ PCl_5$$

However, the pentafluoride is best prepared by halide exchange,

$$2\ PCl_5 + 5\ CaF_2 \rightarrow 2\ PF_3 + 5\ CaCl_2$$

The pentaiodide is unknown, however, selected physical properties are given for the others in Table 13.14.

| Compound | Melting point (°C) | Boiling point (°C) |
|---|---|---|
| $PF_5$ | -93.78 | -84.5 |
| $PCl_5$ | 166.8 | 160 (subl.) |
| $PBr_5$ | 100 | 106.0 |

**Table 13.14: Selected physical properties of phosphorous pentahalides.**

The structures of the phosphorous pentahalides are all trigonal bipyramidal in the gas phase (Figure 13.55). Phosphorous pentafluoride maintains the trigonal bipyramidal structure in solid state, but the chloride and bromide are ionic solids, $[PCl_4]^+[PCl_6]^-$ and $[PBr_4]Br$ (Figure 13.56), respectively. The tetrahedral $[PCl_4]^+$ ion is also formed with the reaction of $PCl_5$ with other chloride ion acceptors,

$$PCl_5 + TiCl_4 \rightarrow [PCl_4]^+[Ti_2Cl_9]^-$$

$$PCl_5 + NbCl_5 \rightarrow [PCl_4]^+[NbCl_6]^-$$

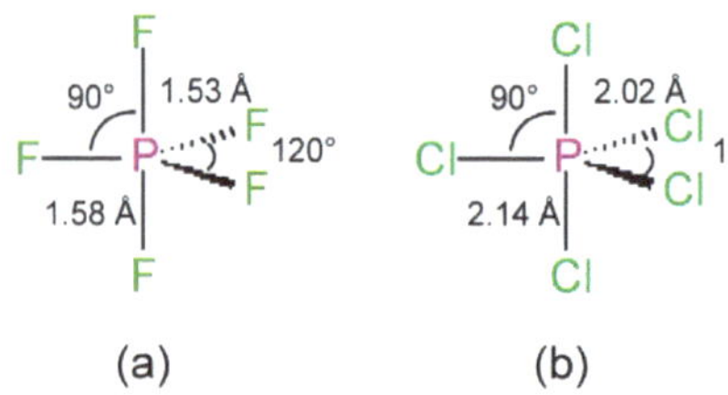


**Figure 13.55: The structures of (a) $PF_5$ and (b) $PCl_5$.**

**Figure 13.56: The crystal structure of phosphorous pentabromide.**

All of the pentahalides undergo thermal decomposition,

$$PX_5 \rightleftharpoons PX_3 + X_2$$

Careful hydrolysis of the pentahalides yields the oxide of the appropriate tri-halide,

$$PX_5 + H_2O \rightarrow O{=}PX_3 + HX$$

while excess hydrolysis yields phosphoric acid,

$$PX_5 + 4\,H_2O \rightarrow H_3PO_4 + 5\,HX$$

# Chapter 14: Group 16 Oxygen and the Chalcogens

## The elements

The Group 16 elements have a particular name chalcogenes. Table 14.1 lists the derivation of the names of the halogens.

| Element | Symbol | Name |
|---|---|---|
| Oxygen | O | Greek *oxys* (sharp, from the taste of acids) and *genes* (producer) |
| Sulfur (sulphur) | S | From the Latin *sulphurium* |
| Selenium | Se | Greek *selene* meaning Moon |
| Tellurium | Te | Latin *tellus* meaning earth |
| Polonium | Po | Named after Poland, Latin *Polonia* |

**Table 14.1: Derivation of the names of each of the Group 16(VI) elements.**

In Latin, the word is variously written *sulpur*, *sulphur*, and *sulfur*. It is an original Latin name and not a classical Greek loan, so the *ph* variant does not denote the Greek letter φ. Sulfur in Greek is thion, whence comes the prefix *thio-* to donate a sulfur derivative, e.g., a thioketone, $R_2C=S$. The simplification of the Latin words *p* or *ph* to an *f* appears to have taken place towards the end of the classical period. The element has traditionally been spelled sulphur in the United Kingdom, India, Malaysia, South Africa, Australia, Ireland, and Canada, but sulfur in the United States. IUPAC adopted the spelling sulfur in 1990, as did the Royal Society of Chemistry Nomenclature Committee in 1992.

## Discovery

*Oxygen*

The 2$^{nd}$ century BC Greek writer, Philo of Byzantium, observed that inverting a jar over a burning candle and surrounding the jar's neck with water resulted in some water rising into the neck. He incorrectly ascribed this to the idea that part of the air in the vessel were converted into the element re and thus were able to escape through pores in the glass. Much later Leonardo da Vinci (Figure 14.1) suggested that this effect was actually due to a portion of air being consumed during combustion.

**Figure 14.1: Italian painter, sculptor, architect, musician, scientist, mathematician, engineer, inventor, anatomist, geologist, cartographer, botanist and writer Leonardo di ser Piero da Vinci (1452 - 1519).**

By the late 17th century, Robert Boyle (Figure 7.2) showed that air is necessary for combustion. His work was expanded by English chemist John Mayow (Figure 14.2) by showing that re requires only a part of air that he called spiritus *nitroaereus* or just *nitroaereus*.

**Figure 14.2: English chemist, physician, and physiologist John Mayow FRS (1641 - 1679).**

The reactive nature of *nitroaereus* was implied by Mayow from his observation that antimony (Sb) increased in weight when heated in air. He also suggested that the lungs separate nitroaereus from air and pass it into the blood and that animal heat and muscle movement result from the reaction of nitroaereus with certain substances in the body; both concepts that were proven to be correct. Robert Hooke (Figure 14.3), Ole Borch (Figure 14.4), Mikhail Lomonosov (Figure 14.5), and Pierre Bayen (Figure 14.6) all produced oxygen in experiments in the 17th and the 18th century but none of them recognized it as an element, probably since the prevalence at that time of the phlogiston, and their attempts to fit their experimental observations to that theory.

**Figure 14.3: Portrait of English natural philosopher, architect Robert Hooke FRS (1635 - 1703).**

**Figure 14.4: Danish scientist, physician, grammarian, and poet Ole Borch (1626 - 1690).**

**Figure 14.5: Russian scientist and writer Mikhail Vasilyevich Lomonosov (1711 - 1765).**

**Figure 14.6: French chemist and pharmacist Pierre Bayen (1725 - 1798).**

The phlogiston theory was postulated in 1667 by the German alchemist J. J. Becher (Figure 14.7) and modified in 1731 by the chemist Georg Ernst Stahl (Figure 14.8). Phlogiston theory stated that all combustible materials were made of two parts. One part, called *phlogiston*, was given off when the substance containing it was burned, while the *dephlogisticated* component was thought to be its true form, or *calx*. Highly combustible materials that leave little residue (e.g., wood) were thought to mostly comprise of phlogiston, while non-combustible substances that corrode (e.g., iron) contained very little phlogiston. Air did not play a role in phlogiston theory, instead, it was based on observations of what happens when something burns, that most common objects appear to become lighter and seem to lose something in the process. However, one observation that overturned phlogiston theory was that metals, gain weight in rusting when they were supposedly losing phlogiston!

**Figure 14.7: German physician, alchemist, and adventurer Johann Joachim Becher (1635 - 1682).**

**Figure 14.8: German chemist, physician, and philosopher Georg Ernst Stahl (1659 - 1734).**

Oxygen was first discovered by Carl Wilhelm Scheele (Figure 9.5) by heating mercuric oxide (HgO). Scheele called the gas re air because it was the only known supporter of combustion. He wrote an account of this discovery in a manuscript (*Treatise on Air and Fire*) submitted in 1775. Unfortunately for Scheele his work was not published until 1777. In August 1774, an experiment conducted by Joseph Priestley (Figure 13.2) showed that exposure of mercuric oxide (HgO) to sunlight inside a glass tube, liberated a gas he named dephlogisticated air. Priestley noted that candles burned brighter in this gas. He even went as far as breathing the gas himself, after which he wrote: "The feeling of it to my lungs was not sensibly different from that of common air, but I fancied that my breast felt peculiarly light and easy for some time afterwards." Priestley published his findings in 1775. Because he published his findings first, Priestley is usually given credit for the discovery of what became known as oxygen.

Interestingly, Lavoisier (Figure 7.4) claimed to have discovered this new substance independently. However, Priestley visited Lavoisier in October 1774 and told him about his experiment and how he liberated the new gas. Furthermore, Scheele also posted a letter to Lavoisier on September 30, 1774 that described his own discovery. Lavoisier never acknowledged receiving it; however, a copy of the letter was found in Scheele's belongings after his death.

Raoul Pictet (Figure 14.7) showed that by the evaporation of liquid sulfur dioxide ($SO_2$), carbon dioxide could be liquefied, which in turn was evaporated to cool oxygen gas enough to liquefy it. Pictet reported his results on December 22, 1877. Two days later, Louis Cailletet (Figure 14.8) announced his own method of liquefying oxygen. In both cases only a few drops could be produced, making analysis difficult. In 1891 James Dewar (Figure 7.5) was able to produce enough liquid oxygen to study. However, it was the process developed independently by Carl von Linde (Figure 13.25) and William Hampson (1854 - 1926).

**Figure 14.7: Swiss chemist and physicist Raoul Pierre Pictet (1846 - 1929).**

**Figure 14.8: French physicist Louis Paul Cailletet (1832 - 1913).**

*Sulfur*

Sulfur was known in ancient times and is referred to in the Bible. English translations of the Bible commonly referred to burning sulfur as *brimstone*, giving rise to the name of *fire-and-brimstone sermons*, in which listeners are reminded of the fate of eternal damnation that await the unbelieving and unrepentant. It is from this part of the Bible that Hell is implied to *smell of sulfur* (likely due to its association with volcanic activity). Sulfur ointments were used in ancient Egypt, while it was used for fumigation in Greece. A natural form of sulfur known as *shiliuhuang* was known in China since the 6$^{th}$ century BC. However, it was not until 1777 that Lavoisier (Figure 7.4) convinced the scientific community that sulfur was an element and not a compound.

*Selenium*

The element was discovered in 1817 by Berzelius (Figure 8.3), who found the element associated with tellurium. It was discovered as a byproduct of sulfuric acid production.

*Tellurium*

Tellurium was discovered in the 18$^{th}$ century in gold ore from the mines in Zlatna, Transylvania. In 1782 Müller von Reichenstein (Figure 14.9), the Hungarian chief inspector of mines in Transylvania, concluded that the ore was bismuth sulfide. However, the following year, he reported that this was erroneous and that the ore contained mostly gold and an unknown metal very similar to antimony. After three years of work Müller determined the specific gravity of the mineral and noted the radish-like smell of the white smoke evolved when the new metal was heated. Nevertheless, he was not able to identify this metal and gave it the names *aurum paradoxium* and *metallum problematicum*, as it did not show the properties predicted for the expected antimony.

**Figure 14.9: A stamp showing Hungarian mineralogist Franz-Joseph Müller von Reichenstein (1742 - 1825).**

In 1789 Kitaibel (Figure 14.10) also discovered the element independently in an ore from Deutsch-Pilsen which had been regarded as argentiferous molybdenite, but later he gave the credit to Müller. In 1798, the name was chosen by Klaproth (Figure 14.11) who earlier isolated it from the mineral calaverite.

**Figure 14.10: Hungarian botanist and chemist Pál Kitaibel (1757 - 1817).**

**Figure 14.11: Figure. German chemist Martin Heinrich Klaproth (1743 1817).**

*Polonium*

Temporarily called *radium F*, polonium was discovered by Marie Curie (Figure 9.7) and her husband Pierre Curie (Figure 9.8) in 1898, but it was later named after Marie Curie's native land of Poland. At the time Poland was not an independent country, but partitioned under Russian, Prussian, and Austrian. It was Curie's hope that naming the element after her native land would publicize its lack of independence. Polonium was the first chemical element named to highlight a political controversy.

## Abundance

The abundance of the chalcogenes is given in Table 14.2.

| Element | Terrestrial abundance (ppm) |
|---|---|
| O | $47\times10^4$ (Earth's crust), constituent of water, $21\times10^4$ (atmosphere) |
| S | 260 (Earth's crust), 870 (sea water), $10^{-3}$ (atmosphere) |
| Se | 0.05 (Earth's crust), 5 (soil), $0.2\times10^{-3}$ (sea water) |
| Te | $5\times10^{-3}$ (Earth's crust), 0.03 (soil), $0.15\times10^{-6}$ (sea water) |
| Po | Trace (Earth's crust) |

**Table 14.2: Abundance of Group 16 elements.**

## Isotopes

The naturally abundant isotopes of the Group 16 elements are listed in Table 14.3. All of the isotopes of polonium are radioactive.

| Isotope | Natural abundance (%) |
|---|---|
| Oxygen-16 | 99.76 |
| Oxygen-17 | 0.039 |
| Oxygen-18 | 0.201 |
| Sulfur-32 | 95.02 |
| Sulfur-33 | 0.75 |
| Sulfur-34 | 4.21 |
| Sulfur-36 | 0.02 |
| Selenium-74 | 0.87 |
| Selenium-76 | 9.36 |
| Selenium-77 | 7.63 |
| Selenium-78 | 23.78 |
| Selenium-80 | 49.61 |
| Tellurium-120 | 0.09 |
| Tellurium-122 | 2.55 |
| Tellurium-123 | 0.89 |
| Tellurium-124 | 4.74 |
| Tellurium-125 | 7.07 |
| Tellurium-126 | 18.84 |
| Tellurium-128 | 31.74 |
| Tellurium-130 | 34.08 |

**Table 14.3: Abundance of the non-synthetic isotopes of the Group 16 elements.**

There are 38 known nuclear isomers of tellurium with atomic masses that range from 105 to 142. Tellurium is the lightest element known to undergo alpha decay, with isotopes $^{106}Te$ to $^{110}Te$ being able to undergo this mode of decay.

*Cigarettes: it is not only the smoke that kills, but also the radioactivity*

Ever since the early 1960s, the presence of polonium-210 in tobacco smoke has been known. The world's biggest tobacco firms spent over 40 years trying to find ways to remove the polonium-210 without success: even to this day. However, they also never published the results, keeping the facts of the radioactive hazards from the consumer. Radioactive polonium-210 is contained in phosphate fertilizers and is absorbed by the roots of plants (such as tobacco) and stored in its tissues. Tobacco plants fertilized by rock phosphates contain polonium-210, which emits *alpha* radiation estimated to cause about 11,700 lung cancer deaths annually worldwide.

# Industrial production of the elements

*Sulfur*

Elemental sulfur is found near hot springs and volcanic regions in many parts of the world. Volcanic deposits are mined in Indonesia, Chile, and Japan. Significant deposits of sulfur also exist in salt domes along the coast of the Gulf of Mexico, and in eastern Europe and western Asia. The sulfur in these deposits is believed to come from the action of anaerobic bacteria on sulfate minerals. However, fossil-based sulfur deposits from salt domes are the basis for commercial production in the United States, Poland, Russia, Turkmenistan, and Ukraine. Sulfur is mainly extracted from natural sources by two processes:

- The Sicilian process.
- The Frasch process.

First used in Sicily from where it takes its name, the Sicilian process was used in ancient times to get sulfur from rocks present in volcanic regions. The sulfur deposits are piled and stacked in brick kilns built on sloping hillsides, and with airspaces between them (Figure 14.12). Then powdered sulfur is put on top of the sulfur deposit and ignited. As the sulfur burns, the heat melts the sulfur deposits, causing the molten sulfur to ow down the sloping hillside. The molten sulfur can then be collected in wooden buckets. The sulfur produced by the Sicilian process must be purified by distillation.

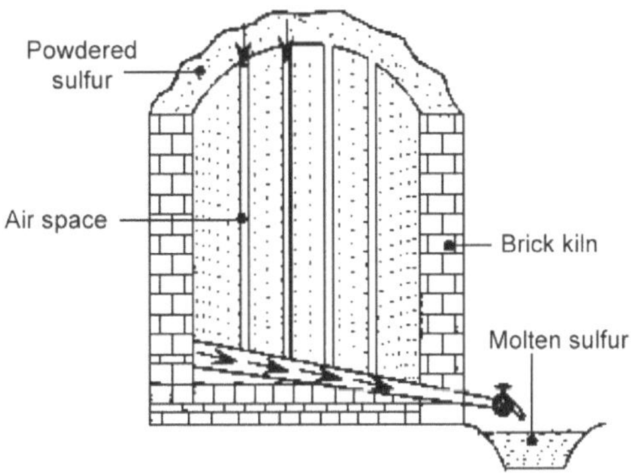


**Figure 14.12: Extraction of sulfur by the Sicilian process.**

In 1867, sulfur was discovered in the caprock of a salt dome in Louisiana; however, it was beneath quicksand, which prevented mining. In 1894 Herman

Frasch (Figure 14.13), devised a method of sulfur removal using pipes to by-pass the quicksand. The process proved successful, but the high cost of fuel needed to heat the water made the process uneconomic until the 1901 discovery of the Spindletop oil field in Texas (Figure 14.14) provided cheap fuel oil to the region.

**Figure 14.13: German-born American chemist Herman Frasch (1851 - 1914).**

**Figure 14.14: Spindletop oil field in Beaumont, Texas.**

In the Frasch process three concentric pipes to extract sulfur at high purity directly out of the ground (Figure 14.15). Superheated steam (160 °C) is pumped down the outermost pipe, which melts the sulfur. Hot compressed air is pumped down the innermost pipe, which serves to create foam and pressure. The resulting molten sulfur foam is then expelled through the middle pipe. The Frasch process produces sulfur with 99.5% purity, which needs no further purification.

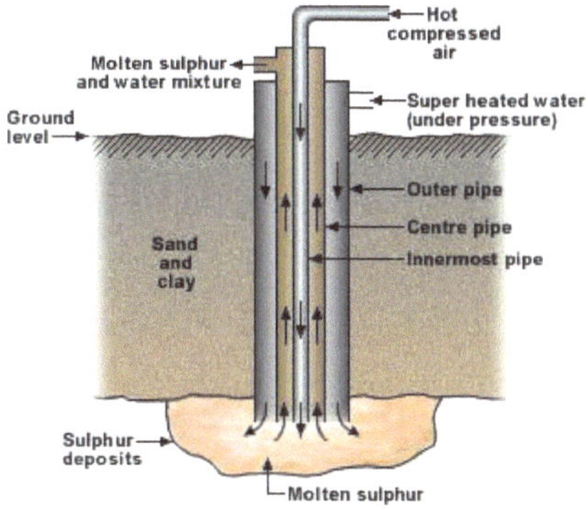


**Figure 14.15: Schematic diagram of the Frasch process.**

Most of the world's sulfur was obtained using the Frasch process until the late 20$^{th}$ century, when sulfur recovered from petroleum sources (recovered sulfur) became more commonplace.

*Selenium*

Elemental selenium is a rare mineral, and most elemental selenium comes as a byproduct of refining copper or producing sulfuric acid. Isolation of selenium begins by oxidation with sodium carbonate to produce selenium dioxide ($SeO_2$). The selenium dioxide is then mixed with water and the solution is acidified to form selenous acid (oxidation step). Selenous acid ($SeO(OH)_2$) is bubbled with sulfur dioxide (reduction step) to give elemental selenium.

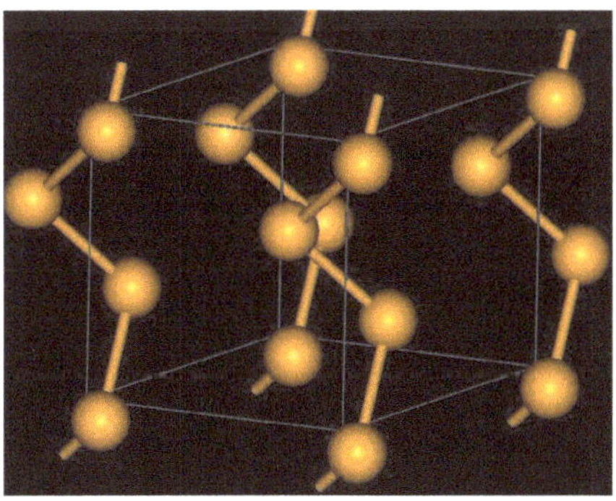

**Figure 14.16: The structure of trigonal selenium.**

Elemental selenium produced by chemical reactions appears as the amorphous red form. When the red form is rapidly melted, it forms the black, vitreous form. The most thermodynamically stable and dense form of selenium is the electrically conductive gray (trigonal) form, which is composed of long helical chains of selenium atoms (Figure 14.16). The conductivity of this form is notably light sensitive. Selenium also exists in three different deep-red crystalline monoclinic forms, which is composed of $Se_8$ molecules, similar to many allotropes of sulfur.

*Tellurium*

The principal source of tellurium is from anode sludges produced during the electrolytic refining of copper. Treatment of 500 tons of copper ore typically yields 1 lb (0.45 kg) of tellurium. The anode sludges contain the selenides and tellurides of the noble metals in compounds with the formula $M_2Se$ or $M_2Te$ (M = Cu, Ag, Au). At temperatures of 500 °C the anode sludges are roasted with sodium carbonate ($Na_2CO_3$) under air. The metals are reduced to the metals, while the tellurium is converted to sodium tellurite,

$$M_2Te + O_2 + Na_2CO_3 \rightarrow Na_2TeO_3 + 2\,M + CO_2$$

Tellurites can be extracted from the mixture with water and are normally present as hydrotellurites $HTeO^{3-}$ in solution. Selenates are also formed during this process, but they can be separated by adding sulfuric acid. The hydrotellurites are converted into the insoluble tellurium dioxide while the selenites stay in solution,

$$HTeO_3^- + OH^- + H_2SO_4 \rightarrow TeO_2 + 2\,SO_4^{2-} + 2\,H_2O$$

The reduction to the metal is done either by electrolysis or by reacting the tellurium dioxide with sulfur dioxide in sulfuric acid,

$$TeO_2 + 2\,SO_2 + 2\,H_2O \rightarrow Te + SO_4^{2-} + 4\,H^+$$

## Physical properties

The physical properties of the Group 16 elements encompass a gas ($O_2$), a non-metallic solid ($S_8$), and metals (Se, Te, Po), Table 14.4.

| Element | Melting point (°C) | Boiling point (°C) | Density (g/cm$^3$) |
|---|---|---|---|
| O | -218.79 | -182.95 | 1.429 g/L |
| S | 115.21 | 444.6 | 1.819 |
| Se | 221 | 685 | 4.81 (gray), 4.39 (alpha), 4.28 (vitreous) |
| Te | 449.51 | 988 | 6.24 (solid), 5.70 (liquid) |
| Po | 254 | 962 | 9.196 (alpha), 9.398 (beta) |

**Table 14.4: Selected physical properties of the Group 16 elements.**

The lighter Group 16 elements form $X_2$ dimers in the vapor phase. Sulfur also forms higher allotropes in the vapor phase (e.g., $S_8$ and $S_6$), while selenium and tellurium forms atomic vapor in addition to the dimmers. Unlike dihydrogen, however, the bonding is associated with the molecular orbital combination of the two $\pi$-orbitals (Figure 14.17). All of the dimeric $X_2$ molecules are paramagnetic.

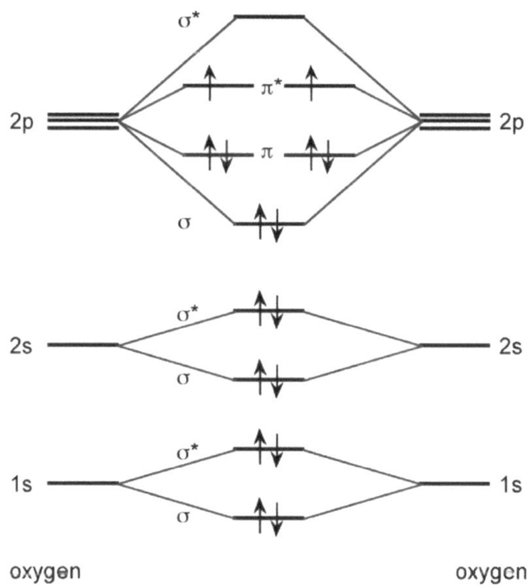


**Figure 14.17: Molecular orbital diagram for the formation of $O_2$.**

*Solid state*

While sulfur forms over 30 allotropes, the common form of sulfur is cyclooctasulfur ($S_8$) has three main allotropes: S$\alpha$, S$\beta$, and S$\gamma$. The orthorhombic form

(S$\alpha$) is more stable up to 95 °C, while the $\beta$-form is the thermodynamic form. The lone pairs of electrons make the S-S-S bend (108°), resulting in $S_8$ having the shape of a crown (Figure 14.18). When sulfur melts the $S_8$ molecules break up. When suddenly cooled, long chain molecules are formed in the plastic sulfur that, behave as rubber. Plastic sulfur transforms into rhombic sulfur over time.

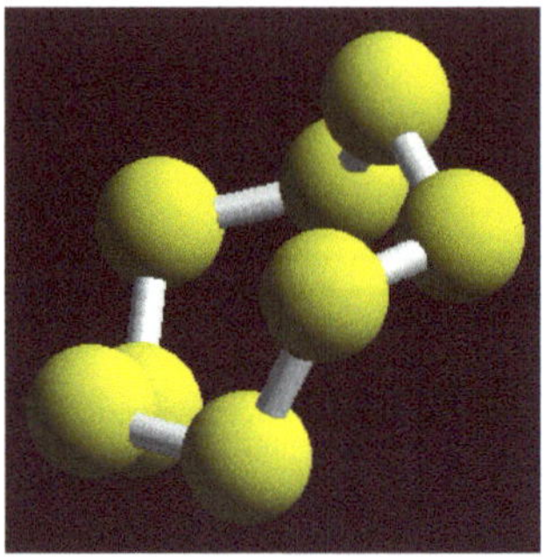

**Figure 14.18: The structure of $S_8$.**

Elemental selenium produced in chemical reactions appears as the amorphous red form. When rapidly melted, it forms the black, vitreous form, which is usually sold industrially as beads. The most thermo- dynamically stable and dense form of selenium is the electrically conductive gray (trigonal) form, which is composed of long helical chains of selenium atoms (Figure 14.16). The conductivity of gray selenium is light sensitive and is hence used in photocopiers. Selenium also exists in three different deep-red crystalline monoclinic forms, which are composed of $Se_8$ molecules, similar to many allotropes of sulfur. Unlike sulfur, however, selenium does not undergo the changes in viscosity when heated.

Tellurium is a crystalline metal with a trigonal structure ($a$ = 4.4572 Å, $c$ = 5.929 Å). Polonium has a simple cubic structure in it's a form ($a$ = 3.359 Å)

## Compounds of the Group 16 elements.

The chemistry of the Group 16 elements is dominated by the stability of the -2 oxidation state and the noble gas configuration of the $X^{2-}$ anion.

## Oxidation state

The electronegativity of oxygen (3.5) results in it having predominantly -2 oxidation state, however, sulfur, selenium and tellurium all for compounds with higher oxidation states, especially with oxygen (Table 14.5).

| Element | -2 | -1 | +4 | +6 |
|---|---|---|---|---|
| O | $Na_2O$, $H_2O$ | $H_2O_2$ | - | - |
| S | $H_2S$ | $H_2S_2$ | $SO_2$ | $H_2SO_4$, $SO_3$ |
| Se | $H_2Se$ | $H_2Se_2$ | $SeO_2$ | $SeF_4$ |
| Te | $H_2Te$ | $^tBu_2Te_2$ | $TeO_2$ | $Te(OH)_6$ |

**Table 14.5: Examples of oxidation states in compounds of the Group 16 elements.**

## Catenation

Catenation is the ability of a chemical element to form a long chain-like structure via a series of covalent bonds. Although oxygen shows this property only in the existence of ozone, sulfur is second only to carbon in exhibiting this mode of combination; the chalcogens beyond sulfur show it to diminishing degrees, polonium having no tendency to catenate.

When aqueous metal sulfide salts are heated with elemental sulfur a range of polysulfide ions are formed,

$$S^{2-} + n\,S \rightarrow S_{(n+1)}^{2-}$$

When alkali polysulfides dissolve in polar solvents (e.g., DMF or DMSO) a deep blue solution is formed. The absorption ($\lambda$ = 610 nm) is associated with the radical anion, $S_3^-$. While, polyselenides and polytellurides are less common, the $Se_3^-$ and $Te_3^-$ ions are known.

The term polysulfide often refers to a class of polymers with alternating chains of several sulfur atoms and hydrocarbon substituents. The general formula is $R_2Sn$, where n ranges from 2 - 10. For the selenium and tellurium analogs the extent of catenation is far more limited.

# Ozone

Ozone ($O_3$) is an allotrope of oxygen that is much less stable than the diatomic molecule ($O_2$). Ground-level ozone is an air pollutant with harmful effects on the respiratory system, while the ozone layer in the upper atmosphere filters potentially damaging ultraviolet light from reaching the Earth's surface.

The structure of ozone is bent, with $C_{2v}$ symmetry, similar to water (Figure 14.19a). The central oxygen has $sp^2$ hybridization with one lone pair. As a consequence of the bent structure, and the resonance hybridization (Figure 14.19b) ozone is a polar molecule (dipole moment = 0.5337 D).

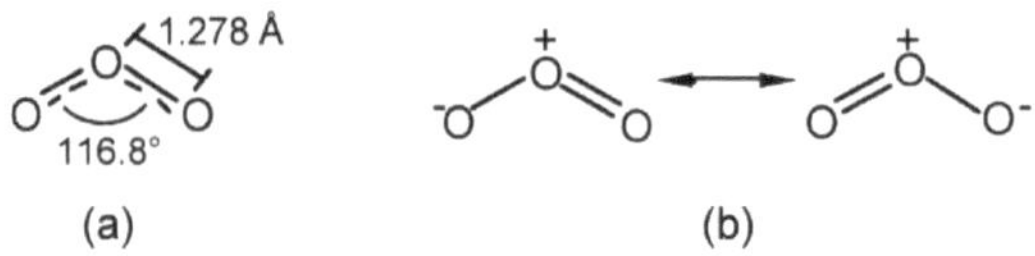


**Figure 14.19: The (a) structural parameters of ozone and (b) its resonance hybridization.**

Ozone is made by the exposure of oxygen ($O_2$) to an electric discharge. Ozone has a characteristic smell can be commonly smelled after a lightening-strike; in fact, the name ozone comes from the Greek *ozein* meaning to smell. In the laboratory, ozone can also be produced by electrolysis using graphite rod cathode, a platinum wire anode, and sulfuric acid (3 M) electrolyte. The half-cell reactions are as follows:

$$3\,H_2O \rightarrow O_3 + 6\,H^+ + 6\,e^- \qquad (\Delta E_o = -1.53\text{ V})$$

$$6\,H^+ + 6\,e^- \rightarrow 3\,H_2 \qquad (\Delta E_o = 0\text{ V})$$

$$2\,H_2O \rightarrow O_2 + 4\,H^+ + 4\,e^- \qquad (\Delta E_o = -1.23\text{ V})$$

Ozone is also produced through photolysis of oxygen both in the laboratory and the atmosphere.

$$O_2 \xrightarrow{h\nu} 2\,O\cdot$$

$$O\cdot + O_2 \rightarrow O_3$$

Ozone is a very strong oxidizing agent, and will readily oxidize a range of materials, e.g.,

$$O_3 + CO \rightarrow CO_2 + O_2$$

$$O_3 + 2\,I^- + H_2O \rightarrow O_2 + 2\,OH^- + I_2$$

It will also oxidize metals (except gold, platinum, and iridium) to their highest oxidation state, e.g.,

$$2\,Cu^+_{(aq)} + 2\,H_3O^+_{(aq)} + O_{3(g)} \rightarrow 2\,Cu^{2+}_{(aq)} + 3\,H_2O_{(l)} + O_{2(g)}$$

Metal ozanides, which contain the ozonide anion ($O_3^-$) are explosive and must be stored at cryogenic temperatures. Ozonides for all the alkali metals are known. $KO_3$, $RbO_3$, and $CsO_3$ can be prepared from their respective superoxides.

$$KO_2 + O_3 \rightarrow KO_3 + O_2$$

*Ozone as a modulator of life on Earth*

The Earth's atmosphere acts as a source of $O_2$ and a repository of $CO_2$, but it also acts as a shield for life. First, nearly all meteorites burn up on entry because of the high temperatures generated by the friction of the atmosphere. Second, the atmosphere acts as a shield for high energy UV radiation.

Although UV radiation converts 7-dehydrocholesterol into vitamin $D_3$ in the skin (Figure 14.20), and is therefore useful, high energy UV destroys living cells. In fact, the darkening we call a suntan is actually the body's mechanism for preventing further UV damage. Sun burn and skin cancer are caused by relatively weak UV light that reaches the Earth's surface, without the atmosphere we would be exposed to high energy UV that would be a hazard to all life on Earth. The Earth's sunscreen is ozone ($O_3$), and without ozone in the upper atmosphere there would be no life on Earth.

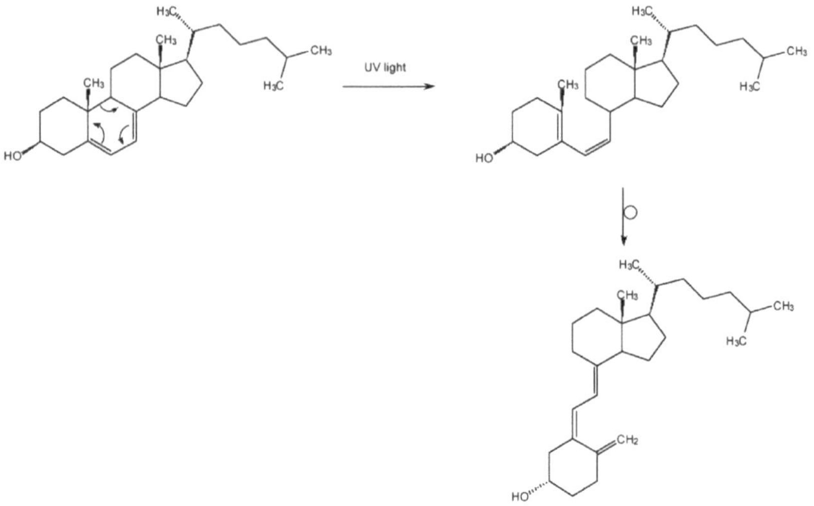


**Figure 14.20: The conversion of 7-dehydrocholesterol (a derivative of cholesterol) to pre-vitamin $D_3$ by photolysis by ultraviolet light, and its subsequent isomerization to vitamin $D_3$.**

The ozone layer is located in the lower portion of the stratosphere from approximately 10 km to 50 km above Earth, though the thickness varies seasonally and geographically. This layer contains over 91% of the ozone in Earth's atmosphere and absorbs 93-99% of the sun's high frequency ultraviolet light. The ozone decomposes photolytically to $O_2$ and molecular oxygen,

$$O_3 + h\nu \rightarrow O_2 + O$$

and it is this reaction that accounts for the UV protection of the atmosphere. Ozone is naturally regenerated by the exothermic reaction of the molecular oxygen with $O_2$,

$$O + O_2 \rightarrow O_3$$

The balance between ozone formation and destruction is thus an important mechanism for the protection of living organisms on the planet. While the ozone layer had been relatively constant on Earth for millions of years, the last 70 have seen a dramatic change including the increase in the polar hole in the ozone layer. The ozone hole is defined geographically as the area where the total ozone concentration is less than 220 Dobson Units. One Dobson unit refers to a layer of ozone that would be 10 μm that is $1 \times 10^{-5}$ m thick under standard temperature (25 °C) and pressure (1 atmosphere).

The ozone hole has steadily grown in size and length of existence over the past two and half decades. At present the size of ozone hole over Antarctica is estimated to be about 30 million sq.km (Figure 14.21).

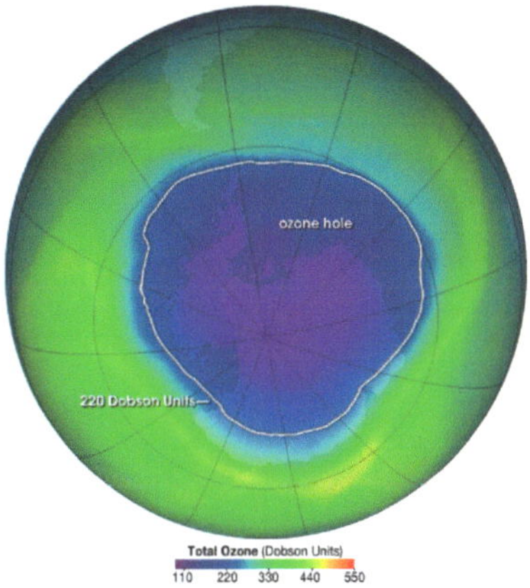


**Figure 14.21: The concentration profile of ozone over Antarctica.**

## Water: the fuel for the medieval industrial revolution

Despite the greatest industrial complex of the Roman Empire being the imperial grain mill at Barbegal near Arles in what is now southern France (Figure 14.22), and the knowledge of gearing used for the mill, the waterwheel (the power source of the Barbegal's power) was little used in the ancient world. This was probably due to the high slave population obviating the need for labor saving devices. However, it may also be that because the Roman Empire was very centralized, they could provide our on a large scale from a few highly mechanized locations. Upon the fall of the Roman Empire, the knowledge of waterpower would have been lost were it not for the writing departments of churches and monasteries that continued to operate through the subsequent Dark Ages.

**Figure 14.22: The Barbegal mill as it stands today.**

The Barbegal mill is probably the first example of industrial mass production. It consisted of eight pairs of waterwheels positioned on a 65-foot slope (Figure 14.23).

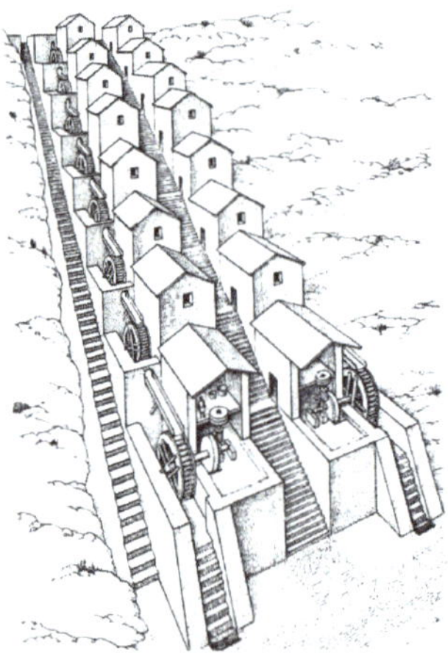

**Figure 14.23: An artist's representation of the overshot waterwheels that are thought to have driven the 16 mills at Barbegal.**

The wheels were turned by water that fell from a reservoir, which in turn was fed by a magnificent aqueduct. The sixteen wheels each powered two grindstones using a set of gear that allowed the horizontal shaft from the wheel to turn a vertical shaft on which the grindstones were positioned. Grain for the mill was imported from as far as Egypt, and the flour production was eight times than that required for the local population of 10,000, resulting in an export business. Unfortunately, upon the fall of Rome the technology of waterpower almost ceased since the small city-states set up had no need for industrial complexes.

There are two general designs of waterwheel. The first is powered by water falling from above the wheel and is called an overshot wheel. The alternative design, the undershot wheel, relied on water owing on a river or pond such that the current moved the paddles at the bottom of the wheel.

Despite its fall from use, the waterwheel was not forgotten. Due to the writings of 14$^{th}$ century BC engineers, such as Vitruvius (Figure 14.23), whose texts had been preserved in the libraries in churches and monasteries across Europe, the waterwheel powered mills made a resurgence between the fifth and the tenth century. In most cases mills were owned by the church, since they had the knowledge (from ancient texts) to construct the mills and, possibly just as important, the literacy to develop the accounting system for their pro table use. The church would lease mills to farmers on a time usage, and they would take payment in flour. The owners of waterwheels and their mills were the first barons of industry post the Dark Ages. In fact, the Saxon word for an aristocrat is Lord, which means loaf giver, suggests the importance of the waterwheel. By the end of the tenth century the waterwheel was in widespread use across Europe. The Domesday Book (the nationwide census carried out by the Normans after the invasion of England) listed nearly six thousand grain mills in 1089.

**Figure 9.34: A depiction of Vitruvius (right) presenting De Architectura to the Emperor Augustus.**

Despite the large number of waterwheels, the gearing used up until the 9$^{th}$ century was little different to that described by Vitruvius and used at Barbegal. Then around 890 AD the monastery of St Gall a new device was attached to the waterwheel. Instead of gearing to transfer power from horizontal to vertical rotation, a piece of wood was set into the shaft driven by the waterwheel (Figure 14.24).

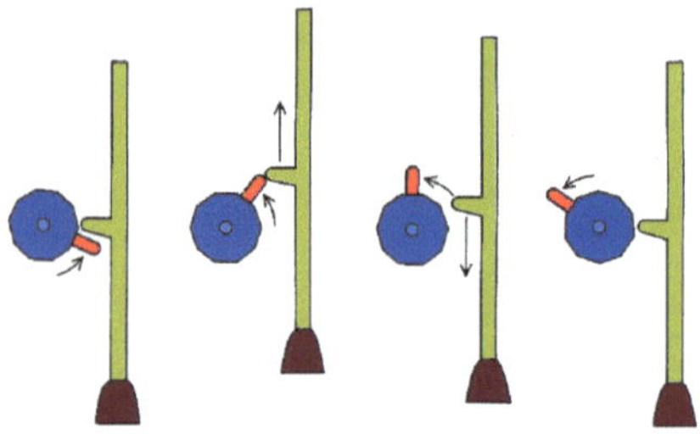

**Figure 14.24: Schematic representation of the transformation of rotary motion into linear motion can be achieved by having a cam on the axle of the wheel**

What had been created was a cam, since as the shaft turns the protruding piece of wood its anything in its way. For example, the first recorded use, by the monks at St Gall, was to crush malt for beer; however, the cam could be made to trip a hammer with every rotation (pounding), or to act on a crank to turn a rotary motion into a horizontal back-and-forward motion (cutting), or to push down a level and activate a suction pump (raising water from a well), or operate a bellows (for a metal forge). The range of motions meant that waterpower could now be used for a wide range of industries. By the end of the tenth century there were waterwheels powering forge hammers, oil and silk mills, sugar cane crusher, tanning mills pounding leather, grinding stones, ore crushing mills (Figure 14.25), and as fulling mills for the rapidly expanding trans-European textile industry.

**Figure 14.25: The cam principle as applied to a rock-crushing mill illustrated by Georgius Agricola in his book De Re Metallica (1556).**

As a consequence of the waterwheel and the cam, the period between the tenth and fourteenth centuries has come to be known as the Medieval Industrial Revolution. It is interesting to note that water played a key role in the driving force during the next Industrial Revolution four hundred years later the steam engine.

## Hydrogen peroxide

Hydrogen peroxide ($H_2O_2$) is a very pale blue liquid but appears colorless in dilute solution. It is prepared by the oxidation of anthraquinol, e.g.,

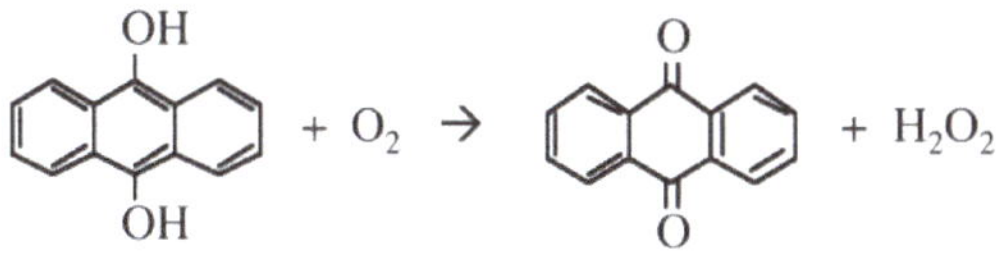


The hydrogen peroxide is extracted with water from the anthraquinone solution and the 20 - 40% solution is purified by solvent extraction. An alternative process involves the oxidation of isopropanol in either the vapor or liquid phase at 100 °C and ca. 15 atm,

$$Me_2C(H)OH + O_2 \rightarrow Me_2C(OOH)OH \rightarrow Me_2C{=}O + H_2O_2$$

The products are separated by fractional distillation. In the gas phase $H_2O_2$ adopts a gauche conformation (Figure 14.26), but there is only a low barrier to rotation about the O-O bond.

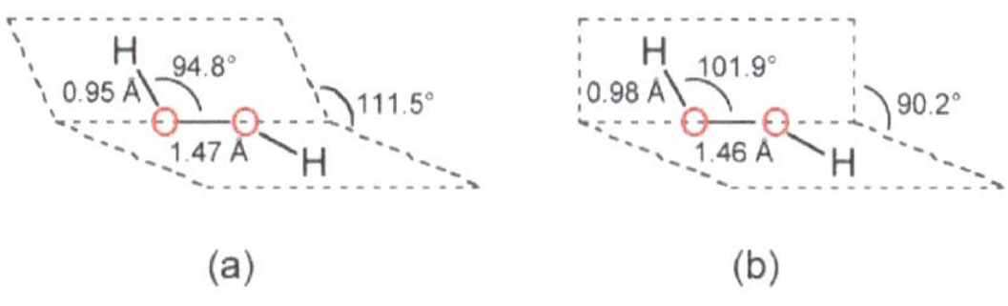


**Figure 14.26: Structure of hydrogen peroxide in (a) the vapor phase and (b) the solid (crystal) phase.**

Hydrogen peroxide is a liquid at standard temperature and pressure (25 °C, 1 atm) due to the presence of strong hydrogen bonding similar to found in water. In fact, the liquid range for $H_2O_2$ (Melting point = -0.43 °C, Boiling point = 150.2 °C) is actually broader than water, and it is slightly more viscous than water. Hydrogen peroxide has a density of 1.44 g/cm$^3$ and is $10^6$ times less basic that water.

As with water, $H_2O_2$ is a good solvent because of its polar nature and broad liquid temperature range, however, it is dangerous in its pure state due to its facile ($\Delta H$ = -99 kJ/mol) auto decomposition,

$$2\,H_2O_2 \rightarrow 2\,H_2O + O_2$$

as well as its strong oxidizing nature.

Hydrogen peroxide is usually sold as 3 - 12% solution for home use; however, laboratory and certain industrial applications require 30% solutions.

Hydrogen peroxide should be stored in a cool, dry, well-ventilated area and away from any flammable or combustible substances. It should be stored in a container composed of non- reactive materials such as stainless steel or glass (other materials including some plastics and aluminum alloys may also be suitable). Because it breaks down quickly when exposed to light, it should be stored in an opaque container, and pharmaceutical formulations typically come in brown bottles that filter out light.

Aqueous solution of $H_2O_2$ are weakly acidic ($K = 1.5\times10^{-12}$),

$$H_2O_2 + H_2O \rightleftharpoons HO_2^- + H_3O^+$$

however, there is no exchange of oxygen atoms between $H_2O_2$ and $H_2O$ in the liquid phase.

As expected, hydrogen peroxide is a strong oxidizing agent,

$$H_2O_2 + 2\,HI \rightarrow I_2 + 2\,H_2O$$

but, it can also act as a reducing agent,

$$5\,H_2O_2 + 2\,MnO_4 + 6\,H^+ \rightarrow 2\,Mn^{2+} + 8\,H_2O + 5\,O_2\uparrow$$

# Hydrogen peroxide providing a lift for 007

In the pre-credit sequence of the 1965 film Thunderball, James Bond 007 (played by Sean Connery) uses a Jetpack to escape from two gunmen after killing Jacques Bouvar, SPECTRE Agent No. 6 (Figure 14.38). The Jetpack was also used in the Thunderball posters, being the "Look Up" part of the "Look Up! Look Down! Look Out!" tagline.

**Figure 14.27: James Bond 007 escaping from gunmen using a Jetpack (Copyright: Eon Productions).**

The Jetpack was actually a real fully functional device, the Bell Rocket Belt. It was designed for use in the US Army but was rejected because of its short flying time of 21-22 seconds. Powered by hydrogen peroxide ($H_2O_2$), it could fly about 250 m and reach a maximum altitude of 18 m, going 55 km/h. Despite its impracticality in the real world, the Jetpack made a spectacular debut in Thunderball. Although Sean Connery is seen in close-up during the takeoff and landings (Figure 14.28), the main flight was actually piloted by Gordon Yeager (Figure 14.29) and Bill Suitor (Figure 14.30).

**Figure 14.28: Close-up shot of Sean Connery as James Bond landing the Jetpack. Note that the harness is not correctly attached, indicating this was a set-up shot for the film rather than a real flight. Presumably the crotch strap would have spoiled the lines of his suit. (Copyright: Eon Productions).**

**Figure 14.29: American pilot Gordon Yeager (1927 2005).**

**Figure 14.30: American rocket pack test pilot Bill Suitor (1944 - ).**

The Bell Rocket Belt is a low-power rocket propulsion device that allows an individual to safely travel or leap over small distances. All subsequent rocket packs were based on the construction design, developed in 1960-1969 by Wendell Moore, which has two major parts:

- Rigid glass-plastic corset (Figure 14.31a), strapped to the pilot (Figure 14.31b). The corset has a tubular metallic frame on the back, on which are fixed three gas cylinders: two with liquid hydrogen peroxide (Figure 14.31c), and one with compressed nitrogen (Figure 14.31d). When the pilot is on the ground, the corset distributes the weight of the pack to the pilot's back.
- The rocket engine, able to move on a ball and socket joint (Figure 14.31e) in the upper part of the corset. The rocket engine consists of a gas generator (Figure 14.31f) and two pipes (Figure 14.31g) rigidly

connected with it, which end with jet nozzles with controlled tips (Figure 9.43h). The engine is rigidly connected to two levers, which are passed under the pilot's hands. Using these levers, the pilot inclines the engine forward or back and to the sides. On the right lever is the thrust control turning handle (Figure 14.31i), connected with a cable to the regulator valve (Figure 14.31j) to supply fuel to the engine. On the left lever is the steering handle, which controlled the tips of jet nozzles.

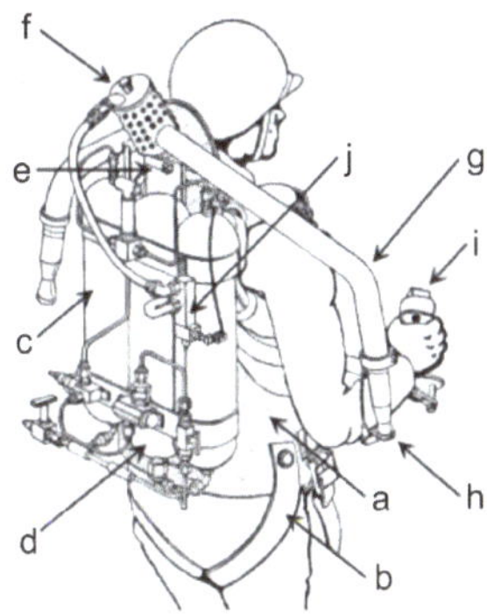


**Figure 14.31: Diagram of the Bell rocket pack. Adapted from US Patent 3,243,144 (1966).**

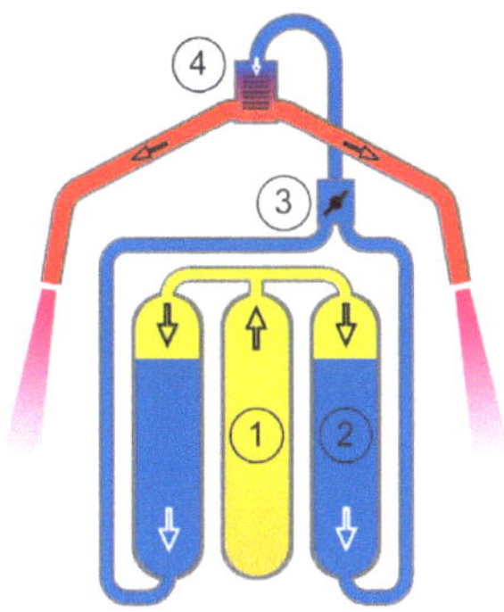


**Figure 14.32: The operating principle of rocket engine.**

The operating principle of the Jetpack is shown in Figure 14.32. The hydrogen peroxide cylinder and compressed nitrogen cylinder are each at a pressure of ca. 40 atm or 4 MPa). To operate the pilot turns the engine thrust control handle and opens the regulator valve (3 in Figure 14.32). Compressed nitrogen (1 in Figure 14.32) displaces liquid hydrogen peroxide (2 in Figure 14.32), which enters the gas generator (4 in Figure 14.32). In the gas generator, the hydrogen peroxide contacts the catalyst (consisting of thin silver plates, covered with a layer of samarium nitrate) and is decomposed. The resulting hot high-pressure mixture of steam and gas enters two pipes (covered with a layer of heat insulator), which emerge from the gas generator. The hot gas enters the jet nozzles, where first they are accelerated, and then expand, acquiring supersonic speed and creating reactive thrust. Pressing on one of two levers, the pilot deflects the nozzles back, and the pack flies forward or back.

## Comparison of sulfur to oxygen

### Size

Table 14.6 summarizes the comparative sizes of oxygen and sulfur.

| Element | Atomic radius (Å) | Covalent radius (Å) | Ionic radius (Å) | van der Waal radius (Å) |
|---|---|---|---|---|
| Oxygen | 0.48 | 0.66 | 1.40 | 1.52 |
| Sulfur | 0.88 | 1.05 | 1.84 | 1.80 |

**Table 14.6: Comparison of physical characteristics for oxygen and sulfur.**

### Electronegativity

Sulfur is less electronegative than oxygen (2.4 and 3.5, respectively) and as a consequence bonds to sulfur are less polar than the corresponding bonds to oxygen. One significant result in that with a less polar S-H bond the subsequent hydrogen bonding is weaker than observed with O-H analogs. A further consequence of the lower electronegativity is that the S-O bond is polar.

### Bonds formed

Sulfur forms a range of bonding types. As with oxygen the -2 oxidation state prevalent. For example, sulfur forms analogs of ethers, i.e., thioethers R-S-R. However, unlike oxygen, sulfur can form more than two covalent (non-dative) bonds, i.e., in compounds such as $SF_4$ and $SF_6$.

Such hypervalent compounds were originally thought be due to the inclusion of low energy d orbitals in hybrids (e.g., $sp^3d^2$ for $SF_6$); however, a better picture involves a combination of s and p orbitals in bonding (Figure 14.33). Any involvement of the d orbitals is limited to the polarization of the p orbitals rather than direct hydridization. In this regard $SF_6$ represents the archetypal hypervalent molecule. Finally, sulfur can form multiple bonds, e.g., $Me_2S=O$.

## Catenation

Catenation is de ned as the ability of a chemical element to form a long chain-like structure via a series of covalent bonds. Oxygen's extent of catenation is limited to ozone ($O_3$) and peroxides (e.g., R-O-O-R). In contrast, the chemistry of sulfur is rich in the formation of multiple S-S bonds.

While elemental sulfur exists as a diatomic molecule (i.e., $S_2$) in the gas phase at high temperatures, sulfur vapor consists of a mixture of oligomers ($S_3$ to $S_8$) as a temperature dependent equilibrium. In the solid state the formation of Sn dominates, and sulfur exists as a range of polymorphs in which extended S-S bonding occurs in either rings of 6 to 20 atoms (e.g., Figure 14.34) or chains (catenasulfur).

The higher level of catenation for sulfur is due to the greater strength of a S-S bond (226 kJ/mol) as compared to the O-O bond (142 kJ/mol). In general, the homoleptic bond strength is expected to decrease going down a period of the Periodic Table. The reason for the unexpected weakness of the O-O bond is that the electronegative oxygen atoms repel each other and thus weaken the bond.

# Chalcogenide hydrides

## Dihydrides

The hydrides of sulfur, selenium and tellurium are all extremely toxic gases with repulsive smells. Hydrogen sulfide ($H_2S$) is very toxic, in fact it is more than 5× as toxic as HCN (Table 14.7). Hydrogen sulfide is considered a broad-spectrum poison, meaning that it can poison several different systems in the body, although the nervous system is most affected. It forms a complex bond with iron in the mitochondrial cytochrome enzymes, thereby blocking oxygen from binding and stopping cellular respiration. Exposure to low concentrations can result in eye irritation, a sore throat and cough, nausea, shortness of breath, and fluid in the lungs. Long-term, low-level exposure may result in fatigue, loss of appetite, headaches, irritability, poor memory, and dizziness.

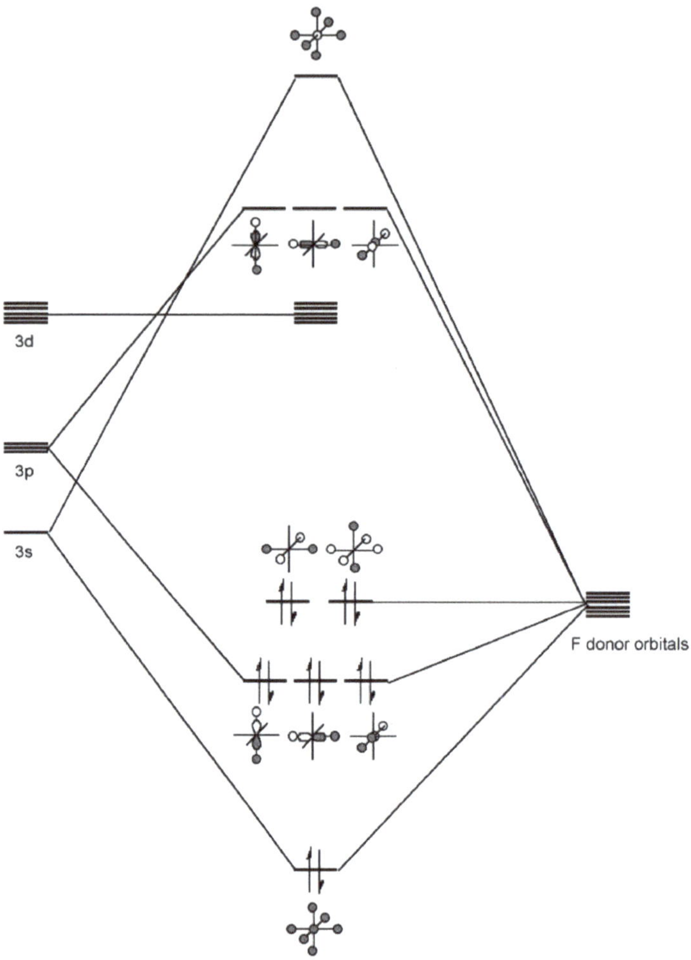


**Figure 14.33: Molecular orbital diagram for $SF_6$.**

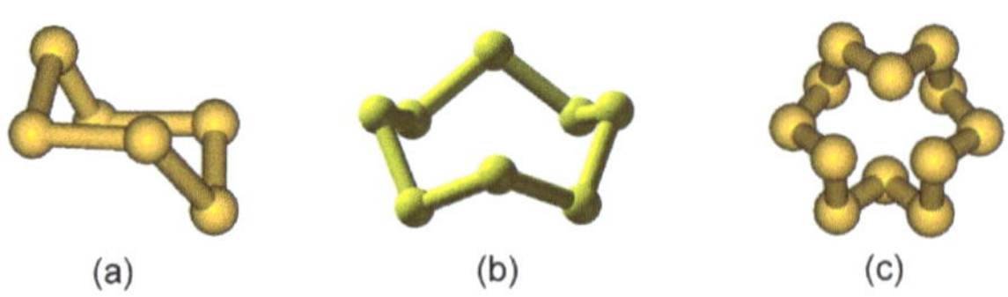


**Figure 14.34: Structures of two polymorphs of sulfur: (a) cyclohexasulfur $S_6$, (b) cyclooctasulfur $S_8$, and (b) cyclododecasulfur $S_{12}$.**

| Concentration (ppm) | Biological effect |
|---|---|
| 0.00047 | Threshold. |
| 10 - 20 | Borderline concentration for eye irritation. |
| 50 - 100 | Eye damage. |
| 100 - 150 | Olfactory nerve is paralyzed, and the sense of smell disappears, often together with awareness of danger. |
| 320 - 530 | Pulmonary edema with the possibility of death. |
| 530 - 1000 | Stimulation of the central nervous system and rapid breathing, leading to loss of breathing. |
| 800 | Lethal concentration for 50% of humans for 5 minutes exposure ($LC_{50}$). |
| +1000 | immediate collapse with loss of breathing, even after inhalation of a single breath. |

**Table 14.7: Toxicity levels for hydrogen sulfide.**

Each of the hydrides is prepared by the reaction of acid on a metal chalcogenide, e.g.,

$$Fe + S \rightarrow FeS$$

$$FeS + 2\,HCl \rightarrow H_2S \uparrow + FeCl_2$$

The unstable $H_2Po$ has been prepared by the reaction of HCl on Po metal. The thermal stability and bond strength of the dihydrides follows the trend:

$$H_2S > H_2Se > H_2Te > H_2Po$$

While $H_2Se$ is thermodynamically stable to 280 °C, $H_2Te$ and $H_2Po$ are thermodynamically unstable. All the dihydrides behave as weak acids in water. Thus, dissolution of $H_2S$ is water results in the formation of the conjugate bases,

$$H_2S + H_2O \rightleftharpoons H_3O^+ + SH^-$$

$$SH^- + H_2O \rightleftharpoons H_3O^+ + S^{2-}$$

with dissociation constants of $10^{-7}$ and $10^{-17}$, respectively.

## Sulfanes

The propensity of sulfur for catenation means that while the hydrides of oxygen are limited to water ($H_2O$) and hydrogen peroxide ($H_2O_2$), the compounds $H_2S_n$ where n = 2 - 6 may all be isolated. Higher homologs are also known, but only as mixtures. All of the sulfanes are yellow liquids whose viscosity increases with increased chain length.

A mixture of lower sulfanes is prepared by the reaction of sodium sulfides ($Na_2Sn$) with HCl,

$$Na_2S_2 + 2\ HCl \rightarrow 2\ NaCl + H_2S_n$$

From this mixture the compounds $H_2S_n$ where n = 2 - 5 are purified by fractional distillation. However, higher sulfanes are made by the reaction of either $H_2S$ or $H_2S_2$ with sulfur chlorides,

$$2\ H_2S + S_nCl_2 \rightarrow 2\ HCl + H_2S_{n+2}$$

$$2\ H_2S_2 + S_nCl_2 \rightarrow 2\ HCl + H_2S_{n+4}$$

# Oxides and oxyacids of sulfur

Alternative spellings of sulfurous and sulfuric acids are based upon the traditional UK spelling of sulphur, i.e., sulphurous and sulphuric acid.

## Sulfur dioxide and sulfurous acid solutions

The combustion of sulfur results in the formation of gaseous sulfur dioxide,

$$S + O_2 \rightarrow SO_2$$

The bent structure of $SO_2$ is shown in Figure 14.35, and as a consequence of the $sp^2$ hybridization the molecule is polar.

The modest boiling temperature of $SO_2$ (-10 °C) means that it is readily liquefied and easily kept as a liquid at room temperature under a slight pressure. The liquid is associated by dipole-dipole attractions due to the polar nature of $SO_2$. Liquid $SO_2$ is a good solvent due to the polarity of the molecule; as a consequence, it readily solubilizes polar compounds and salts. It is also convenient since it is easy to remove from reaction products by evaporation.

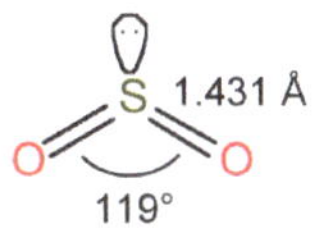


**Figure 14.35: The structure of sulfur dioxide.**

Sulfur dioxide is soluble in water forming aqueous solutions where most of the $SO_2$ is maintained as a hydrogen-bonded hydrate, in a similar manner to that observed for aqueous solutions of carbon dioxide. At equilibrium in neutral water (no added base) a small fraction reacts, to give a mixture of bisulfite ($HSO_3^-$, Figure 14.36a) and sulfite ($SO_4^-$, Figure 14.36b),

$$SO_{2(aq)} + H_2O \rightleftharpoons HOSO_2^-{}_{(aq)} \rightleftharpoons SO_3^{2-}{}_{(aq)}$$

$$\downarrow$$

$$H\text{-}SO_3^-$$

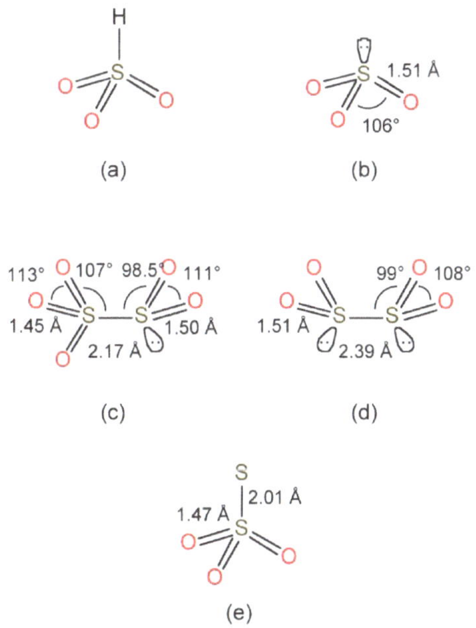


**Figure 14.36: The structures of the (a) bisulfite, (b) sulfite, (c) disulfite, (d) dithionite, and (e) thiosulfate anions.**

The free acid does not to exist. Bisulfite undergoes a further equilibrium,

$$2\,HSO_3^- \rightleftharpoons S_2O_5^{2-} + H_2O$$

to form disulfite, whose structure is shown in Figure 14.36c.

Salts of these anions are known, and complexes of the sulfite ion are known (Figure 14.37), while $SO_2$ itself can act as a ligand to heavy metals. The bisulfite ion has strong reducing properties, e.g.,

$$2\,Fe^{3+} + SO_3^{2-} + 2\,OH^- \rightarrow 2\,Fe^{2+} + H_2O + SO_4^{2-}$$

$$2\,MnO_4^+ + 5\,SO_3^{2-} + 6\,H^+ \rightarrow 2\,Mn^{2+} + 3\,H_2O + 5\,SO_4^{2-}$$

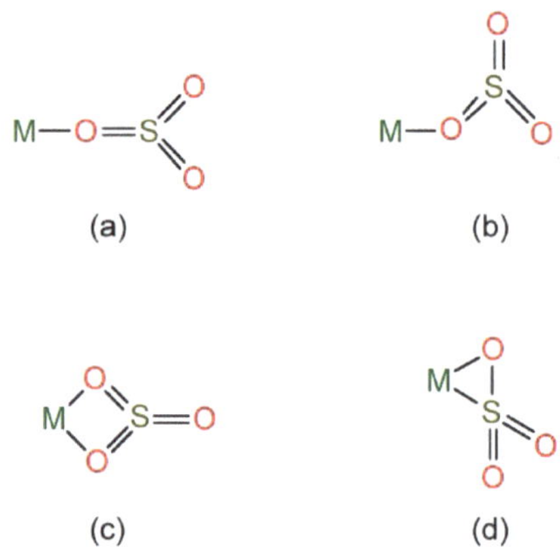


**Figure 14.37: The structural modes of sulfite coordination.**

Bisulfite is also reduced by zinc in the presence of additional $SO_2$,

$$SO_3^{2-} + SO_2 \xrightarrow{Zn} S_2O_4^{2-}$$

to form the highly reducing dithionite anion (Figure 14.36d). Reaction of bisulfite with elemental sulfur yields the thiosulfate anion (Figure 14.36e),

$$SO_3^{2-} + S \rightarrow S_2O_3^{2-}$$

# Sulfur trioxide and sulfuric acid

Oxidation of sulfur dioxide in the presence of a catalyst (e.g., platinum) yields sulfur trioxide,

$$2\ SO_2 + O_2 \xrightarrow{Pt} 2\ SO_3$$

which may be condensed to a liquid at room temperature (Boiling point = 45 °C). Liquid $SO_3$ exists as a mixture of monomer and trimers (Figure 14.38a and b), while as a solid (Melting point = 16.9 °C) it forms polymers (Figure 14.38c).

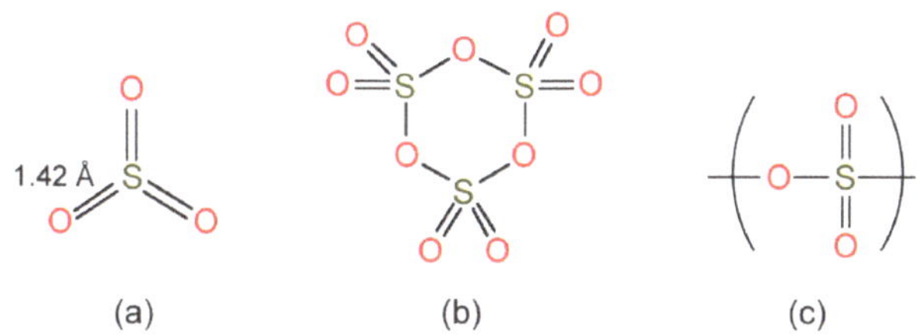


**Figure 14.38: The structure of (a) monomeric, (b) trimeric, and (c) polymeric sulfur trioxide.**

The reaction of $SO_3$ with water results in the formation of sulfuric acid, $H_2SO_3$, as a viscous, hydrogen bonded liquid. Sulfuric acid is a strong protic acid, which in dilute solutions (in water) reacts as a dibasic acid forming bi-sulfate ($HSO_4^-$) and sulfate ($SO_4^{2-}$) anions.

$$H_2SO_4 \underset{+\,H^+}{\overset{-\,H^+}{\rightleftharpoons}} HSO_4^- \underset{+\,H^+}{\overset{-\,H^+}{\rightleftharpoons}} SO_4^{2-}$$

A large number of salts are known for both anions. In addition, sulfate is known to act as a monodentate or bidentate ligand in coordination complexes.

The dissolution of $SO_3$ in concentrated sulfuric acid yields very corrosive, fuming sulfuric acid, which contains some pyrosulfuric acid,

$$H_2SO_4 + SO_3 \rightleftharpoons H_2S_2O_7$$

The corrosive properties of sulfuric acid are accentuated by its highly exothermic reaction with water. Burns from sulfuric acid are potentially more serious than those of comparable strong acids (e.g., hydrochloric acid), as there is additional tissue damage due to dehydration and particularly secondary thermal damage due to the heat liberated by the reaction with water.

## Sulfur as a source of atmospheric pollution and acid rain

Sulfur dioxide is formed as a pollutant during the combustion of sulfur containing fuels, in particular coal. While the emission of $SO_2$ itself leads to concerns it is its conversion to sulfuric acid in the form of acid rain that has been of concern for several decades. The pathway for the formation of sulfuric acid in the atmosphere is dependent on whether the reaction occurs in dry atmosphere or in clouds and rain.

*Gaseous reactions in a dry atmosphere*

In the dry atmosphere, gaseous sulfur dioxide reacts with the hydroxide radical (formed by the photochemical decomposition of ozone,

$$O_3 + h\nu \rightarrow O^* + O_2$$

$$O^* + H_2O \rightarrow 2\ HO^{\cdot}$$

in the presence of a non-reactive gas molecule such as nitrogen,

$$O^* + H_2O \rightarrow 2\ HO^{\cdot}$$

The sulfurous acid thus formed reacts with oxygen to generate sulfur trioxide,

$$HSO_3 + O_2 \rightarrow HO_2 + SO_3$$

which reacts with water to form sulfuric acid,

$$SO_3 + H_2O \rightarrow H_2SO_4$$

Measurements indicate that the conversion rate of $SO_2$ to $H_2SO_4$ is 4% per hour on a clear sunny day, but the rate is slower during the winter.

*Liquid phase reactions in clouds and rain*

In the liquid phase $SO_2$ reacts directly with water,

$$2\ SO_2 + 2\ H_2O \rightarrow SO_3^- + HSO_3^- + 3\ H^+$$

The bisulfite ($HSO_3^-$) is oxidized by hydrogen peroxide forming a forming bisulfate ($HSO_4^-$) solution,

$$HSO_3^- + H_2O_2 \rightarrow HSO_4^- + H_2O$$

Water soluble hydrogen peroxide is formed by the oxidation of water,

$$HO_2 + HO_2 \rightarrow H_2O_2 + O_2$$

The $HO_2^{\cdot}$ radical is formed by the photolysis of organic carbonyl compounds, e.g., formaldehyde,

$$H_2C{=}O + h\nu \rightarrow H^{\cdot} + HCO^{\cdot}$$

$$HCO + O_2 \rightarrow HO_2 + CO$$

The conversion rate is independent of pH is very fast: almost 100% per hour in summer. However, the conversion is limited by the supply of hydrogen peroxide, which is often present in much lower levels than $SO_2$. Thus, a reduction in sulfur dioxide emissions does not always correlate with a reduction of wet acid deposition.

## Sulfur halides

### Sulfur hexafluoride

Sulfur hexafluoride ($SF_6$) is a gas at standard temperature and pressure (25 °C, 1 atm). The most common synthesis involves the direct reaction of sulfur with fluorine yields $SF_6$.

$$S + 3\ F_2 \rightarrow SF_6$$

It should be noted that while $SF_6$ is highly stable, $SCl_6$ is not formed. The explanation of this difference may be explained by a consideration of the Born-Haber cycle shown in Figure 14.39. A similar cycle may be calculated for $SCl_6$; however, a higher dissociation energy for $Cl_2$ and a lower S-Cl bond energy (Table 14.8) provide the rational for why $SCl_6$ is not formed.

| Bond dissociation energy | kJ/mol | Bond energy | kJ/mol |
|---|---|---|---|
| D(F-F) | 158 | E(S-F) | 362 |
| D(Cl-Cl) | 262 | E(S-Cl) | 235 |

**Table 14.8: Comparison of diatomic bond dissociation and S-X bond energy for the fluorine and chlorine analogs.**

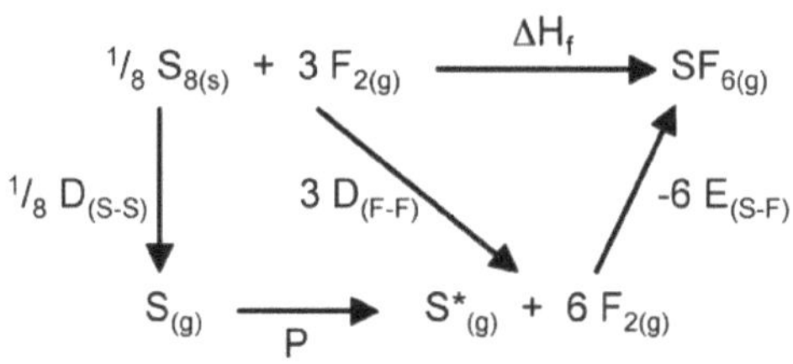


**Figure 14.39: Born-Haber cycle for the formation ($\Delta H_f$) of $SF_6$: where $D_{(X-Y)}$ = dissociation energy for X-Y bond, $E_{(S-F)}$ = S-F bond energy, and S* indicates 6 coordinate sulfur.**

The S-F bond length (1.56 Å) is very short and consistent with π-bonding in addition to σ-bonding. Like $SiF_6^{2-}$, $SF_6$ is an example of a hypervalent molecule (Figure 14.40).

Sulfur hexafluoride is an unreactive, non-toxic compound. Its inert nature provides one of its applications, as a spark suppressor. The hexafluoride is generally resistant to chemical attack, e.g., no reaction is observed with potassium hydroxide (KOH) at 500 °C. The low reactivity is due to $SF_6$ being kinetically inert due to:

- Coordination saturation precluding associative reactions with nucleophiles.
- Strong S-F bond (360 kJ/mol) limiting dissociative reactions.

Thermodynamically $SF_6$ should react with water (ΔH = -460 kJ/mol), but the rate factors are too great. Sulfur hexafluoride can be reduced with sodium in liquid ammonia,

$$SF_6 + 8\,Na \rightarrow Na_2S + 6\,NaF$$

or with $LiAlH_4$. In each of these reactions the mechanism involves the formation of a radical,

$$SF_6 + e^- \rightarrow SF_6^{\cdot-}$$

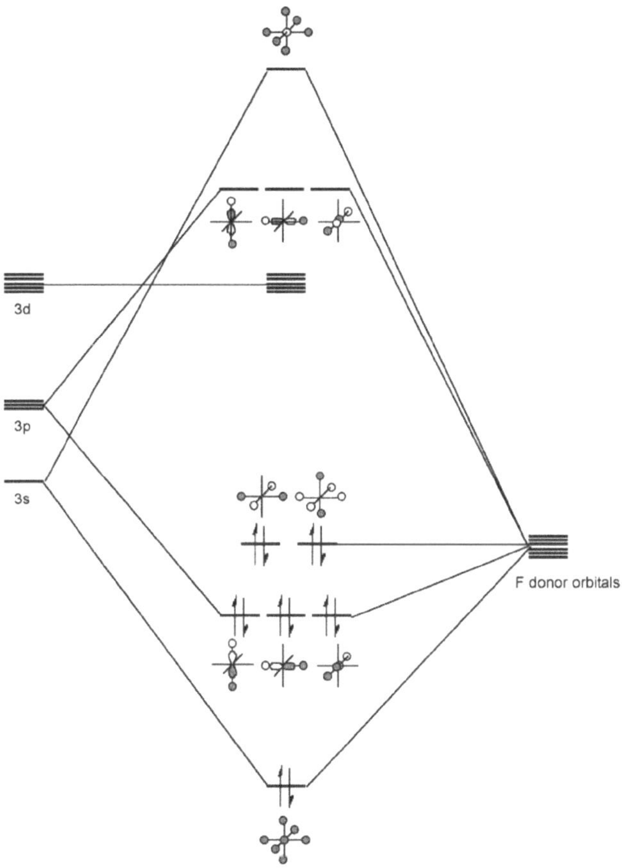


**Figure 14.40: Molecular orbital bonding in SF6.**

The reaction with sulfur trioxide yields $SO_2F_2$, however, the reactions with carbon or $CS_2$ only occur at elevated temperatures (500 °C) and pressure (4000 atm).

$$SF_6 + 2\,SO_3 \xrightarrow{250\ °C} 3\,SO_2F_2$$

Although the hexachloride is unknown, it is possible to isolate the monochloride derivative ($SF_5Cl$) by the oxidative addition of Cl-F across $SF_4$.

$$SF_4 + ClF \rightarrow SF_5Cl$$

Sulfur monochloride pentafluoride is a gas (Boiling point = -21 °C), but unlike $SF_6$ it is fairly reactive due to the polarization of the S-Cl bond (Figure 14.41), and as a consequence it reacts with water,

$$SF_5Cl + 3\,H_2O \rightarrow SO_3 + 5\,HF + HCl$$

$$\overset{\delta+}{Cl}—\overset{\delta-}{SF_5}$$

**Figure 14.41: Polarization of the S-Cl bond in $SF_5Cl$.**

## Sulfur pentafluoride

Although $SF_5$ does not exist as a stable molecule, the gaseous dimmer $S_2F_{10}$ (Boiling point = 29 °C) may be isolated from the photochemical hydrogen reduction of $SF_5Cl$,

$$2\,SF_5Cl + H_2 \xrightarrow{h\nu} S_2F_{10} + 2\,HCl$$

While the sulfur is octahedral in $S_2F_{10}$ (Figure 14.42a) the S-S bond is weak and long (2.21 Å versus an expected 2.08 Å for a single S-S bond). Despite the apparently weak S-S bond, $S_2F_{10}$ shows almost no reactivity at room temperature; however, the S-S bond undergoes homoleptic cleavage at high temperatures. The resultant $SF_5^{\cdot}$ radicals disproportionate to give highly reactive fluoride radicals,

$$2\,SF_5^{\cdot} \rightarrow 2\,SF_4 + 2\,F^{\cdot}$$

which is the source of the highly oxidative properties of $S_2F_{10}$.

The $SF_5^{\cdot}$ fragment is stabilized by the addition of an alkyl radical, and thus, there are a large number of $RSF_5$ derivatives known. Unlike, the chloride analog, these are very stable.

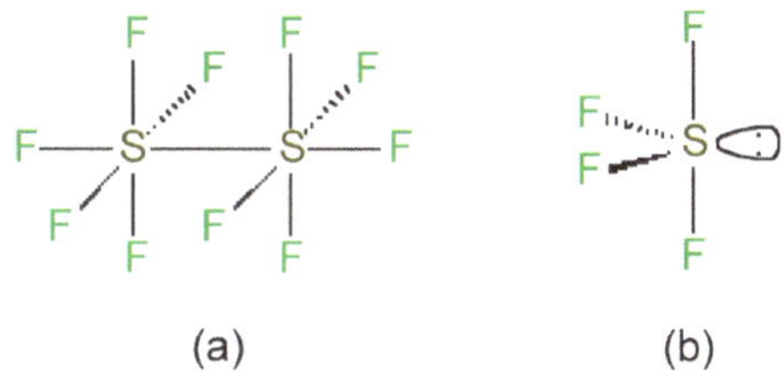


**Figure 14.42: Structures of (a) $S_2F_{10}$ and (b) $SF_4$.**

## Sulfur tetrafluoride

Sulfur tetrafluoride ($SF_4$) is prepared from sulfur dichloride and sodium fluoride in acetonitrile solution at 70 - 80 °C.

$$3\ SCl_2 + 4\ NaF \rightarrow SF_4 + S_2Cl_2 + 4\ NaCl$$

The structure of $SF_4$ (and its substituted derivatives $RSF_3$) is based upon a trigonal bipyramidal structure with one of the equatorial sites being occupied by a lone pair (Figure 14.42b). Unlike the hexafluoride, sulfur tetrachloride is a highly reactive compound. It hydrolyzes readily and is a useful fluorinating agent (Figure 14.43).

$$SF_4 + 2\ H_2O \rightarrow SO_2 + 4\ HF$$

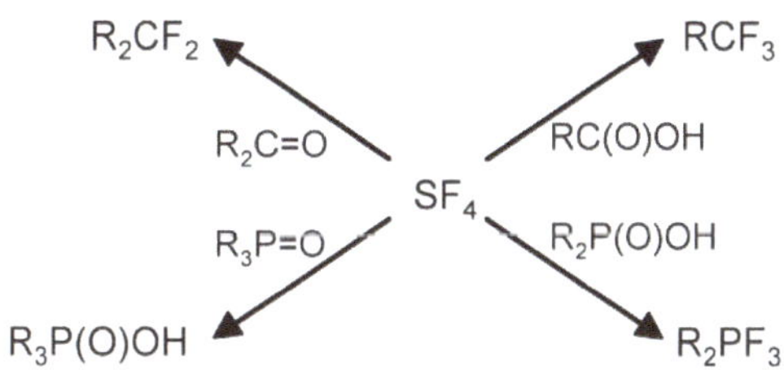


**Figure 14.43: Examples of the use of $SF_4$ as a fluorinating agent.**

# Sulfur chlorides

The chlorination of molten sulfur yields the fowl smelling disulfur dichloride ($S_2Cl_2$). If the reaction is carried out with a catalyst such as $FeCl_3$, $SnI_4$ or $I_2$, an equilibrium mixture containing sulfur dichloride ($SCl_2$) is formed. However, the dichloride dissociates readily,

$$2\ SCl_2 \rightleftharpoons S_2Cl_2 + Cl_2$$

although it can be isolated as a dark red liquid if it distilled in the presence of $PCl_5$. The reaction of chlorine at -80 °C with $SCl_2$ or $S_2Cl_2$ allows for the formation of $SCl_4$ as a yellow crystalline compound which dissociates above -31 °C. Sulfur chlorides are readily hydrolyzed. Sulfur chlorides are used to dissolve sulfur (giving species up to $S_{100}Cl_2$) for the vulcanization of rubber.

In the vapor phase $S_2Cl_2$ has $C_2$ symmetry (Figure 14.44a) while that of $SCl_2$ has $C_{2v}$ symmetry (Figure 14.44b).

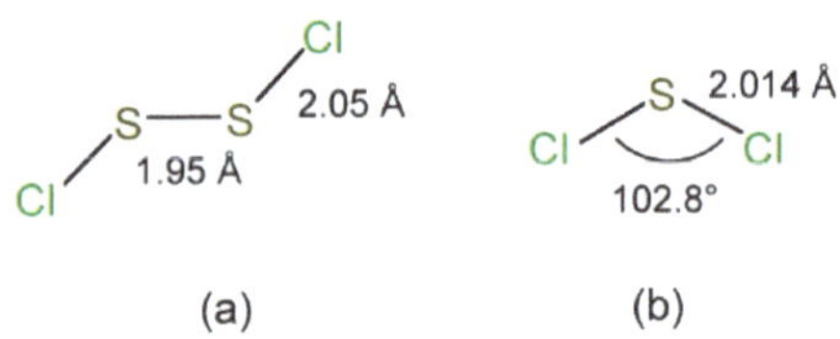


**Figure 14.44: Structures of (a) $S_2Cl_2$ and (b) $SCl_2$.**

# Chapter 15: Group 17 The Halogens

## The elements

The Group 17 elements have a particular name: the *halogens* meaning 'born of salt'. This is due to the formation of salts when they form compounds with a metal. Table 10.1 lists the derivation of the names of the halogens.

| Element | Symbol | Name |
|---|---|---|
| Fluorine | F | Latin *uere* meaning to ow |
| Chlorine | Cl | Greek *khlôros* meaning pale green |
| Bromine | Br | Greek *brómos* meaning stench |
| Iodine | I | Greek *odes* meaning violet or purple |
| Astatine | At | Greek *astatos*, meaning unstable |

**Table 15.1: Derivation of the names of each of the halogens.**

## Discovery

*Fluorine*

The mineral fluorspar (also known as fluorite) consists mainly of calcium fluoride and was described in 1530 by Agricola (Figure 11.2) for its use as a ux. Fluxes are used to promote the fusion of metals or minerals, and it was from this use that fluorine derived its name. In 1670 Heinrich Schwanhard found that when he mixed fluorspar with an acid the fumes produced (hydrogen fluoride) etched the glasses he was wearing. Despite many researchers investigating the chemistry of hydrogen fluoride (HF) the elemental form of fluorine was not isolated until 1886 when Henri Moissan (Figure 9.16) studied the electrolysis of a solution of potassium hydrogen di fluoride ($KHF_2$) in liquid hydrogen fluoride (HF). The mixture was needed because hydrogen fluoride is a non-conductor. The device was built with platinum/iridium electrodes in a platinum holder and the apparatus was cooled to -50 °C.

The generation of elemental fluorine from hydrofluoric acid proved to be exceptionally dangerous, killing or blinding several scientists who attempted early experiments on this halogen. The victims became known as 'fluorine martyrs'.

*Chlorine*

Archaeological evidence has shown that sodium chloride (known as table salt) has been used as early as 3000 BC and brine (the saturated water solution) as early as 6000 BC. It is thought that hydrochloric acid was probably known to alchemist Abū Mūsā Jābir ibn Hayyān (Figure 11.1) around 800 AD, while aqua regia (a mixture of nitric acid and hydrochloric acid) began to be used to dissolve gold sometime before 1400 AD. Upon dissolving gold in aqua regia, chlorine gas is released along with other nauseating and irritating gases.

Chlorine was first prepared and studied in 1774 by Carl Wilhelm Scheele (Figure 9.5), and therefore he is credited for its discovery despite his failing to establish chlorine as an element, mistakenly thinking that it was the oxide obtained from the hydrochloric acid. Regardless of what he believed, Scheele did isolate chlorine by reacting $MnO_2$ (as the mineral pyrolusite) with HCl,

$$4\,HCl + MnO_2 \rightarrow MnCl_2 + 2\,H_2O + Cl_2$$

*Bromine*

Bromine was discovered independently by two chemists Antoine Balard (Figure 15.1) in 1825 and Carl Jacob Löwig (Figure 15.2) in 1826.

**Figure 15.1: French chemist Antoine Jérôme Balard (1802 - 1876).**

Balard found bromide salts in the ash of seaweed from the salt marshes of Montpellier. The seaweed was used to produce iodine, but also contained bromine. Balard distilled the bromine from a solution of seaweed ash saturated with chlorine. The properties of the resulting substance resembled that of an intermediate of chlorine and iodine; with those results he tried to prove that the substance was iodine monochloride (ICl), but after failing to do so he was sure that he had found a new element and named it *muride*, derived from the Latin word *muria* for brine.

**Figure 15.2: German chemist Carl Jacob Löwig (1803 - 1890).**

In contrast, Löwig isolated bromine from a mineral water spring in Bad Kreuznach. Löwig used a solution of the mineral salt saturated with chlorine and extracted the bromine with $Et_2O$. After evaporation a brown liquid remained. Unfortunately, the publication of his results was delayed and Balard published first.

*Iodine*

Iodine was discovered by Bernard Courtois (Figure 15.3) in 1811 when he was destroying the waste material from the production of saltpeter ($KNO_3$) during gunpowder production. Saltpeter produced from French niter beds required sodium carbonate ($Na_2CO_3$), which could be isolated from seaweed washed up on the coasts of Normandy and Brittany. To isolate the sodium carbonate, seaweed was burned, and the ash then washed with water; the remaining waste was destroyed by the addition of sulfuric acid ($H_2SO_4$). After adding too much acid, Courtois observed a cloud of purple vapor that crystallized on cold surfaces making dark crystals. Courtois suspected that this was a new element but lacked the money to pursue his observations. In supplying samples to his friends, Charles Desormes and Nicolas Clément, he hoped his research was to continue. On 29 November 1813, Dersormes and Clément made public Courtois's discovery, describing the substance to a meeting of the Imperial Institute of France.

**Figure 15.3: French chemist Bernard Courtois (1777 - 1838).**

*Astatine*

The existence of *eka-iodine* had been predicted by Mendeleev (Figure 11.7), but astatine was first synthesized in 1940 by Corson (Figure 15.4), MacKenzie (Figure 15.5), and Segrè (Figure 15.6) at the University of California, Berkeley by bombarding bismuth with alpha particles.

**Figure 15.4: Physicist and President of Cornell University, Dale R. Corson (1914 - 2012).**

**Figure 15.5: Physicist Kenneth Ross MacKenzie (1912 - 2002).**

**Figure 15.6: Italian physicist Emilio Segrè (1905 - 1989).**

## Abundance

The abundance of the halogens is given in Table 15.2.

| Element | Terrestrial abundance (ppm) |
|---|---|
| F | 950 (Earth's crust), 330 (soil), 1.3 (sea water), $6\times10^{-4}$ (atmosphere) |
| Cl | 130 (Earth's crust), 50 2000 (soil), $1.8\times10^{4}$ (sea water) |
| Br | 0.4 (Earth's crust), 5 40 (soil), 65 (sea water) |
| I | 0.14 (Earth's crust), 3 (soil), 0.06 (sea water), $60\times10^{-3}$ (atmosphere) |
| At | Trace in some minerals |

**Table 15.2: Abundance of the halogens.**

## Isotopes

The naturally abundant isotopes of the halogens are listed in Table 10.3. All 33 isotopes of astatine are radioactive.

| Isotope | Natural abundance (%) |
|---|---|
| Fluorine-19 | 100 |
| Chlorine-35 | 75.77 |
| Chlorine-36 | trace |
| Chlorine-37 | 24.23 |
| Bromine-79 | 50.69 |
| Bromine-81 | 49.31 |
| Iodine-127 | 100% |

**Table 15.3: Abundance of the major isotopes of the halogens.**

While $^{19}F$ is the only naturally abundant isotope of fluorine, the synthetic isotope, $^{18}F$, has half-life of about 110 minutes, and is commercially an important source of positrons for positron emission tomography (PET). PET is a nuclear medicine imaging technique that produces a 3-D image of processes within the body. The system detects pairs of γ-rays emitted indirectly by a positron-emitting radionuclide (tracer), which is introduced into the body on a biologically active molecule.

Trace amounts of radioactive $^{36}Cl$ exist in the environment at about $7\times10^{-11}\%$. $^{36}Cl$ is produced in the atmosphere by the interaction of cosmic rays with $^{36}Ar$. In the ground $^{36}Cl$ is generated through neutron capture by $^{35}Cl$ or *muon* (an elemental particle similar to an electron) capture by $^{40}Ca$. $^{36}Cl$ decays with a half-life of 308,000 years making it suitable for geologic dating in the range of 60,000 to 1 million years. However, due to the large amounts of $^{36}Cl$ produced by irradiation of seawater during atmospheric detonations of nuclear weapons between 1952 and 1958, it is also sued as an event marker for 1950s water in soil and ground water.

Iodine has 37 isotopes of iodine, but only one, $^{127}I$, is stable. Of the radioactive isotopes, $^{129}I$ (half-life 15.7 million years) is used for radiometric dating of the first 85 million years of solar system evolution. $^{129}I$ is also a product of uranium and plutonium fission, and as a consequence of nuclear fuel reprocessing and atmospheric nuclear weapons tests, the natural signal has been swamped. As a consequence, it can now be used as a tracer of nuclear waste dispersion into the environment. $^{129}I$ was used in rainwater studies to track fission products following the Chernobyl disaster.

Due to preferential uptake of iodine by the thyroid, isotopes with short half-lives such as $^{131}I$ can be used for thyroid ablation, a procedure in which radioactive iodine is administered intravenously or orally following a diagnostic scan. The lower energy isotopes $^{123}I$ and $^{125}I$ are used as tracers to evaluate the anatomic and physiologic function of the thyroid.

## Industrial production.

Industrial production of fluorine involves the electrolysis of hydrogen fluoride (HF) in the presence of potassium fluoride (KF) during which fluorine gas is formed at the anode and hydrogen gas is formed at the cathode (Figure 15.7). The potassium fluoride (KF) is converted to potassium bifluoride ($KHF_2$),

$$HF + KF \rightarrow KHF_2$$

which is the electrolyte and intermediate to the fluorine and hydrogen,

$$2\ KHF_2 \rightarrow 2\ KF + F_2 + H_2$$

**Figure 15.7: The fluorine cell room at F2 Chemicals Ltd, Preston, UK.**

The HF is formed as a byproduct of the production of phosphoric acid, since phosphate-containing minerals contain significant amounts of calcium fluorides, which upon treatment with sulfuric acid release hydrogen fluoride,

$$CaF_2 + H_2SO_4 \rightarrow 2\ HF + CaSO_4$$

Chlorine is generally manufactured by electrolysis of a sodium chloride solution (brine). The production of chlorine results in the co-products caustic soda (sodium hydroxide, NaOH) and hydrogen gas ($H_2$). Chlorine can also be produced by the electrolysis of a solution of potassium chloride, in which case the co-products are hydrogen and caustic potash (potassium hydroxide). There are three industrial methods for the extraction of chlorine by electrolysis of chloride solutions, all proceeding by the same reaction at the cathode and anode,

$$2\,H^+_{(aq)} + 2\,e^- \rightarrow H_{2(g)}$$

$$2\,Cl^-_{(aq)} \rightarrow Cl_{2(g)} + 2\,e^-$$

which lead to the overall reaction where M = Na or K.

$$2\,MCl + 2H_2O \rightarrow Cl_2 + H_2 + 2\,MOH$$

Bromine exists exclusively as bromide salts in the Earth's crust, however, due to leaching, bromide salts have accumulated in sea water, but at a lower concentration than chloride. The majority of bromine is isolated from bromine-rich brines, which are treated with chlorine gas, flushing through with air. In this treatment, bromide anions are oxidized to bromine by the chlorine gas,

$$2\,Br^- + Cl_2 \rightarrow 2\,Cl^- + Br_2$$

Two major sources of iodine are used for commercial production: the caliche (a hardened sedimentary deposit of calcium carbonate found in Chile) and the iodine containing brines of gas and oil fields in Japan and the United States. The caliche contains sodium nitrate ($NaNO_3$); in which traces of sodium iodate ($NaIO_3$) and sodium iodide (NaI) are found. During the production of sodium nitrate the sodium iodate and iodide is extracted. Iodine sourced from brine involves the acidification with sulfuric acid to form hydrogen iodide (HI), which is then oxidized to iodine with chlorine,

$$2\,HI_{(aq)} + Cl_2 \rightarrow I_{2(aq)} + 2\,HCl_{(aq)}$$

The aqueous iodine solution is concentrated by passing air through the solution causing the iodine to evaporate. The iodine solution is then re-reduced with sulfur dioxide,

$$I_2 + 2\,H_2O + SO_2 \rightarrow 2\,HI + H_2SO_4$$

The dry hydrogen iodide (HI) is reacted with chlorine to precipitate the iodine,

$$2\,HI + Cl_2 \rightarrow I_2\downarrow + 2\,HCl$$

## Physical properties

The physical properties of the halogens (Table 15.4) encompasses gases ($F_2$ and $Cl_2$), a liquid ($Br_2$), a non-metallic solid ($I_2$), and a metallic metal (At).

| Element | Melting point (°C) | Boiling point (°C) | Density (g/cm$^3$) |
|---|---|---|---|
| F | -219.62 | -188.12 | $1.7\times10^{-3}$ @ 0 °C, 101 kPa |
| Cl | -101.5 | -34.04 | $3.2\times10^{-3}$ @ 0 °C, 101 kPa |
| Br | -7.2 | 58.8 | 3.1028 (liquid) |
| I | 113.7 | 184.3 | 4.933 |
| At | 302 | 337 | ca. 7 |

**Table 15.4: Selected physical properties of the halogens.**

## Reactivity

All the halogens are highly reactive and are as a consequence of the stability of the $X^-$ ion are strong oxidizing agents (Table 15.5).

| Reduction | Reduction potential (V) |
|---|---|
| $F_2 + 2\,e^- \rightarrow 2\,F^-$ | 2.87 |
| $Cl_2 + 2\,e^- \rightarrow 2\,Cl^-$ | 1.36 |
| $Br_2 + 2\,e^- \rightarrow 2\,Br^-$ | 1.07 |
| $I_2 + 2\,e^- \rightarrow 2\,I^-$ | 0.53 |

**Table 15.5: Electrochemical reduction potential for halogens.**

Elemental fluorine (fluorine gas) is a highly toxic, corrosive oxidant, which can cause ignition of organic material. Fluorine gas has a characteristic pungent odor that is detectable in concentrations as low as 20 ppb. As it is so reactive, all materials of construction must be carefully selected, and metal surfaces must be passivated. In high concentrations, soluble fluoride salts are also toxic and skin or eye contact with high concentrations of many fluoride salts is dangerous.

## Use of chlorine as a weapon

Chlorine gas, also known as *bertholite*, was first used as a weapon in World War I by Germany on April 22, 1915 in the Second Battle of Ypres. At around 5:00 pm on April 22, 1915, the German Army released one hundred and sixty-eight tons of chlorine gas over a 4 mile front against French and colonial Moroccan and Algerian troops of the French 45$^{th}$ and 78$^{th}$ divisions (Figure 10.13). The attack involved a massive logistical effort, as German troops hauled 5730 cylinders of chlorine gas, weighing ninety pounds each, to the front by hand. The German soldiers also opened the cylinders by hand, relying on the prevailing winds to carry the gas towards enemy lines. Because of this method of dispersal, a large number of German soldiers were injured or killed in the process of carrying out the attack. Approximately 6,000 French and colonial troops died within ten minutes at Ypres, primarily from asphyxiation and subsequent tissue damage in the lungs. Many more were blinded. Chlorine gas forms hydrochloric acid when combined with water, destroying moist tissues such as lungs and eyes. The chlorine gas, being denser than air, quickly filled the trenches, forcing the troops to climb out into heavy enemy fire.

**Figure 15.8: The gas attack of the Second Battle of Ypres.**

As described by the soldiers it had a distinctive smell of a mixture between pepper and pineapple. It also tasted metallic and stung the back of the throat and chest. The damage done by chlorine gas can be prevented by a gas mask, or other filtration method, making the fatalities from a chlorine gas attack much lower than those of other chemical weapons. The use as a weapon was pioneered by Fritz Haber (Figure 13,22) of the Kaiser Wilhelm Institute in Berlin, in collaboration with the German chemical conglomerate IG Farben, who developed methods for discharging chlorine gas against an entrenched enemy. It is alleged that Haber's role in the use of chlorine as a deadly weapon drove his wife, Clara Immerwahr, to suicide. After its first use, chlorine was used by both sides as a chemical weapon (Figure 15.9), but it was soon replaced by the more deadly gases phosgene and mustard gas.

**Figure 15.9: A poison gas attack, in World War I.**

## Structure

*Vapor phase*

All the halogens form $X_2$ dimers in the vapor phase in an analogous manner to hydrogen. Unlike dihydrogen, however, the bonding is associated with the molecular orbital combination of the two p-orbitals (Figure 15.10). The bond lengths and energies are given in Table 15.6.

| Element | Bond length (Å) | Energy (kJ/mol) |
|---|---|---|
| $F_2$ | 1.42 | 158 |
| $Cl_2$ | 1.99 | 243 |
| $Br_2$ | 2.29 | 193 |
| $I_2$ | 2.66 | 151 |

**Table 15.6: Bond lengths and energies for halogens.**

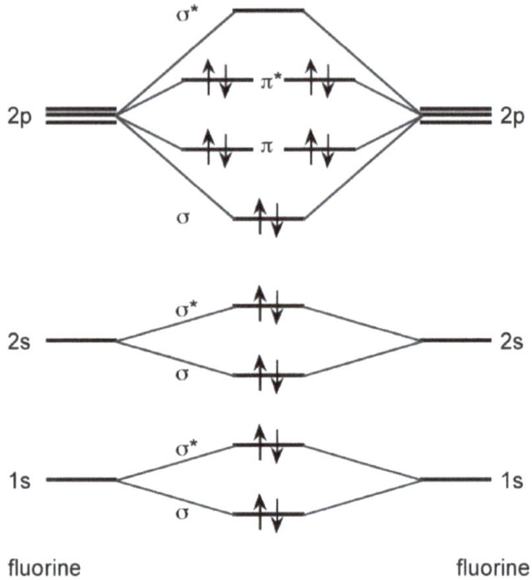


**Figure 15.10: Molecular orbital diagram for the formation of $F_2$.**

*Solid state*

Iodine crystallizes in the orthorhombic space group *Cmca* (Figure 15.11). In the solid state, $I_2$ molecules still contain short I-I bond (2.70 Å).

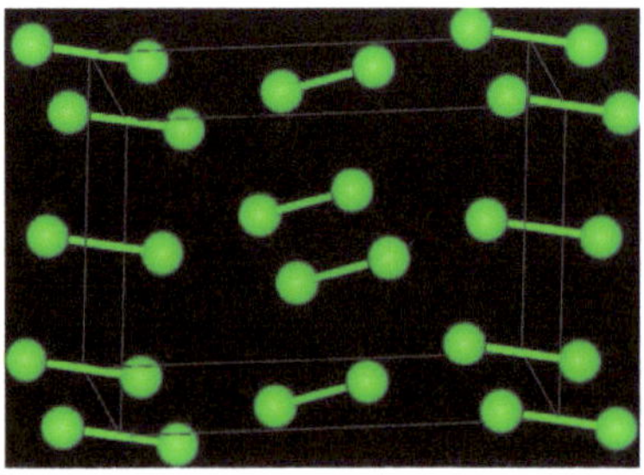

**Figure 15.11: The solid-state structure of $I_2$.**

## Compounds of the halogens.

The chemistry of the halogens is dominated by the stability of the -1 oxidation state and the noble gas configuration of the $X^-$ anion.

# Oxidation state

The use of oxidation state for fluorine is almost meaningless since as the most electronegative element, fluorine exists in the oxidation state of -1 in all its compounds, except elemental fluorine ($F_2$) where the oxidation state is zero by definition. Despite the general acceptance that the halogen elements form the associated halide anion ($X^-$), compounds with oxidation states of +1, +3, +4, +5, and +7 are common for chlorine, bromine, and iodine (Table 10.7).

| Element | -1 | +1 | +3 | +4 | +5 | +7 |
|---|---|---|---|---|---|---|
| Cl | HCl | ClF | $ClF_3$, $HClO_2$ | $ClO_2$ | $ClF_5$, $ClO_3^-$ | $HClO_4$ |
| Br | HBr | BrCl | $BrF_3$ | $Br_2O_4$ | $BrF_5$, $BrO_3^-$ | $BrO_4^-$ |
| I | HI | ICl | $IF_3$, $ICl_3$ | $I_2O_4$ | $IO_3^-$ | $IO_4^-$ |

**Table 15.7: Examples of multiple oxidation states for the halogens.**

# Compounds of Fluorine

Elemental fluorine ($F_2$) is the most reactive element. Fluorine combines directly with all other elements, except nitrogen and the lighter noble gases. It also reacts with many compounds forming fluorides, and many organic compounds in flame and burn in the gas. The highly reactive nature is due to the weak F-F bond (thermodynamically unstable), which provides a low activation energy to reactions (kinetically unstable). The $\Delta G$ for reactions is often large due to the strength of the resulting X-F bonds. The weak F-F bond (158 kJ/mol) is due to the small size (0.5 Å) and high nuclear charge of fluorine that result in a small overlap of the bonding orbitals and a repulsion between the non-bonding orbitals (lone pairs) on the two fluorine atoms.

*Ionic salts*

The ease of formation of $F^-$ anion is due to the high electron affinity of fluorine (-322 kJ/mol). Since the fluoride ion is small (1.33 Å) and the least polarizable anion (i.e., hard) it is stable in ionic lattices with metal cations in a high oxidation state (high charge), e.g., $MnF_4$ and $CrF_5$. In general, the highest oxidation states for any metal are found with the fluoride salts. The large ionization energies needed to produce the cations are recovered by the high lattice energies.

*Covalent compounds*

The high electronegativity of fluorine means that it forms a single electron pair bond polar bond with a high ionic character. The polar nature of the bond means that there is a large inductive effect within a molecule. For example, perfluoroethanol ($CF_3CF_2OH$) has an acidity comparable to acetic acid.

The high strength of X-F bonds (Table 15.8) is also due to the high ionic character (up to 50%) that results in a high activation energy for bond breaking. In contrast, the low polarizability of the fluorine means that the intermolecular van der Waals bonds are very weak. Thus, even with very high molecular weights the boiling point can be very low, e.g., $WF_6$, Boiling point = 17 °C, $M_w$ = 297.84 g/mol.

| Bond | Bond energy (kJ/mol) |
|---|---|
| C-F | 486 |
| N-F | 272 |
| P-F | 490 |

**Table 15.8: Typical bond energies for X-F bonds.**

A wide range of fluoride complexes may be prepared from both metal ($FeF_6^{3-}$, $RuF_6^-$, $PtF_6^{2-}$, and $SnF_6^{2-}$) and non-metal ($BF_4^-$, $SiF_6^{2-}$, and $PF^-$) fluorides. While many fluorides are salts, when the metal is in its higher oxidation states (e.g., $OsF_6$ and $WF_6$), the formation of an ionic lattice with the appropriate cation (i.e., $Os^{6+}$ and $W^{6+}$ respectively) is energetically unfavorable.

*Hydrogen fluoride*

Hydrogen fluoride (HF) is converted to highly corrosive hydrofluoric acid upon contact with moisture. Pure hydrogen fluoride must be handled in metal or polythene vessels, while aqueous solutions will readily etch and dissolve standard laboratory glassware requiring the use of fluorinated polymer (e.g., Teflon®) containers.

Hydrogen fluoride is synthesized by the reaction of a fluoride salt with a concentrated acid,

$$CaF_2 + H_2SO_4 \rightarrow CaSO_4 + 2\,HF\uparrow$$

The HF vapor may be condensed, and then subsequently purified by distillation.

The H-F bonding in hydrogen fluoride involves an electron pair bond with a high degree of ionic character. This results in a very polar H-F bond and a large dipole moment (1.86 D).

In the vapor phase, hydrogen fluoride is monomeric above 80 °C, but at lower temperatures it associates into oligomers and small polymers, e.g., cyclic $(HF)_6$, as a consequence of strong intermolecular hydrogen bonds. As a pure liquid (Melting point = -83 °C, Boiling point = 19.5 °C) hydrogen fluoride is extensively associated by strong hydrogen bonding to form zig-zag polymers (Figure 15.12).

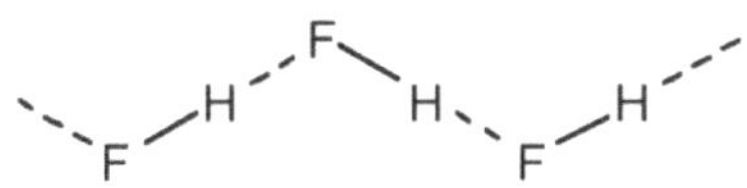


**Figure 15.12: The hydrogen bonding in HF.**

Hydrogen fluoride has a high dielectric constant (84.2) and as such is a good solvent for polar molecules. However, it is not a good solvent for salts (even fluorides) because it does not solvate cations too well. Despite this, it is useful as a solvent because it is non-oxidizing and easy to evaporate off products.

In a similar manner to water, hydrogen fluoride self-ionizes,

$$2\,HF \rightleftharpoons H_2F^+ + F^-$$

Salts of $H_2F^+$ are known and the $F^-$ anion is further solvated by the HF to form a series of salts,

$$F^- + n\,HF \rightleftharpoons HF_2^- + H_2F_3^- + H_3F_4^- \text{ etc.}$$

The complex anion $HF_2^-$ is also formed in aqueous solutions of hydrogen fluoride (pK = 0.7).

Hydrogen fluoride is actually a weak acid in aqueous solution with a low pK = 3.5. In fact, HF is a weaker acid that the other halogen analogs:

$$HF < HCl < HBr < HI$$

This trend is despite the fluorine being more electronegative than the other halogens but is consistent with the strength of the H-F bond (568 kJ/mol).

Hydrogen fluoride is used as a non-oxidizing acid for the hydrolysis of proteins and acid catalyzed condensation reaction. The stability of its salt ($HF_2^-$) allow for the study of very strong acids,

$$2\,HF + R_2C{=}O \rightarrow R_2C{=}OH^+ + HF_2^-$$

$$2\,HF + H_2O \rightarrow H_3O^+ + HF_2^-$$

The acidity of HF may be increased sufficiently by the addition of a fluoride acceptor (e.g., $SbF_5$) to facilitate the reaction with a weak base such as benzene,

$$C_6H_6 + HF + SbF_5 \rightarrow C_6H_7^+ + SbF_6^-$$

Finally, hydrogen fluoride can be used in the synthesis of other fluorine-containing compounds:

$$6\,HF + KCl + PCl_5 \rightarrow K[PF_6] + 6\,HCl \uparrow$$

$HF + SO_3 \rightarrow$ 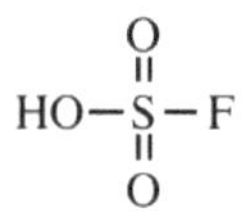

*Organic fluorine compounds*

Organic compounds in which some or all of the hydrogen atoms are replaced by fluorine have unique (and often important) properties. The high stability of fluorocarbon compounds is a consequence of the C-F bond energy (486 kJ/mol) in comparison with that of C-H (415 kJ/mol); however, while kinetically stable, fluorocarbons are not necessarily particularly thermodynamically stable.

Replacement of hydrogen with fluorine results in an increased density; since the small size of fluorine means that the minimal distortion or structural change occurs as a result of the substitution. As with metal salts, the weak inter-molecular forces mean that completely fluorinated organic compounds have low boiling points. One attribute of the low inter-molecular forces is the low coefficient of friction for fluoropolymers such as polytetrafluoroethylene, commonly known as Teflon® (Figure 15.13).

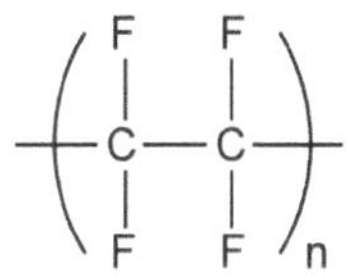

**Figure 15.13: The structure of polytetrafluoroethylene (Teflon®).**

*Synthetic routes to fluorocarbon compounds*

The simplest route to a fluorocarbon compounds involves the direct replacement of another halogen by a metal fluoride,

$$R\text{-}Cl + MF \rightleftharpoons R\text{-}F + MCl$$

The driving force for this reaction depends on the free-energy difference of MF and MCl, which is related to the difference in lattice energies. Thus, the larger the metal cation, the more favored the reaction. In this regard, AgF and CsF are the most effective fluorination agent.

Anhydrous hydrogen fluoride (HF) reacts with chlorocarbon compounds in the presence of a catalyst such as $SbCl_5$ or $CrF_4$,

$$2\ CCl_4 + 3\ HF \rightarrow CCl_2F_2 + CCl_3F + 3\ HCl$$

However, elevated temperatures (50 - 150 °C) and high pressures (50 - 500 psi) are required. The direct replacement of hydrogen with fluorine is possible if the reaction is carried out under dilute conditions in the presence of a catalyst,

$$C_6H_6 + 9\ F_2 \xrightarrow{Cu} C_6F_{12} + 6\ HF$$

Sulfur tetra fluoride ($SF_4$) is a particularly selective fluorination agent. It can be used to convert ketones to difluoro compounds,

$$2\ R_2C{=}O + SF_4 \rightarrow R_2CF_2 + SO_2$$

# Compounds of Chlorine

*Comparison to fluorine*

To appreciate the chemistry of chlorine in comparison to that of fluorine it is necessary to appreciate the differences and trends between the elements. As may be seen in Table 15.9, chloride is significantly larger than fluorine. In addition, while chlorine is an electronegative element its electronegativity is significantly less than that of fluorine, resulting in less polar bonding.

| Element | Ionic radius (Å) | Covalent radius (Å) | van der Waal radius (Å) | Electronegativity |
|---|---|---|---|---|
| Fluorine | 1.33 | 0.64 | 1.47 | -4.1 |
| Chlorine | 1.81 | 0.99 | 1.75 | -2.9 |

**Table 15.9: Comparison of physical characteristics for fluorine and chlorine.**

The X-Cl bond is an electron pair covalent bond with a highly polar nature. In this regard, chlorine is similar to fluorine. However, there are two key features with regard to chlorine's bonding that differentiates it from fluorine:

- Unlike fluorine, chlorine can form multiple covalent bonds, e.g., $ClO_4^-$ and $ClF_3$.
- Unlike fluorine, chlorine can form $\pi$-bonds with oxygen, i.e., Cl=O.

The chloride ion ($Cl^-$) forms salts with ionic lattices (e.g., NaCl) but also forms a wide range of complexes, e.g., $[Fe(H_2O)_5Cl]^+$ and $[RhCl_6]^{3-}$. Chloride also acts as a bridging ligand in which one, two or three chlorides can bridge two metal centers (Figure 15.14).

Chloride (and bromide) bridges are usually bent, whereas fluoride bridges can be either linear or bent. As an example, $BeF_2$ and $BeCl_2$ are isostructural, consisting of infinite chains with bent bridges (Figure 15.15). In contrast, transition metal pentahalides show different structures depending on the identity of the halide. This, $TaCl_5$ dimerizes with bent bridges (Figure 15.16a), while $TaF_5$ forms a cyclic tetramer with linear fluoride bridges (Figure 15.16b).

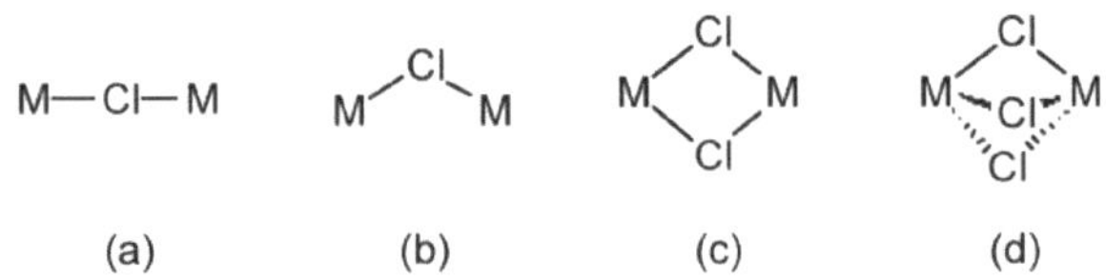


**Figure 15.14: Structures of common halide bridging moieties.**

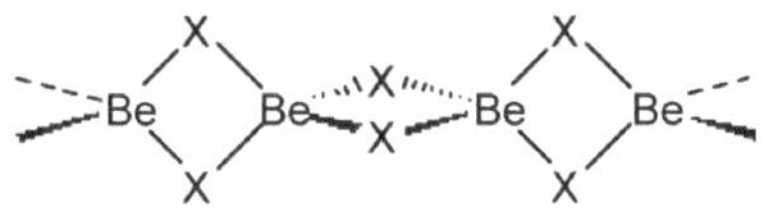

**Figure 15.15: The solid-state structure of $BeX_2$ (X = F or Cl).**

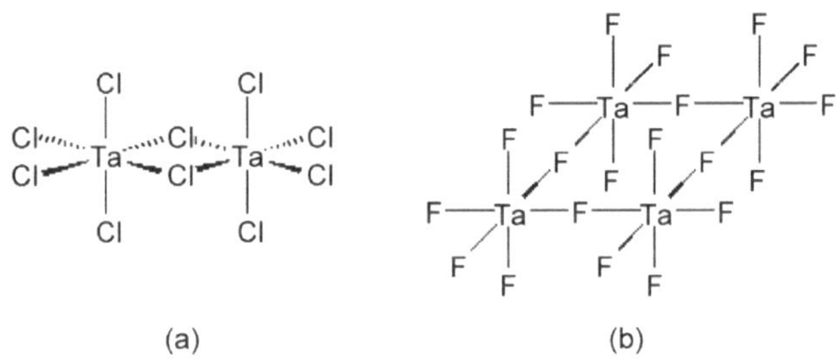


**Figure 15.16: The solid-state structure of (a) $TaCl_5$ and (b) $TaF_5$.**

*The halide bridge*

The bridging halide bonds can be described by both Lewis and molecular orbital (MO) theory. In a simple picture, the lone pair of a terminal halide can be thought to act as a Lewis base donor ligand to the second Lewis acidic metal center. Indeed, some bridging halides are asymmetric consistent with this view; however, the symmetrical ones can be described by a resonance form. From a molecular orbital point of view, the bridging halide is represented by a combination of two metal centered orbitals with two halogen orbitals.

*Hydrogen chloride*

Hydrogen chloride (HCl) is prepared by the reaction of concentrated sulfuric acid ($H_2SO_4$) with either NaCl or concentrated HCl solution.

Hydrogen chloride is a polar molecule with a dipole of 1.08 D. However, the lower polarity as compared to that of hydrogen fluoride (1.91 D) is consistent with the physical and chemical properties. Hydrogen chloride is a gas at room temperature (Melting point = -114.25 °C, Boiling point = -85.09 °C), and its low boiling point is consistent with weak hydrogen bonding in the liquid state. While self-ionization,

$$3\ HCl \rightleftharpoons H_2Cl^+ + HCl_2^-$$

is very small, liquid HCl dissolves some inorganic compounds to give conducting solutions,

$$R_3N + 2\ HCl \rightleftharpoons R_3NH^+ + HCl_2^-$$

Hydrogen chloride is soluble (and reacts) in water,

$$HCl + H_2O \rightleftharpoons H_3O^+ + Cl^-$$

The $pK_a$ of the reaction (-7.0) is larger than observed for fluorine (3.2) and as such HCl is a stronger acid than HF.

*Oxides of chlorine*

Chlorine forms a series of oxides (Table 15.10, Figure 15.17) in which the chlorine has the formal oxidation states +1, +4, +6, and +7. The physical properties of the oxides are summarized in Table 10.10. While, the oxides of chlorine are not very stable (in fact several are shock sensitive and are prone to explode) the conjugate oxyacids are stable.

| Compound | Melting point (°C) | Boiling point (°C) |
|---|---|---|
| $Cl_2O$ | -116 | 4 |
| $ClO_2$ | -5.9 | 10 |
| $Cl_2O_4$ | -117 | 44.5 |
| $Cl_2O_6$ | 3.5 | unstable |
| $Cl_2O_7$ | -91.5 | 82 |

**Table 15.10: Physical properties of the oxides of chlorine.**

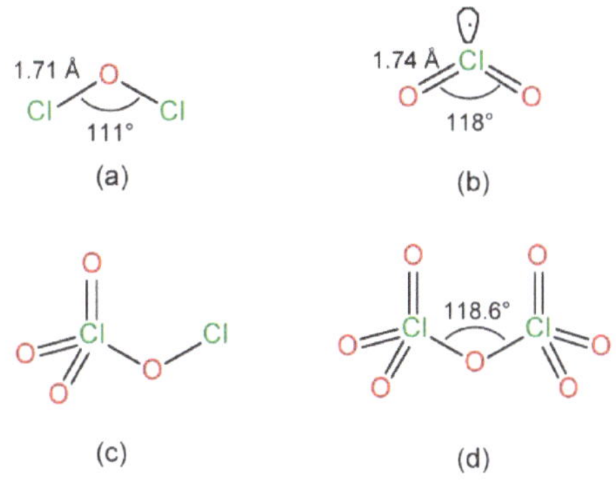


**Figure 15.17: The structure of (a) $Cl_2O$, (b) $ClO_2$, (c) $Cl_2O_4$, and (d) $Cl_2O_7$.**

Dichlorine monoxide ($Cl_2O$, Figure 15.17a) is a yellowish-red gas that is prepared by the reaction of chlorine with mercury oxide or with a solution of chlorine in $CCl_4$.

$$2\,Cl_2 + 2\,HgO \rightarrow HgCl_2.HgO + Cl_2O$$

When heated or subject to a spark, $Cl_2O$ explodes to $Cl_2$ and $O_2$. Dichlorine monoxide reacts with water to form an orange-yellow solution of hypochlorous acid,

$$H_2O_{(g)} + Cl_2O_{(g)} \rightleftharpoons 2\,HOCl_{(g)}$$

Chlorine dioxide ($ClO_2$) is a yellowish gas at room temperature and is commonly used in industry as an oxidizing agent. The best synthesis of $ClO_2$ involves the reduction of potassium chlorate ($KClO_3$) by oxalic acid at 90 °C, since the $ClO_2$ formed acts as a diluent for the highly explosive $ClO_2$. On an industrial scale $ClO_2$ is made by the exothermic reaction of sodium chlorate with $SO_2$ in sulfuric acid,

$$2\,NaClO_3 + SO_2 + H_2SO_4 \rightarrow 2\,ClO_2 + 2\,NaHSO_4$$

The photolysis of $ClO_2$ yields a dark brown solid with the formula $Cl_2O_3$; however, its facile explosive decomposition precludes study.

The structure of $ClO_2$ (Figure 15.17b) is equivalent to $SO_2$ with one extra electron, resulting in a paramagnetic unpaired electron species. Unusually, despite the unpaired electron configuration, $ClO_2$ shows no tendency to dimerize. This is unlike the analogous $NO_2$ molecule.

Dichlorine tetraoxide ($Cl_2O_4$) is commonly called chlorine perchlorate as a consequence of its structure (Figure 15.17c). Dichlorine hexaoxide ($Cl_2O_6$) is an unstable red oil that has the ionic structure, $[ClO_2]^+[ClO_4]^-$, in the solid state.

Dichlorine heptoxide ($Cl_2O_7$) is a relatively stable oil, that is prepared by the dehydration of perchloric acid at -10 °C,

$$HClO_4 \xrightarrow[- H_2O]{+ P_4O_5} Cl_2O_7$$

followed by vacuum distillation. The structure of $Cl_2O_7$ (Figure 15.17d) has been determined by gas phase electron diffraction. The reaction of $Cl_2O_7$ with alcohols and amines yields alkyl perchlorates ($ROClO_3$) and amine perchlorates ($R_2NClO_3$), respectively.

*Fluorides of chlorine*

Given the isolobal relationship between the halogens it is not surprising that the mixed dihalogens can be prepared, e.g., ClF, ICl, and BrCl. Chlorine fluoride is a highly reactive gas (Boiling point = -100.1 °C) that is a powerful fluorinating agent, and is prepared by the oxidation of chlorine by chlorine trifluoride,

$$KCl + 3 F_2 \xrightarrow{200\ °C} ClF_5 + KF$$

The higher electronegativity of fluorine as compared to chlorine (Table 15.9), and the ability of chlorine to form more than one bond, means that higher fluorides of chlorine are also known, i.e., $ClF_3$ and $ClF_5$. Chlorine trifluoride ($CF_3$, Boiling point = 11.75 °C) is a useful fluorinating agent, that is prepared by the high temperature reaction of elemental chlorine and fluorine, is a useful fluorinating age.

$$Cl_2 + ClF_3 \rightarrow 3\ ClF$$

The gaseous pentafluoride ($ClF_5$, Boiling point = -31.1 °C) is prepared by the reaction of potassium chloride with fluorine,

$$Cl_2 + 3\ F_2 \xrightarrow{200\ °C} 2\ ClF_3$$

The structure of $ClF_3$ is T-shaped with two lone pairs on chlorine (Figure 15.18a), while that of $ClF_5$ is square pyramidal with a single lone pair on chlorine (Figure 15.18b).

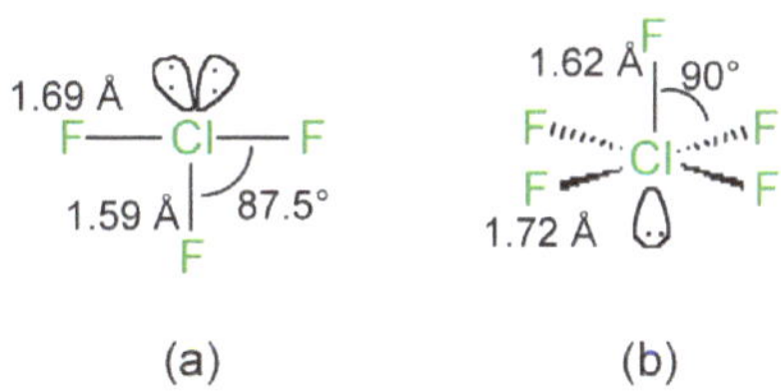


**Figure 15.18: The structures of (a) $ClF_3$ and (b) $ClF_5$.**

In general, the halogen fluorides are very reactive; explosive reactions occur with organic compounds. They are all powerful fluorinating agents when diluted with nitrogen, and the order of reactivity follows:

$$ClF_3 > BrF_3 > BrF_5 > IF_7 > ClF > IF_5 > BrF$$

Like most halogen fluorides, ClF, $ClF_3$ and $ClF_5$ all react with strong bases (e.g., alkali metal fluorides) to form anions,

$$ClF + CsF \rightarrow Cs^+ + ClF_2^-$$

$$ClF_3 + CsF \rightarrow Cs^+ + ClF_4^-$$

and strong acids (e.g., $AsF_5$ and $SbF_5$) to form cations,

$$2\ ClF + AsF_5 \rightarrow FCl_2^+\ AsF_6^-$$

$ClF_3 + AsF_5 \rightarrow ClF_2^+ + AsF_6^-$

$ClF_5 + SbF_5 \rightarrow ClF_4^+ + SbF_6^-$

*Oxyacids of chlorine*

Table 15.11 lists the various oxyacids of chlorine. The relative strengths increase with the number of oxygen atoms since the more there are, the greater is the extent to which the negative charge on the resulting anion can be delocalized.

| **Oxyacid** | **Formula** | **pK$_a$** |
|---|---|---|
| Hypochlorous | HOCl | 7.5 |
| Chlorous | $HClO_2$ | 1.9 |
| Chloric | $HClO_3$ | -2 |
| Perchloric | $HClO_4$ | -10 |

**Table 15.11: Relative acidity of oxyacids of chlorine.**

*Hypochlorous acid*

Hypochlorous acid (HOCl) can be made pure in the gas phase,

$$H_2O_{(g)} + Cl_2O_{(g)} \rightleftharpoons 2\ HOCl_{(g)}$$

while strong acid solutions can be made from $Cl_2O$. In contrast, dilute aqueous solutions are obtained with a suspension of mercury oxide to remove the chloride,

$$2\ Cl_2 + 2\ HgO_{(s)} + H_2O \rightarrow 2\ HOCl + HgO.HgCl_2 \downarrow$$

Solutions of the anion, $OCl^-$, are obtained by electrolysis of brine solution; allowing the products to mix at low temperature,

$$Cl_2 + 2\ OH^- \rightarrow ClO^- + Cl^- + H_2O$$

The anion (hypochlorite) is a good oxidant,

$$ClO^- + NH_3 \rightarrow NH_2Cl + OH^-$$

$ClO^- + 2\,I^- + 2\,H^+ \rightarrow I_2 + Cl^- + H_2O$

but can undergo disproportionation slowly at 25 °C, but fast above 80 °C.

$3\,ClO^- \rightarrow 2\,Cl^- + ClO_3^-$

*Chlorous acid*

Chlorous acid (HOClO) is prepared by the reaction of $ClO_2$ with base,

$2\,ClO_2 + 2\,OH^- \rightarrow ClO_2^- + ClO_3^- + H_2O$

followed by the precipitation of the $ClO_2^-$ salt with barium chloride,

$2\,ClO_2^- + BaCl_2 \rightarrow 2\,Cl^- + Ba(ClO_2)_2 \downarrow$

The barium salt is dried and then reacted with a calculated amount of $H_2SO_4$,

$Ba(ClO_2)_2 + H_2SO_4 \rightarrow Ba(SO_4) + 2\,HClO_2$

The pure acid is unknown since it is too unstable, however, salts can be prepared directly, e.g.,

$2\,ClO_2 + Na_2O_2 \rightarrow 2\,NaClO_2 + O_2$

The anion ($ClO_2^-$) is stable in alkaline solutions but in acid solutions decomposition occurs,

$5\,HClO_2 \rightarrow 4\,ClO_2 + Cl^- + H^+ + 2\,H_2O$

As with hypochlorite, the chlorite anion is a strong oxidant,

$ClO_2^- + 4\,I^- + 4\,H^+ \rightarrow 2\,I_2 + 2\,H_2O + Cl^-$

*Chloric acid*

The chloric anion ($ClO_3^-$) is made from the reaction of chlorine gas with hot alkali (80 °C) or by the electrolysis of hot NaCl solution.

$$3\,Cl_2 + 6\,OH^- \rightarrow ClO_3^- + 5\,Cl^- + 3\,H_2O$$

To obtain a solution of the acid, $ClO_3^-$ is precipitated as the barium salt,

$$2\,ClO_3^- + BaCl_2 \rightarrow 2\,Cl^- + Ba(ClO_3)_2 \downarrow$$

which is removed, dried, and suspended in water and treated with a calculated amount of $H_2SO_4$,

$$Ba(ClO_3)_2 + H_2SO_4 \rightarrow Ba(SO_4) + 2\,HClO_3$$

The free acid cannot be isolated and a maximum concentration of only 40% can be obtained in water.

The $ClO_3^-$ anion is pyramidal both in solid salts and in solution, and many salts are known; however, those with organic cations are explosive. The anion is a strong oxidizing agent,

$$ClO_3^- + 6\,I^- + 6\,H^+ \rightarrow 3\,I_2 + 3\,H_2O + Cl^-$$

$$ClO_3^- + 3\,NO_2^- \rightarrow 3\,NO_3^- + Cl^-$$

and it disproportionates slowly in solution,

$$4\,ClO_3^- \rightarrow 3\,ClO_4^- + Cl^-$$

*Perchloric acid*

Perchloric acid is a very dangerous liquid that will explode if traces of metal ions are present. It is also a very strong oxidizing agent that will convert organic compounds to $CO_2$ and $H_2O$.

The perchlorate anion ($ClO_4^-$) is best made by electrolytic oxidation of chlorate in aqueous solution,

$$ClO_3^- + H_2O \rightarrow ClO_4^- + 2\,H^+ + 2\,e^-$$

Fractional distillation can concentrate the solution to 72.5% which is a constant boiling mixture. This concentration is moderately safe to use, however, 100% perchloric acid may be obtained by dehydration with $H_2SO_4$.

Perchloric acid is a very strong acid that is fully ionized in aqueous solution, such that the salt $[H_3O][ClO_4]$ can be isolated. Many other perchlorate salts are known, but those with organic cations are explosive. Perchlorate salts of metals are often used when studying complex formation in aqueous solution, because $ClO_4^-$ is a very weak ligand ($PF_6^-$ is better) and unlikely to form complexes itself. However, perchlorate does complex with +3 and +4 cations.

## Compounds of bromine

*Bromine trifluoride as a solvent*

Bromine trifluoride is a toxic, colorless, and corrosive liquid with a pungent choking smell that is soluble in sulfuric acid but explodes on contact with water and organic compounds. Vapors severely irritate and may burn the eyes, skin, and respiratory system. The liquid burns all human tissue and causes severe damage.

Bromine trifluoride ($BrF_3$) has a liquid range similar to water (Melting point = 8.8 °C and Boiling point = 127 °C), and like water it auto ionizes,

$$BrF_3 \rightarrow BrF_2^+ + BrF_4^-$$

The products, like those of water's self-ionization, are an acid ($BrF_2^+$) and a base ($BrF_4^-$). However, unlike water, $BrF_3$ reacts with fluoride acids and bases not proton acids and bases. Thus, in $BrF_3$ a base is a salt that provides $F^-$, i.e., potassium fluoride (KF) is a base in $BrF_3$ solution in the same manner as potassium hydroxide (KOH) is a base in water. The product from the reaction of a fluoride salt with $BrF_3$ is the formation of the conjugate base, $BrF_4^-$,

$$AgF + BrF_3 \rightarrow Ag^+ + BrF_4^-$$

Other examples of this type of reaction include:

$$KF + BrF_3 \rightarrow K^+ + BrF_4^-$$

$$NOF + BrF_3 \rightarrow NO^+ + BrF_4^-$$

By analogy, an acid in $BrF_3$ solution acts as a fluoride ($F^-$) acceptor, i.e., a Lewis acid,

$$SbF_5 + BrF_3 \rightarrow SbF_6^- + BrF_2^+$$

*Bromine trifluoride as a fluorinating agent*

Bromine trifluoride is a strong fluorinating agent that is able to convert a metal (e.g., vanadium) to its associated fluoride compound, (i.e., $VF_5$). A wide range of salts and oxides may be converted to fluorides with the metal in a high oxidation state. However, it should be noted that BeO, MgO, and $Al_2O_3$ form oxo fluorides rather than the fluoride.

The reaction of silver with $BrF_3$ yields the mono fluoride, while the same reaction with gold yields the trifluoride,

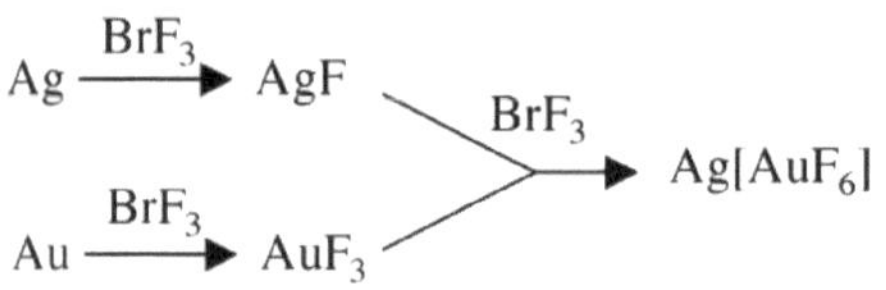


If the reactions are combined in $BrF_3$ solution a mixed metal fluoride salt is formed. A similar reaction occurs with NOCl and $V_2O_5$,

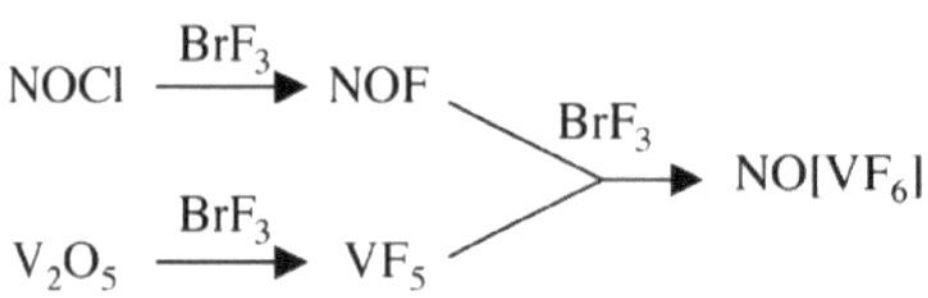

# Chapter 16: Group 18 The Noble Gases

## The elements

The Group 18 elements have a particular name: Noble gases. Noble gas is translated from the German noun *Edelgas*, first used in 1898 by Hugo Erdmann (Figure 16.1) (1862 - 1910) to indicate their extremely low level of reactivity. The noble gases were often also called the inert gases, however, since noble gas compounds are now known this name is no longer used. Table 11.1 lists the derivation of the names of the Noble gases.

**Figure 16.1: German chemist Hugo Wilhelm Traugott Erdmann (1862 - 1910).**

| Element | Symbol | Name |
|---|---|---|
| Helium | He | Greek *helios* meaning the Sun |
| Neon | Ne | From the Greek meaning *new one* |
| Argon | Ar | From the Greek meaning *inactive* |
| Krypton | Kr | From the Greek kryptos meaning the *hidden one* |
| Xenon | Xe | From the Greek xenos], meaning *foreigner*, *stranger*, or *guest* |
| Radon | Rn | From its radioactive nature |

**Table 16.1: Derivation of the names of each of the Group 18 (VIII) elements.**

## Discovery

*Helium*

The first evidence of helium was the observation by astronomer Pierre Janssen (Figure 16.2) on August 18, 1868 as a bright yellow line with a wavelength of 587.49 nm in the spectrum of the chromosphere of the Sun. On October 20 of the same year, English astronomer Norman Lockyer (Figure 16.3) observed a yellow line in the solar spectrum, which he named the *D3 Fraunhofer* line because it was near the known *D1* and *D2* lines of sodium.

He concluded that it was caused by an element in the Sun unknown on Earth. Lockyer and Edward Frankland (Figure 10.6) named the element with the Greek word for the Sun, *helios*.

**Figure 16.2: Portrait of French astronomer Pierre Jules César Janssen (1824 - 1907).**

**Figure 16.3: English scientist and astronomer Sir Joseph Norman Lockyer, FRS (1836 - 1920).**

On March 26, 1895 British chemist Sir William Ramsay (Figure 16.4) isolated helium on Earth by treating the mineral cleveite (a radioactive mineral containing uranium and found in Norway) with mineral acids.

**Figure 16.4: Scottish chemist Sir William Ramsay, KCB FRSE (1852 - 1916).**

*Neon*

Neon was discovered in 1898 by Sir William Ramsay (Figure 16.4) and Morris Travers (Figure 16.5). When Ramsay chilled a sample of air until it became a liquid, then warmed the liquid and captured the gases as they boiled off. After nitrogen, oxygen, and argon, the three gases that boiled off were krypton, xenon, and neon.

**Figure 16.5: English chemist and founding director of the Indian Institute of Science, Morris William Travers (1872 - 1961).**

*Argon*

In 1785 Henry Cavendish (Figure 7.3) suspected that argon was present in air but it was not isolated until 1894 by Lord Rayleigh (Figure 16.6) and Sir William Ramsay (Figure 16.4) in an experiment in which they removed all of the oxygen, carbon dioxide, water and nitrogen from a sample of clean air.

**Figure 16.6: English physicist John William Strutt, 3rd Baron Rayleigh, OM (1842 - 1919).**

*Krypton*

Krypton was discovered in 1898 by Sir William Ramsay (Figure 16.4) and Morris Travers (Figure 16.5) in residue left from evaporating nearly all components of liquid air.

In 1960, an international agreement defined the meter (m) in terms of wavelength of light emitted by the $^{86}Kr$ isotope (wavelength of 605.78 nm). This agreement replaced the standard meter located in Paris, which was a metal bar made of a Pt-Ir alloy and was itself replaced by a definition based on the speed of light, a fundamental physical constant. In October 1983, the *Bureau International des Poids et Mesures* defined the meter as the distance that light travels in a vacuum during $^{1}/_{299,792,458}$ s.

*Xenon*

Xenon was discovered by William Ramsay (Figure 16.4) and Morris Travers (Figure 16.5) on July 12, 1898, shortly after their discovery of krypton and neon.

*Radon*

Radon was the fifth radioactive element to be discovered after uranium, thorium, radium and polonium. Discovered in 1900 by Friedrich Dorn (Figure 11.8) after he noticed that radium compounds emanate a radioactive gas that he named Radium Emanation (*Ra Em*). Prior to these experiments, in 1899, Marie and Pierre Curie (Figures 16.8) observed that the gas emitted by radium remained radioactive for a month. Later that year, Ernest Rutherford (Figure 16.9) noticed variations when trying to measure radiation from thorium oxide. In 1901, he demonstrated that the emanations are radioactive, but credited the Curies for the discovery of the element.

**Figure 16.7: German physicist Friedrich Ernst Dorn (1848 - 1916).**

**Figure 16.8: Pierre Curie (1859 - 1906) and Marie Skaodowska-Curie (1867 - 1934) in their Paris laboratory.**

**Figure 16.9: British-New Zealand chemist and physicist Ernest Rutherford, 1st Baron Rutherford of Nelson, OM, FRS (1871 - 1937).**

## Abundance

The abundance of the Noble gases is given in Table 16.2.

| Element | Terrestrial abundance (ppm) |
|---|---|
| He | $8\times10^{-3}$ (Earth's crust), $4\times10^{6}$ (sea water), 5 (atmosphere) |
| Ne | $70\times10^{-3}$ (Earth's crust), 0.2 (sea water), 18 (atmosphere) |
| Ar | 1.2 (Earth's crust), 0.45 (sea water), $0.93\times10^{4}$ (atmosphere) |
| Kr | $10\times10^{-6}$ (Earth's crust), $80\times10^{-6}$ (sea water), 1 (atmosphere) |
| Xe | $2\times10^{-6}$ (Earth's crust), $100\times10^{-6}$ (sea water), $90\times10^{-3}$ (atmosphere) |

**Table 16.2: Abundance of Group 18 elements.**

## Isotopes

The naturally abundant isotopes of the Group 18 elements are listed in Table 16.3. All of the isotopes of radon are radioactive.

Unlike most elements, helium's isotopic abundance varies greatly by origin, due to the different formation processes. The most common isotope, $^{4}He$, is produced on Earth by a decay of heavier radioactive elements. It was also formed in enormous quantities during the Big Bang.

Naturally occurring $^{40}K$ with a half-life of $1.25 \times 10^{9}$ years, decays to stable $^{40}Ar$ (11.2%) by electron capture and positron emission, and also to stable

$^{40}$Ca (88.8%) via beta decay. These properties and ratios are used to determine the age of rocks.

| Isotope | Natural abundance (%) |
|---|---|
| Helium-3 | 0.000137 |
| Helium-4 | 99.999863 |
| Neon-20 | 90.48 |
| Neon-21 | 0.27 |
| Neon-22 | 9.25 |
| Argon-36 | 0.337 |
| Argon-86 | 0.063 |
| Argon-40 | 99.600 |
| Krypton-78 | 0.35 |
| Krypton-80 | 2.25 |
| Krypton-81 | trace |
| Krypton-82 | 11.6 |
| Krypton-83 | 11.5 |
| Krypton-84 | 57 |
| Krypton-86 | 17.3 |
| Xenon-124 | 0.095 |
| Xenon-126 | 0.089 |
| Xenon-128 | 1.91 |
| Xenon-129 | 26.4 |
| Xenon-130 | 4.07 |
| Xenon-131 | 21.2 |
| Xenon-132 | 26.9 |
| Xenon-134 | 10.4 |
| Xenon-136 | 8.86 |
| Radon-222 | trace |

**Table 16.3: Abundance of the non-synthetic isotopes of the Group 18 elements.**

With a half-life of 230,000 years $^{81}$Kr is used for dating 50,000 - 800,000 year old groundwater. $^{85}$Kr is an inert radioactive noble gas with a half-life of 10.76 years. It is produced in nuclear bomb testing and nuclear reactors. $^{85}$Kr is released during the reprocessing of fuel rods from nuclear reactors.

## Industrial production of the elements

Helium is extracted by fractional distillation from natural gas, which contains up to 7% helium. Since helium has a lower boiling point than any other element, low temperature and high pressure are used to liquefy nearly all the other gases. The resulting helium gas is purified by successive exposures to lowering temperatures. A final purification step with activated charcoal results in 99.995% pure Grade-A helium.

Argon is produced industrially by the fractional distillation of liquid air, a process that separates liquid nitrogen, which boils at 77.3 K, from argon, which boils at 87.3 K and oxygen, which boils at 90.2 K. Xenon is obtained commercially as a byproduct of the separation of air into oxygen and nitrogen.

## Physical properties

The physical properties of the Group 18 elements are given in Table 16.4.

| Element | Melting point (°C) | Boiling point (°C) |
|---|---|---|
| He | -272.20 | -268.93 |
| Ne | -248.59 | -246.08 |
| Ar | -189.35 | -185.85 |
| Kr | -157.36 | -153.22 |
| Xe | -111.7 | -108.12 |
| Rn | -71.15 | -61.85 |

**Table 11.4: Selected physical properties of the Group 18 elements. All of the Noble gases show characteristic spectral lines (Figure 16.10 - Figure 16.14).**

**Figure 16.10: Spectral lines of helium (He).**

**Figure 16.11: Spectral lines of neon (Ne).**

**Figure 16.12: Spectral lines of argon (Ar).**

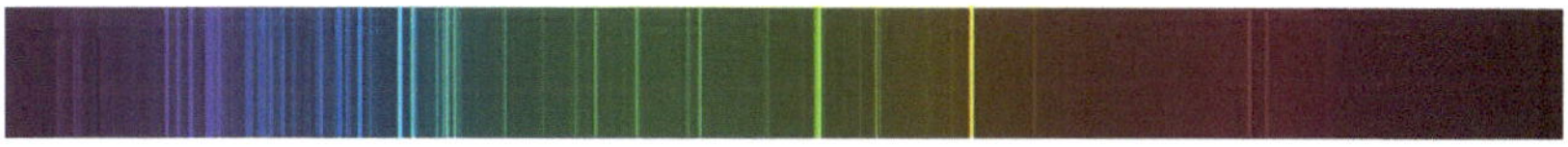

**Figure 16.13: Spectral lines of krypton (Kr).**

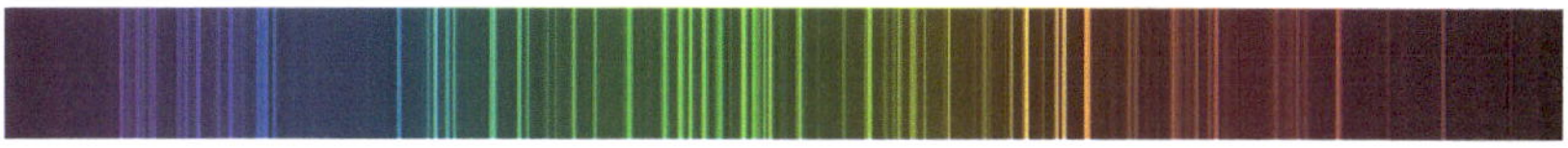

**Figure 16.14: Spectral lines of xenon (Xe).**

## Compounds of the Group 18 elements.

Only a few hundred noble gas compounds have been formed. Neutral compounds of helium and neon have not been formed, while xenon, krypton, and argon have shown only minor reactivity. The reactivity follows the order:

$$\text{Ne} < \text{He} < \text{Ar} < \text{Kr} < \text{Xe} < \text{Rn}$$

Xenon compounds are the most numerous of the noble gas compounds. Oxidation states of +2, +4, +6, and +8 with electronegative elements, e.g., $XeF_2$, $XeF_4$, $XeF_6$, $XeO_4$, and $Na_4XeO_6$. Compounds of xenon bound to boron, hydrogen, bromine, iodine, beryllium, sulfur, titanium, copper, and silver have also been observed but only at low temperatures in noble gas matrices, or in supersonic noble gas jets.

Although radon is more reactive than xenon it should form chemical bonds more easily than xenon, however, due to the high radioactivity and short half-life of radon isotopes, only a few fluorides and oxides of radon have been formed.

Krypton is less reactive than xenon, and oxidation states are generally limited to +2, $KrF_2$. Compounds in which krypton forms a bond to nitrogen and oxygen are only stable below -60 °C and -90 °C, respectively. Krypton atoms chemically bound to other nonmetals (hydrogen, chlorine, carbon) as well as some late transition metals (copper, silver, gold), but only at low temperatures in noble gas matrices, or in supersonic noble gas jets. Similar conditions were used to obtain the first compounds of argon.

Noble gases also form non-covalent compounds, for example clathrates that consist of an atom trapped within cavities of crystal lattices of organic and inorganic compounds. Noble gases can form endohedral fullerene compounds, in which the noble gas atom is trapped inside a fullerene molecule (Figure 16.15).

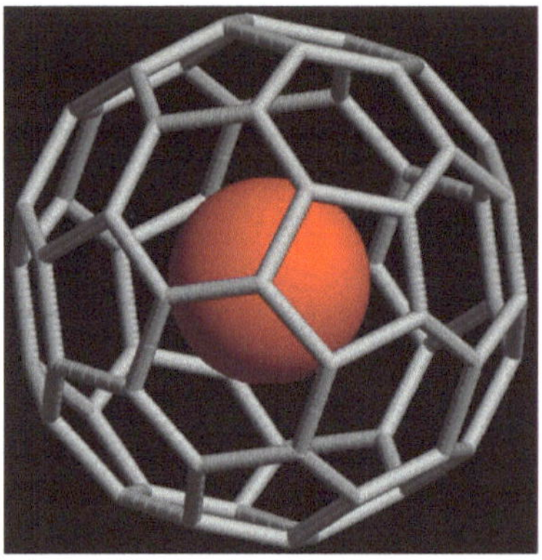

**Figure 16.15: A skeletal structure of He@$C_{60}$.**

# Bibliography

The following is intended to provide further reading.

A. C. Adams and C. D. Capio, Edge profiles in the plasma etching of polycrystalline silicon. *J. Electrochem. Soc.*, 1981, **128**, 366.

H. Adkins, The selective activation of alumina for decarboxylation or for dehydration. *J. Am. Chem. Soc.*, 1922, **44**, 2175.

*Advances in Ceramics*, Eds. J. A. Mangels and G. L. Messing, American Ceramic Society, Westville, OH, 1984, Vol. 9.

G. Agricola, *De Re Metallica*, Dover Publications, UK (1950).

J. J. Allen, C. E. Hamilton, and A. R. Barron, Molecular Structures of RN(H)Py (R = 2,4,6-$Me_3C_6H_2$, 2,6-$Et_2C_6H_3$, $Ph_3C$), and the Copper Complex [Cu{(2,4,6-$Me_3C_6H_2$)N(H)Py}$_2$]$BF_4$. *J. Chem. Cryst.*, 2009, **39**, 573.

R. E. Anderson and A. R. Barron, Effect of carbon nanomaterials on calcium carbonate crystallization. *Main Group Chem.*, 2005, **4**, 279.

R. E. Anderson and A. R. Barron, Solubilization of single-wall carbon nanotubes in organic solvents without sidewall functionalization. *J. Nanosci. Nanotechnol.*, 2007, **7**, 3436.

M. Aucott, M. McLinden, and M. Winka, Release of mercury from broken fluorescent bulbs. *J. Air. Waste Manag. Assoc.*, 2003, **53**, 143.

S. M. Bachilo, M. S. Strano, C. Kittrell, R. H. Hauge, R. E. Smalley, and R. B. Weisman, Structure-assigned optical spectra of single-walled carbon nanotubes. *Science*, 2002, **298**, 2361.

H. Bader and N. M. Smyth, Grignard reactions of polyhalocarbons. II. In situ Grignard reactions of carbon tetrahalides with organochlorosilanes. *J. Org. Chem.*, 1964, **29**, 953.

A. R. Barron, Oxide, chalcogenide and related clusters of aluminum, gallium and indium. *Comm. Inorg. Chem.*, 1993, **14**, 123.

A. R. Barron, The chemistry behind explosives and the application of nano technology. *The Detonator*, 2009, **36**, 60.

A. R. Barron, M. B. Hursthouse, M. Motevalli, and G. Wilkinson, Transition metal polyhydride anions: a new synthetic route. X-ray crystal structure of $\{[(Me_3P)_3WH_5]^-Li^+\}_4$. *J. Chem. Soc., Chem. Commun.*, 1986, 81.

D. Barisic, S. Lulic, and P. Miletic, Radium and uranium in phosphate fertilizers and their impact on the radioactivity of waters. *Water Res.*, 1992, **26**, 607.

H. A. Bent, Distribution of atomic s character in molecules and its chemical implications. *J. Chem. Educ.*, 1960, **37**, 616.

H. A. Bent, An appraisal of valence-bond structures and hybridization in compounds of the first-row elements. *Chem. Rev.*, 1961, **61**, 275.

R. S. Berry, Correlation of rates of intramolecular tunneling processes, with application to some group V compounds. *J. Chem. Phys.*, 1960, **32**, 933.

D. S. Bethune, C. H. Klang, M. S. deVries, G. Gorman, R. Savoy, J. Vazquez, and R. Beyers, Cobalt-catalysed growth of carbon nanotubes with single-atomic-layer walls. *Nature*, 1993, **363**, 605.

V. Biringuccio, The Pirotechnia of Vannoccio Biringuccio: *The Classic Sixteenth-Century Treatise on Metals and Metallurgy*, Dover Publications (1990).

M. Bishop, N. Shahid, J. Yang, and A. R. Barron, Determination of the mode and efficacy of the cross-linking of guar by borate using MAS $^{11}$B NMR of borate cross-linked guar in combination with solution $^{11}$B NMR of model systems. *Dalton Trans.*, 2004, 2621.

K. M. Bissett and T. M. Gilbert, Pericyclic reactions between aminoboranes $R_2B=NR_2$' and alkenes: [4+2] vs [2+2] transition states. *Organometallics*, 2004, **23**, 850.

P. Blake, P. D. Brimicombe, R. R. Nair, T. J. Booth, D. Jiang, F. Schedin, L. A. Ponomarenko, S. V. Morozov, H. F. Gleeson, E. W. Hill, A. K. Geim, and K. S. Novoselov, Graphene-based liquid crystal device. *Nano Lett.*, 2008, **8**, 1704.

P. M. Boorman and D. Potts, Group V chalcogenide complexes of boron trihalides. *Can. J. Chem.*, 1974, **52**, 2016.

A. S. Borovik and A. R. Barron, Reaction of olefins with aluminium chloride stabilized arene-mercury complexes. *Main Group Chem.*, 2005, **4**, 135. 9.

C. S. Branch, S. G. Bott, and A. R. Barron, Group 13 trihalide complexes of 9-fluorenone: a comparison of methods for assigning relative Lewis acidity. J. *Organomet. Chem.*, 2003, **666**, 23.

M. A. Bredig, The crystal structure of calcium cyanamide. *J. Am. Chem. Soc.*, 1942, **64**, 1730.

J, J. Brege, C. Gallaway, and A. R. Barron, Fluorescence quenching of single-walled carbon nanotubes in SDBS surfactant suspension by metal ions: quenching efficiency as a function of metal and nanotube identity. *J. Phys. Chem., C*, 2007, **111**, 17812.

J. M. Buriak, L. K. Cheatham, R. G. Gordon, J. J. Graham, and A. R. Barron, The use of amines to enhance the volatility of barium β-diketonates: chemical vapor deposition of barium oxide thin films. *Euro. J. Solid State Inorg. Chem.*, 1992, **29**, 43.

Carbon Dioxide Capture and Storage: Special Report of the Intergovernmental Panel on Climate Change, Cambridge University Press (2005).

J. Chattopadhyay, A. Mukherjee, C. E. Hamilton, J.-H. Kang, S. Chakraborty, W. Guo, K. F. Kelly, A. R. Barron, and W. E. Billups, Graphite epoxide. *J. Am. Chem. Soc.*, 2008, **130**, 5414.

K. Christe, Chemical synthesis of elemental fluorine. *Inorg. Chem.*, 1986, **25**, 3721.

J. C. Clifton II, Mercury exposure and public health. *Pediatr. Clin. North Am.*, 2007, **54**, 237.

F. A. Cotton and G. Wilkinson, *Advanced Inorganic Chemistry*, 4$^{th}$ Ed. Wiley Interscience (1980).

J. L. Courtneidge, A. G. Davies, P. S. Gregory, D. C. McGuchan, and S. N. Yazdi, $^{199}$Hg hyperfine coupling in the ESR spectra of mercurated arene radical cations: evidence for an alternative mechanism of mercuration. *J. Chem. Soc., Chem. Commun.*, 1987, 1192.

L. C. Damude and P. A. W. Dean, A preparative and $^{1}$H and $^{13}$C NMR spectroscopic study of mercury(II)-arene complexes. *J. Organomet. Chem.*, 1979, **181**, 1.

M. Davies, A Thousand Years of Science and Scientists: 988 to 1988. *Hist. Sc.*, 1995, **33**, 239.

T. G. Duplinsky1 and D. V. Cicchetti, The health status of dentists exposed to mercury from silver amalgam tooth restorations. *Int. J. Stats. Med. Res.*, 2012, **1**, 1.

A. D. Fortes, E. Suard, M. -H. Lemée-Cailleau, C. J. Pickard, and R. J. Needs, Crystal structure of ammonia monohydrate phase II. *J. Am. Chem. Soc.*, 2009, **131**, 13508.

D. F. Gaines, Chemistry of pentaborane(9). *Acc. Chem. Res.*, 1973, **6**, 416.

Z. Ge, J. C. Duchamp, T. Cai, H. W. Gibson, and H. C. Dorn, Purification of endohedral trimetallic nitride fullerenes in a single, facile step. *J. Am. Chem. Soc.*, 2005, **127**, 16292.

S. K. Ghandhi, VLSI fabrication principles: silicon and gallium arsenide. Wiley-Interscience, New York, (1994).

R. J. Gillespie, The electron-pair repulsion model for molecular geometry. *J. Chem. Educ.*, 1970, **47**, 18.

R. J. Gillespie, The VSEPR model revisited. *Chem. Soc. Rev.*, 1992, **21**, 59.

R. J. Gillespie and R. S. Nyholm, Inorganic stereochemistry. *Quart. Rev.*, 1957, **11**, 339.

L. A. Girifalco, M. Hodak, and R. S. Lee, Carbon nanotubes, buckyballs, ropes, and a universal graphitic potential. *Phys. Rev. B*, 2000, **62**, 13104.

J. Goodyear and G. Steigman, The crystal structure of α-$Ga_2S_3$. *Acta Crystallogr.*, 1963, **16**, 946.

G. Graham, S. Richtsmeier, and D. A. Dixon, Electronic structure of the alkyllithium clusters, $(CH_3Li)_n$, n = 1-6, and $(C_2H_5Li)_n$, n = 1-2. *J. Am. Chem. Soc.*, 1980, **102**, 5759.

N. N. Greenwood and A. Earnshaw, *Chemistry of the Elements*, Pergamon Press, Oxford (1984).

T. Guo, P. Nikolaev, A. G. Rinzler, D. Tománek, D. T. Colbert, and R. E. Smalley, Self-assembly of tubular fullerenes. *J. Phys. Chem.*, 1995, **99**, 10694.

R. B. Gupta and J.-J. Shim, *Solubility in Supercritical Carbon Dioxide*, CRC Press (2006).

J. J. Habeeb, A. Osman, and D. G. Tuck, The electrochemical synthesis of neutral and anionic organozinc halides. *J. Organomet. Chem.*, 1980, **185**, 117.

J. H. Hafner, M. J. Bronikowski, B. R. Azamian, P. Nikolaev, A. G. Rinzler, D. T. Colbert, K. A. Smith, and R. E. Smalley, Catalytic growth of single-wall carbon nanotubes from metal particles. *Chem. Phys. Lett.*, 1998, **296**, 195.

M. Y. Han, B. Ozyilmaz, Y. Zhang, and P. Kim, Phys. Energy band-gap engineering of graphene nanoribbons. *Rev. Lett.*, 2008, **98**, 206805.

M. D. Healy, J. T. Leman, and A. R. Barron, $AlMe_2(BHT)NH_3$: An unusually stable organoaluminum-ammonia complex and its extended coordination sphere solvate.*J. Am. Chem. Soc.*, 1991, **113**, 2776.

M. D. Healy, P. E. Laibinis, P. D. Stupik, and A. R. Barron, The reaction of indium(III) chloride with tris(trimethylsilyl)phosphine: a novel route to indium phosphide. *J. Chem. Soc., Chem. Commun.*, 1989, 359.

K. Hedberg and R. Ryan, Molecular structure of gaseous diboron tetrachloride, $B_2Cl_4$. *J. Chem. Phys.*, 1964, **41**, 2214.

Y. Hernandez, V. Nicolosi, M. Lotya, F. M. Blighe, Z. Sun, S. De, I. T. McGovern, B. Holland, M. Byrne, Y. K. Gun'Ko, J. J. Boland, P. Niraj, G. Duesberg, S. Krishnamurthy, R. Goodhue, J. Hutchinson, V. Scardaci, A. C. Ferrari, and J. N. Coleman, High-yield production of graphene by liquid-phase exfoliation of graphite. *Nat. Nanotechnol.*, 2008, **3**, 563.

A. Hirsch, New cages and unusual guests: fullerene chemistry continues to excite. *Angew. Chem. Int. Ed.*, 2001, **40**, 1195.

P. B. Hitchcock, J. M. Keates, and G. A. Lawless, Mercurycyclopentadienyl derivatives are not always fluxional. *J. Am. Chem. Soc.*, 1998, **120**, 599.

V. Q. Ho and T. Sugano, Selective anodic oxidation of silicon in oxygen plasma. *IEEE Trans. Electron Devices*, 1980, **ED-27**, 1436.

H. O. House, Use of lithium organocuprate additions as models for an electron-transfer process. *Acc. Chem. Res.*, 1976, **9**, 59.

C. E. Housecroft, Boranes and metalloboranes: structure, bonding and reactivity, Ellis Horwood, Chichester (1990).

W. S. Hummers and R. E. Offeman, Preparation of graphitic oxide. *J. Am. Chem. Soc.*, 1958, **80**, 1339.

S. Iijima and T. Ichihashi, Single-shell carbon nanotubes of 1-nm diameter. *Nature*, 1993, **363**, 603.

*International Tables for X-ray Crystallography*. Vol. IV; Kynoch Press: Birmingham, UK (1974).

H. R. Jafry, E. A. Whitsitt, and A. R. Barron, Silica coating of vapor grown carbon fibers. *J. Mater. Sci.*, 2007, **42**, 7381.

G. E. Jellison, Jr., L. W. Panek, P. J. Bray, and G. B. Rouse, Jr., Determinations of structure and bonding in vitreous $B_2O_3$ by means of $B^{10}$, $B^{11}$, and $O^{17}$ NMR. *J. Chem. Phys.*, 1977, **66**, 802.

W. B. Jensen, The origin of the oxidation-state concept. *J. Chem. Educ.*, 2007, **84**, 1418.

S. Jensen and A. Jernelöv, Biological methylation of mercury in aquatic organisms. *Nature*, 1969, **223**, 753.

B. F. G. Johnson, in *Comprehensive Inorganic Chemistry*, Pergamon Press, Vol. 4, Chapter 52 (1973).

L. Jiao, L. Zhang, X. Wang, G. Diankov, and H. Dai, Narrow graphene nanoribbons from carbon nanotubes. *Nature*, 2009, **458**, 877.

D. Jones and C. Dunnill, Hydrogen-enriched natural gas as a domestic fuel: An analysis based on flash-back and blow-off limits for domestic natural gas appliances within the UK. *Sustain. Energy Fuels*, 2018, **2**, 710.

S. Kabalkina, V. G. Losev, and N. M. Gasanly, Polymorphism of InS at high pressures. *Solid State Commun.*, 1982, **44**, 1383.

T. F. Kelley, Polonium-210 content of mainstream cigarette smoke. *Science*, 1965, **149**, 537.

S. Kesavan and R. K. Prud'homme, Rheology of guar and (hydroxypropyl) guar crosslinked by borate. *Macromolecules*, 1992, **25**, 2026.

A. Keys, S. G. Bott, and A. R. Barron, Molecular structures of ($^{t}$Bu)Ga($S_2CN^nPr_2$)$_2$ and ($^{i}$PrO)Ga($S_2CNEt_2$)$_2$: An example of an unusual ligand pseudorotation. *J. Chem. Cryst.*, 1998, **28**, 629.

A. Keys, S. G. Bott, and A. R. Barron, MOCVD growth of gallium sulfide using di-*tert*-butyl gallium dithiocarbamate precursors: formation of a metastable phase of GaS. *Chem. Mater.*, 1999, **11**, 3578.

H. W. Kroto, J. R. Heath, S. C. O'Brien, R. F. Curl, and R. E. Smalley, $C_{60}$: Buckminsterfullerene. *Nature*, 1985, **318**, 162.

C. F. Lane, Reduction of organic compounds with diborane. *Chem. Rev.*, 1976, **76**, 773.

J. M. Lehn, *Supramolecular Chemistry: Concepts and Perspectives*. John Wiley & Sons, New Jersey (1995).

J. T. Leman, J. Braddock-Wilking, A. J. Coolong, and A. R. Barron, 1,3-Diaryltriazenido compounds of aluminum. *Inorg. Chem.*, 1993, **32**, 4324.

F. Liang, A. K. Sadana, A. Peera, J. Chattopadhyay, Z. Gu, R. H. Hauge, and W. E. Billups, A convenient route to functionalized carbon nanotubes. *Nano Lett.*, 2004, **4**, 1257.

C. H. Lochmüller and R. W. Souter, Chromatographic resolution of enantiomers: selective review. *J. Chromatogr. A*, 1975, **113**, 283.

A. N. MacInnes, W. M. Cleaver, A. R. Barron, M. B. Power, and A. F. Hepp, Metal–Organic chemical vapour deposition of polycrystalline tetragonal indium sulphide (InS) thin films. *Adv. Mater. Optics. Electron.*, 1992, **1**, 229.

D. McComb, Occupational exposure to mercury in dentistry and dentist mortality.*J. Can. Dent. Assoc.*, 1997, **63**, 372.

R. Minkwitz, A. Kornath, W. Sawodny, and H. Härtner, Über die reaktion von HCl im system HF/$SbF_5$ und die kristallstruktur von $SbCl_3F_2$. *Z. Anorg. Allg. Chem.*, 1994, **620**, 753.

R. S. Mulliken, A new electroaffinity scale; together with data on valence states and on valence ionization potentials and electron affinities. *J. Chem. Phys.*, 1934, **2**, 782.

L. R. Murphy, T. L. Meek, A. L. Allred, and L. C. Allen, Evaluation and test of Pauling's electronegativity scale. *J. Phys. Chem. A* 2000, **104**, 24.

C. B. Murray, D. J. Norris, and M. G. Bawendi, Synthesis and characterization of nearly monodisperse CdE (E = sulfur, selenium, tellurium) semiconductor nanocrystallites. *J. Am. Chem. Soc.*, 1993, **115**, 8706.

K. Nakamoto, M. Margoshes, and R. E. Rundle, Stretching frequencies as a function of distances in hydrogen bonds. *J. Am. Chem. Soc.*, 1955, **77**, 6480.

D. Ogrin and A. R. Barron, Highly oxygenated fullerenes by catalytic epoxidation of $C_{60}$ and single walled carbon nanotubes with methyltrioxorhenium–hydrogen peroxide. *J. Mol. Cat. A: Chem.*, 2006, **244**, 267.

D. Ogrin, J. Chattopadhyay, A. K. Sadana, E. Billups, and A. R. Barron, Epoxidation and deoxygenation of single-walled carbon nanotubes: quantification of epoxide defects. *J. Am. Chem. Soc.*, 2006, **128**, 11322.

P. Paetzold, Iminoboranes. *Adv. Inorg. Chem.*, 1987, **31**, 123.

L. Pauling, The nature of the chemical bond. IV. The energy of single bonds and the relative electronegativity of atoms. *J. Am. Chem. Soc.*, 1932, **54**, 3570.

L. Pauling, The formulas of antimonic acid and the antimonates. *J. Am. Chem. Soc.*, 1933, **55**, 1895.

W. G. Pfann, *Zone Melting*, John Wiley & Sons, New York, (1966).

R. Phillips and C. Dunnill, Zero gap alkaline electrolysis cell design for renewable energy storage as hydrogen gas. *RSC Advances*, 2016, **6**, 100643.

K. S. Pitzer, Relativistic effects on chemical properties. *Acc. Chem. Res.*, 1979, **12**, 271.

M. B. Power, A. R. Barron, S. G. Bott, E. J. Bishop, K. D. Tierce and J. L. Atwood, Acylation and esterification of the aryloxide ligand in $[AlMe(dbmp)_2]$. Crystal structures of [AlMe(dbmp)(bhmap)], Hbhmap and O=C(dbmp)$Bu^t$(Hdbmp = 2,6-di-*tert*-butyl-4-methylphenol, Hbhmap = 3-*tert*-butyl-2-hydroxy-5-methylacetophenone). *J. Chem. Soc., Dalton Trans.*, 1991, 241.

M. B. Power, J. W. Ziller, A. N. Tyler, and A. R. Barron, Interaction of tri-tert-butyl-gallium with elemental sulfur, selenium, and tellurium. *Organometallics*, 1992, **11**, 1055.

R. W. Rudolph, Boranes and heteroboranes: a paradigm for the electron requirements of clusters? *Acc. Chem. Res.*, 1976, **9**, 446.

G. Santiso-Quiñones and I. Krossing, Reference values for the B-X bond lengths of $BI_3$ and $BBr_3$. *Z. Anorg. Allg. Chem.*, 2008, **634**, 704.

M. Saunders, H. A. Jiménez-Vázquez, R. J. Cross, and R. J. Poreda, Stable compounds of helium and neon: He@$C_{60}$ and Ne@$C_{60}$. *Science*, 1993, **259**, 1428.

H. Schneider, K. Okada, and J. Pask, *Mullite and Mullite Ceramics*, Wiley (1994).

*Semiconductors: Group IV Elements and III-V Compounds*, Ed. O. Madelung, Springer-Verlag, Berlin (1991).

A. G. Sharpe, Interhalogen compounds and polyhalides. *Q. Rev. Chem. Soc.*, 1950, **4**, 115.

J. H. Simons, D. M. Yost, and M. T. Rogers, Bromine(III) fluoride. *Inorg. Synth.*, 1950, **3**, 184.

L. Staudenmaier, Verfahren zur darstellung der graphitsäure. *Ber. Dtsch. Chem. Ges.*, 1898, **31**, 1481.

N. Stern, The Stern Review: The Economics of Climate Change, HM Treasury, London.

R. Taylor and O. Kennard, Hydrogen-bond geometry in organic crystals. *Acc. Chem. Res.*, 1984, **17**, 320.

L. R. Thorne, R. D. Suenram, and F. J. Lovas, Microwave spectrum, torsional barrier, and structure of $BH_3NH_3$. *J. Chem. Phys.*, 1983, **78**, 167.

M. van Duin, J. A. Peters, A. P. G. Kieboom, and H. Van Bekkum, Studies on borate esters 1: The pH dependence of the stability of esters of boric acid and borate in aqueous medium as studied by $^{11}$B NMR. *Tetrahedron*, 1984, **40**, 2901.

L. H. van Poppel, S. G. Bott and A. R. Barron, Molecular structures of $Ga(^tBu)_2(OPh)(pyz)\cdot PhOH$ and $[(^tBu)_2Ga(H_2O)(\mu\text{-}OH)Ga(^tBu)_2]_2(\mu\text{-}OC_6H_4O)\cdot 4(2\text{-Mepy})$: intra- and inter-molecular hydrogen bonding to gallium aryloxides. *Polyhedron*, 2002, **21**, 1877.

N.-G. Vannerberg, The crystal structure of calcium carbide II and IV. *Acta Chem. Scand.*, 1962, **16**, 1212.

A. I. Vogel and G. Svehla, *Textbook of Macro and Semimicro Qualitative Inorganic Analysis*, Longman, London (1979).

E. von Frankland, Ueber die Isolirung der organischen Radicale. *Justus Liebigs Ann. Chem.*, 1849, **71**, 171.

J. L. Vossen and W. Kern, Thin film formation. *Phys. Today*, 1980, **33**, 26.

K. Wade, Structural and bonding patterns in cluster chemistry. *Adv. Inorg. Chem. Radiochem.*, 1976, **18**, 1.

H. A. Waldron, Did the Mad Hatter have mercury poisoning? *Br. Med. J.*, 1983, **287**, 24.

T. J. Ward and K. D. Ward, Chiral separations: fundamental review 2010. *Anal. Chem.*, 2010, **82**, 4712.

K. D. Weaver, J. J. Heymann, A. Mehta, P. L. Roulhac, D. S. Anderson, A. J. Nowalk, P. Adhikari, T. A. Mietzner, M. C. Fitzgerald, and A. L. Crumbliss, $Ga^{3+}$ as a mechanistic probe in $Fe^{3+}$ transport: characterization of $Ga^{3+}$ interaction with FbpA. *J. Biol. Inorg. Chem.*, 2008, **13**, 887.

K. Wefers and C. Misra, *Oxides and Hydroxides of Aluminum*, Alcoa Laboratories (1987).

E. Weiss, Die kristallstruktur des dimethylmagnesiums. *J. Organomet. Chem.*, 1964, **2**, 314.

E. Weiss, E. A. C. Lucken, Die kristall- und elektronenstruktur des methyllithiums. *J. Organomet. Chem.*, 1964, **2**, 197.

R. L. Wells, C. G. Pitt, A. T. McPhail, A. P. Purdy, S. R. B. Shafieezad, and Hallock, The use of tris(trimethylsilyl)arsine to prepare gallium arsenide and indium arsenide. *Chem. Mater.*, 1989, **1**, 4.

H. L. Wen and F. S. Yen, Growth characteristics of boehmite-derived ultrafine theta and alpha-alumina particles during phase transformation. *J. Cryst. Growth*, 2000, **208**, 696.

A. R. West, Solid State Chemistry and its Applications, Wiley, New York (1984).

R. West, M. J. Fink, and J. Michl, Tetramesityldisilene, a stable compound containing a silicon-silicon double bond. *Science*, 1981, **214**, 1343.

J. Wu and N. J. Evans, Indications of inflow motions in regions forming massive stars. *Astrophys. J.*, 2003, **592**, L79.

Y. Yamamoto, K. Kinoshita, K. Tamaru, and T. Yamanaka, Redetermination of the crystal structure of calcium cyanamide. *Bull. Chem. Soc. Japan*, 1958, **31**, 501.

J. Yang and A. R. Barron, A new route to fullerene substituted phenylalanine derivatives. *Chem. Commun.*, 2004, 2884.

L. Zeng, L. B. Alemany, C. L. Edwards, and A. R. Barron, Demonstration of covalent sidewall functionalization of single wall carbon nanotubes by NMR spectroscopy: Side chain length dependence on the observation of the sidewall $sp^3$ carbons. *Nano Res.*, 2008, **1**, 72.

*Zonal Isolation, Borate Crosslinked Fluids*, by R. J. Powell and J. M Terracina, Halliburton, Duncan, OK (1998).

www.ingramcontent.com/pod-product-compliance
Lightning Source LLC
LaVergne TN
LVHW051127210926
848482LV00001B/4